D1499983

The Challenge of Remaining Innovative

INNOVATION *and* TECHNOLOGY *in the* WORLD ECONOMY

Editor

MARTIN KENNEY
University of California at Davis/
Berkeley Roundtable on the International Economy

Other titles in the series

The Challenge of Remaining Innovative

Insights from Twentieth-Century American Business

EDITED BY
SALLY H. CLARKE,
NAOMI R. LAMOREAUX,
and STEVEN W. USSELMAN

Stanford Business Books

An Imprint of Stanford University Press
Stanford, California

Stanford University Press
Stanford, California

Printed in the United States of America on acid-free,
archival-quality paper

Library of Congress Cataloging-in-Publication Data

The challenge of remaining innovative : insights from twentieth-
century American business / edited by Sally H. Clarke, Naomi R.
Lamoreaux, and Steven W. Usselman.
    p. cm. — (Innovation and technology in the world economy)
    Includes bibliographical references and index.
    ISBN 978-0-8047-5892-5 (cloth : alk. paper)
    1. Technological innovations—United States—History—20th
century. 2. Research, Industrial—United States—History—20th
century. 3. Business enterprises—Technological innovations—
United States—History—20th century. I. Clarke, Sally H.
II. Lamoreaux, Naomi R. III. Usselman, Steven W. IV. Series.

HC110.T4C47 2009
658.4'063—dc22   2008031532

Typeset by Classic Typography in 10/12.5 Electra

For Lou,
Reader & Advisor

Contents

Contents

List of Figures and Tables

ACKNOWLEDGMENTS

Over the many years this volume has been in the works we have accumulated numerous debts. Many individuals and companies provided vital assistance along the way, including Susan Aaronson, Eric Abrahamson, Franco Amatori, Alan Anderson, Brian Balogh, William Becker, Nancy Berlage, Albert Beveridge, Robert Brugger, Christopher Casteneda, Michael Cavino, Alfred Chandler, Carl Christ, Robert Collins, Mary Butler Davies, Energy Planning, Pat Forster, Jill Friedman, Louis Galambos, Robert Garnet, Williamjames Hull Hoffer, David Hounshell, Elizabeth S. Hughes, Peter Jelavich, Catherine Kerr, Julie Kimmell, Mark Kornbluh, Bill Leslie, Harold Livesay, Harry Marks, William M. McClenahan Jr., Merck & Co., David Milobsky, Susan Morris, Sydney Nathans, Luigi Orsenigo, Fabio Pammolli, Edwin Perkins, Glenn Porter, Ben Primer, The Prologue Group, Daniel M. G. Raff, Leonard Reich, William R. Roberts, Richard Rosenbloom, Doug Rossinow, Margaret Rung, Steven Sass, Kelly Schrum, Mary Schweitzer, Christopher Sellers, David Sicilia, Leo Slater, Frank Smith, John K. Smith, Gabrielle Spiegel, Jeffrey Sturchio, Christopher Tomlins, Sharon Widomski, Mary Yeager, Julian Zelizer, and The Johns Hopkins University Department of History. We would also like to thank the contributors for their rapid responses to our many demands for revisions and information and Martin Kenney, Margo Beth Croupen, and the staff at Stanford University Press for their advice, support, and patience. Finally, we owe Sasha Nichols-Geerdes a deep debt of gratitude for his tireless editorial assistance.

CONTRIBUTORS

Stephen B. Adams is an associate professor of management at the Franklin P. Perdue School of Business, Salisbury University. He is the author of *Mr. Kaiser Goes to Washington: The Rise of a Government Entrepreneur* (University of North Carolina Press, 1997) and *Manufacturing the Future: A History of Western Electric* (Cambridge University Press, 1999, coauthor Orville R. Butler). He is writing a book on the institutional forces in the development of Silicon Valley.

Sally H. Clarke is professor of history at the University of Texas at Austin where she specializes in the study of political economy. She received her Ph.D. from Brown University under the direction of Naomi R. Lamoreaux. She has published *Regulation and the Revolution in United States Farm Productivity* (Cambridge University Press, 1994) and *Trust and Power: Consumers, the Modern Corporation, and the Making of the United States Automobile Market* (Cambridge University Press, 2007). In addition, she has written articles in various journals including the *Journal of Economic History*, *Business History*, the *Journal of Design History*, and *Law and History Review*.

Margaret B. W. Graham is associate professor of strategy and organization at the Desautels Faculty of Management of McGill University. She has divided her career between the history and practice of technology-based innovation, with special emphasis on industrial research and development. Her scholarly work has been based on in-depth archival study, on multiyear fieldwork, and on long-term interaction with companies. She has published widely on the history of innovation in large companies, including *RCA and the VideoDisc: The Business of Research* (Cambridge University Press, 1986), *R&D for Industry: A*

Century of Technical Innovation at Alcoa (Cambridge University Press, 1990, coauthor Bettye Pruitt), and *Corning and the Craft of Innovation* (Oxford University Press, 2001, coauthor Alec Shuldiner). As a member of the senior staff at Xerox Palo Alto Research Center in the early 1990s she helped lead an advanced research organization. She has served on National Research Council committees and has directed studies for the National Science Foundation, including the Machine Tool Industry and National Preparedness, which documented the lack of innovation in a U.S. industry that subsequently disappeared. Her current research interest is in understanding the history and practice of innovation and entrepreneurship from a global perspective.

Naomi R. Lamoreaux is professor of economics, history, and law at the University of California, Los Angeles, a research associate at the National Bureau of Economic Research, and a fellow of the American Academy of Arts and Sciences. She received her Ph.D. in history from The Johns Hopkins University in 1979, studying under Louis Galambos, and then taught at Brown University until she moved to UCLA in 1996. She has written *The Great Merger Movement in American Business, 1895–1904* (Cambridge University Press, 1985) and *Insider Lending: Banks, Personal Connections, and Economic Development in Industrial New England* (Cambridge University Press, 1994), as well as many articles on business, economic, and financial history. She is a past president of the Business History Conference and a recipient of that organization's Harold F. Williamson Prize.

Kenneth J. Lipartito is professor of history at Florida International University and former editor of *Enterprise & Society: The International Journal of Business History*. A specialist on technology, business, and culture, he is author of *A History of the Kennedy Space Center* (University Press of Florida, 2007, coauthor Orville Butler). He is the author or editor of four other books, including *Constructing Corporate America: History, Politics, Culture* (Oxford University Press, 2004) and *The Bell System and Regional Business: The Telephone in the South, 1877–1920* (Johns Hopkins University Press, 1989). His articles have appeared in the *American Historical Review*, the *Journal of Economic History*, *Technology and Culture*, *Industrial and Corporate Change*, and the *Business History Review*. He is a recipient of the Harold F. Williamson Prize from the Business History Conference and the Abbott Payson Usher Prize from the Society for the History of Technology.

Christopher McKenna is a reader in business history and strategy at the Saïd Business School, a fellow of Brasenose College, and a founding member of the Clifford Chance Centre for the Management of Professional Service Firms, all within the University of Oxford. A graduate of Amherst College and The Johns Hopkins University, he is particularly interested in the global evolution of professional firms and their role in the governance of business, nonprofits, and the state. His first book, *The World's Newest Profession: Management Consulting in the Twentieth Century* (Cambridge University Press, 2006), won both the Hagley Prize and the Newcomen-Harvard Book Award for the best book on business history. McKenna is currently writing an international history of white-collar crime, tentatively titled "Partners in Crime."

Paul J. Miranti received his doctorate in history at The Johns Hopkins University in 1985 and is a professor in the Department of Accounting, Business Ethics and Information Systems at Rutgers Business School—Newark and New Brunswick. He is the author of *Accountancy Comes of Age: The Development of an American Profession, 1886–1940* (University of North Carolina Press, 1990); *A History of Corporate Finance* (Cambridge University Press, 1997, coauthor J. B. Baskin); and *The Institute of Accounts: Nineteenth-Century Origins of Accounting Professionalism in the United States* (Routledge, 2004, coauthor Stephen Loeb). He is currently working on a book that analyzes the computer automation of the New York Stock Exchange during the period 1955–1975.

Joseph A. Pratt is the NEH-Cullen Professor of History and Business at the University of Houston. He is an energy historian who has published books on a variety of oil- and business-related topics, including histories of the Amoco Corporation and the National Petroleum Council (written with William Becker and William McClenahan Jr.). His most recent publication is *Energy Metropolis* (University of Pittsburgh Press, 2007, coeditor Martin Melosi), a collection of essays on the environmental history of the Houston area. He is currently working on histories of the offshore oil industry and of the Exxon Corporation from 1975 to 2000.

The late **Kenneth L. Sokoloff** was professor of economics at the University of California, Los Angeles, a research associate at the National Bureau of Economic Research, and a fellow of the American Academy of Arts and Sciences. He received his Ph.D. in economics in 1982 from Harvard University, where he studied under Robert Fogel. He taught at UCLA from 1980 until his death in 2007. A renowned economic historian, Sokoloff published numerous articles on the sources of productivity growth and inventive activity in early American industry, on comparative patent systems, on the role of factor endowments and institutions in the differential growth paths of countries in North and South America, and on the development of follower nations such as Taiwan, Korea, and Mexico.

Introduction

Sally H. Clarke, Naomi R. Lamoreaux,
and Steven W. Usselman

"Can capitalism survive? No, I do not think it can." So declared Joseph Schumpeter, the foremost theorist of entrepreneurial innovation, as he watched governments ratchet up their level of economic intervention during the 1930s. Despite the length and severity of the Great Depression, Schumpeter did not believe that the capitalist system would collapse under the weight of its own failure. Rather, the problem, as he saw it, was capitalism's extraordinary success, which perversely but "inevitably" undermined the social institutions that protected it.[1] At the root of Schumpeter's pessimism was the increasing dominance of large-scale managerial enterprises. To be sure, Schumpeter recognized that the modern bureaucratic corporation was itself a major innovation, one that made possible the production of vast quantities of goods at low prices. By "routinizing" and "mechanizing" technological discovery in their in-house research and development (R&D) laboratories, however, these giants had rendered the individual entrepreneur superfluous. In the process they had fatally weakened the social and political forces that prevented capitalism's enemies on the left from increasing the role of government in the economy. Though perhaps well intentioned, the state's growing interventions to mitigate suffering during downturns would in Schumpeter's assessment have disastrous consequences. By protecting inefficient firms from the competition of more innovative enterprises, government would disrupt the process of creative destruction that was the engine of capitalist economic growth.[2]

With the benefit of hindsight, most scholars who wrote during the decades following World War II rejected Schumpeter's pessimistic view of government's increased role in the economy, embracing instead the idea that government intervention provided a stable macroeconomic environment within which

2

innovation could flourish. Some scholars accepted Schumpeter's idea that the large-scale corporation had become the major source of productivity-enhancing ideas, but they did not share his preoccupation with the stifling effect of bureaucracy on entrepreneurship. Instead, they considered the large corporation to be a source of both economic stability and growth. In *The Visible Hand*, for example, Alfred D. Chandler Jr. proclaimed the superiority of the large, vertically integrated, managerially directed enterprise over the entrepreneurial firm. Not only did large firms coordinate more efficiently the flow of resources through the production process, but "once a managerial hierarchy had been formed and had successfully carried out its function of administrative coordination," he declared boldly, "the hierarchy itself became a source of permanence, power, and continued growth."[3] Chandler's work was so influential during the 1970s and 1980s that the idea that managerial hierarchies had replaced entrepreneurs as the driving force of the economy became a central tenet of business history.[4]

Although there were contrary voices, they had relatively little influence on the direction of scholarship until the boom of the 1980s and 1990s reawakened interest in entrepreneurial enterprises.[5] At the same time, the severe setbacks that many giant firms experienced beginning in the 1970s stimulated a new crop of scholars to contest Chandler's core idea that large-scale enterprises were an inherently superior means of coordinating economic activity. Looking at business history from an evolutionary perspective, these scholars argued instead that the advantages of giant, managerial enterprises were specific to the particular historical context in which they arose. As the economic environment changed during the late twentieth century, large "Chandlerian" firms were not readily able to adapt, and new forms of economic coordination moved to the fore.[6]

This reversal of fortune stimulated a parallel reexamination of the role that large firms played in innovation. Amidst the booming affluence of the postwar decades, observers had taken for granted that large corporations would henceforth account for most important new technological discoveries. The experiences during the last couple of decades of the twentieth century cast considerable doubt upon this assumption. While American giants such as Radio Corporation of America (RCA) and the Big Three automakers lost their leadership in mass producing consumer durables, startup firms dominated the highly innovative segments of computing, telecommunications, and biotechnology. Scholars began, as a consequence, to look more critically at in-house research laboratories, trying to understand exactly how they functioned and whether they truly generated a continuous stream of innovations for their parent companies.[7] Scholars also scrutinized the sources of the innovations that large firms successfully exploited. Some emphasized the role played by suppliers.[8] Others examined how consumers influenced managers' approaches to innovation.[9] Still others showed how large firms, in some industries, formed symbiotic relationships with entrepreneurial enterprises. Rather than displacing them, they benefited from a division of labor in which they commercialized the new technological ideas that their smaller competitors developed.[10]

Our intent in *The Challenge of Remaining Innovative* is to offer a sampling of this new work and advance its research agenda. All of the chapters in the volume take as their starting point Schumpeter's (and Chandler's) idea that large firms had become the most important source of innovation in the economy by the early twentieth century, but they part company with their predecessors by emphasizing the complexities and contingent nature of this leading role. Part I homes in on the emergence and operation of centralized corporate research, examining where and why it originated in the opening decades of the century and how it functioned at two prominent firms thereafter. Part II looks beyond individual firms to consider how companies, large and small, have joined together in more informal ways to foster innovation. Part III turns to government and examines how it has sometimes influenced the course of innovation by altering the structure of interactions among firms and between firms and consumers.

Collectively, the chapters in this volume suggest that in order to meet the challenge of remaining innovative, large firms had to learn to manage their organizational boundaries effectively. They had to learn to tap into external sources of knowledge and creativity without undermining their sense of corporate purpose. They had to learn to preserve the advantages that their core capabilities brought them while at the same time accommodating themselves to the shifting complexities of their markets and to changes in the larger economic and political environment in which they operated. And they had to be able to do all this while making sure that the various internal units of their enterprise worked as a coordinated whole.

This sustained exercise in managing boundaries, both within the enterprise and with the outside world, revolved around the flow of information. Virtually all of the studies in this volume emphasize the centrality of information. They portray firms as continually striving both to create new knowledge about technology and markets and to access knowledge being developed in universities, trade associations, government agencies, and even other firms. They also portray firms as monitoring informed thinking about the approaches to innovation that were likely to prove most effective. As many of the chapters show, moreover, a key goal of all this activity was to reduce the uncertainty inherent in innovation and transform it into manageable risk. The chapter authors conceive of managers as confronting a process fraught with difficulty, gambling on

4   a strategy to maintain their firms' lead but trying at the same time to hedge their bets. To meet the challenge of remaining innovative, managers had to look outward as well as inward, and many found this extremely difficult to do. In the remainder of this introduction we elaborate upon this theme of boundary crossing by connecting it explicitly to the book's three parts. Our aim in the sections that follow is to situate the authors' contributions in a larger conceptual framework. More detailed summaries of the individual chapters precede each section of the collection.

WITHIN FIRMS

In the prologue to this volume (Chapter 1), Naomi Lamoreaux and Kenneth Sokoloff use data on patent assignments to trace the movement of inventors inside large firms. Their evidence suggests that by the early twentieth century, inventors were finding it increasingly difficult to raise sufficient resources to pursue independent careers. As a result, employment in a research laboratory was an increasingly attractive alternative, especially for those who had the formal scientific or engineering training that was the ticket to a good job. But why did these jobs exist in the first place? Why did large firms invest in laboratories that would provide employment to so many engineers and scientists?

Of course, it had long been common for firms to have some technological capabilities in-house. Manufacturers who made capital goods needed to have employees who could modify their products to meet the needs of particular customers. These employees were often an important source of improvements and inventions.[11] Similarly, employees whose job it was to keep machinery in good running order often found ways to make it better. In early telegraph offices, for example, operators had responsibility for maintaining their equipment as well as for transmitting messages. Thomas Edison is only the most famous example of an operator who learned about electricity on the job and got his start as an inventor by devising new or improved ways to send information by wire.[12]

Moreover, when firms were founded to exploit particular inventions, the inventor typically played an ongoing role in the business. Sometimes the inventor took charge of running the firm; sometimes he (or, more rarely, she) left the day-to-day operations to someone whose background was better suited to these tasks. Regardless, the inventor usually continued to work on technological development, sometimes hiring additional staff to assist him. With the formation of the Thomson-Houston Electric Company in 1883, for example, the electrical inventor Elihu Thomson handed responsibility for the company's management over to Charles Coffin, so that he could devote his time to inven-

tion in the company's "Model Room." There Thomson supervised a small 5
team of machinists, draftsmen, and other employees who transformed his new
technological ideas into working machines.[13]

Although Thomson-Houston's Model Room was in some ways an incipient
industrial research lab, it functioned very differently. Facilities such as these
were rooted in the assumptions of what Thomas Hughes has called "the golden
age of the independent inventor."[14] They were extensions of the inventor
around whom the company had formed, and they revolved around his needs.
Thomson and his staff did not work closely with colleagues in manufacturing
or in the field, and no one at the firm assumed the role of a research director
who supervised Thomson's labors or attempted to coordinate them with other
functions and other inventors employed by the company. In providing Thomson and a handful of other prolific inventors with steady support to carry out
their work, Thomson-Houston and its successor General Electric were doing
little more than enhancing the ongoing specialization of inventive activity that
many independents pursued on their own. Of course, in return they expected
exclusive rights to the inventors' patents.[15]

Most firms in the late nineteenth century were not willing to go much farther in developing their in-house R&D capabilities. Although Western Union
probably devoted more resources to technological discovery than any other
enterprise in the late nineteenth century, it typically financed its employees'
inventive activity by helping them set up independent businesses. Managers at
Western Union believed that in a time of rapid technological change such spun-
off companies would stay closer to the technical frontier than groups located
within the telegraph company itself. Western Union financed the consolidation
of the innovative partnership of Elisha Gray and Enos Barton with its own
machine shop, for instance, but the resulting enterprise, Western Electric, operated as an independent firm. Western Electric adopted a similar policy. When
Gray began work on his invention of a harmonic telegraph (essentially the telephone), he resigned as superintendent but continued to work in the company's
facilities as an independent inventor.[16] The position of the American Telephone
& Telegraph Company (AT&T) at this time was even more extreme. As T. D.
Lockwood, head of the company's patent department, explained, "I am fully
convinced that it has never, is not now, and never will pay commercially, to keep
an establishment of professional inventors, or of men whose chief business it is
to invent." Instead, AT&T invested in building the capacity to track and assess
inventions generated in the external world. It was only after Theodore N. Vail
became president in 1907 that the company reversed this policy.[17]

The earliest firms to create true in-house R&D programs featuring research
laboratories often did so for idiosyncratic reasons, proceeding on a small scale

6   until they discovered that the labs brought competitive advantages. During the
         1890s, for example, General Electric (GE) faced increasing competition. Its
         basic (Edison) patents on lightbulbs had expired, and inventors elsewhere were
         developing new, more efficient types of filaments. Charles Steinmetz, a con-
         sulting engineer at GE's Schenectady factory, had been trained in Germany
         and believed that American firms would do well to emulate the R&D labs that
         German firms had pioneered. He convinced the company to support a mod-
         est research initiative (the first allocation was only $15,830) to develop an
         improved incandescent light. Although the lab did not immediately succeed
         in its mission (GE had to purchase the technology from German inventors),
         the experiment nonetheless established the value of an in-house R&D facility.
         The lab had provided important services to other parts of the firm by testing
         materials and resolving technical problems. More important, in the process of
         experimenting with different kinds of filaments, company researchers had filed
         a number of minor patents that turned out to be useful—not only defensively,
         by helping the company protect its product line from infringing competitors,
         but also offensively, as bargaining chips in negotiations with rivals. Sales of duc-
         tile tungsten filament bulbs, which the lab claimed credit for developing,
         accounted for between one-third and two-thirds of GE's profits every year
         between the late 1910s and mid-1930s.[18]
             As Kenneth Lipartito details in Chapter 4 in this volume, AT&T's lab started
         in a similar way. Under competitive pressure from new wireless technologies
         (radio) that threatened its control over local voice communications, the com-
         pany focused its energies on building the capacity to provide long-distance serv-
         ice. This strategy depended on the development of an improved amplifier. At
         the urging of J. J. Carty, the head of the company's Mechanical Department,
         Vail agreed to fund a modest in-house laboratory to develop the necessary tech-
         nology. As was the case at GE, the lab failed in this effort, and AT&T had to
         purchase Lee de Forest's patents. But again the research team proved its use-
         fulness. By solving a number of technical problems with de Forest's inventions
         that had to be overcome for the technology to be commercially practicable, it
         made possible the successful inauguration of coast-to-coast telephone service
         in 1915. Moreover, the lab's accumulation of "a thousand and one little patents"
         (in the words of the company's president) helped to keep competitors at bay.[19]
             The research labs had other advantages as well. The knowledge they gen-
         erated internally helped firms monitor external technological developments
         more effectively, reducing uncertainty and enabling them to make reasoned
         assessments of which patents were important to buy and which inventions
         could be safely ignored.[20] Chapter 3, by Paul Miranti, shows how researchers
         at AT&T joined such technical assessments with novel statistical methods that

made it possible to estimate the returns from investing in trunk lines, automatic switching, and other new technologies. As Lipartito emphasizes in his chapter, this arrangement fit the philosophy of new research director Frank Jewett, who would lead Bell Labs from the 1920s into the 1940s. Jewett insisted on situating the Labs at the center of the corporation, firmly planted between the commercial operations and the manufacturing facilities at Western Electric (by this time part of AT&T). Scientists and engineers in the Labs thus remained closely tied to the commercial and technical problems that most concerned AT&T, without growing too close to the immediate practical concerns that occupied engineers tasked with designing exchanges and manufacturing equipment. Personnel in the Labs used their knowledge of emerging developments in the sciences to evaluate the many components that constituted the phone system, always with an eye for raising standards of performance, both technically and economically.

As the advantages of having in-house R&D facilities became apparent, more and more firms developed them. Between 1919 and 1936, U.S. manufacturing firms established over a thousand industrial research laboratories, roughly half of the total number of such facilities founded prior to 1946. The number of scientists employed in research laboratories increased tenfold between 1920 and 1940, from 2,775 to 27,777, and by 1946 it had jumped to 45,941. Once concentrated in fields such as chemistry and electricity, labs appeared in many sectors of the economy, including heavy industry. Automakers and large metals producers such as U.S. Steel, for instance, launched conspicuous ventures in research during the interwar period. Companies in areas such as petroleum and pharmaceuticals, which had not previously tapped emerging scientific expertise in chemicals and chemical engineering, rapidly built significant research programs. Although the total number of labs proliferated, research facilities at the largest firms accounted for the lion's share of activity. When AT&T formally established Bell Labs in 1925, the research facility already had a staff of 3,600 (including nonscientific personnel) and a budget in excess of $12 million. David Mowery and Nathan Rosenberg calculated for the period 1921 to 1946 that between 45 and 61 percent of all industrial research scientists worked in the top decile of laboratories.[21]

Having expanded and proliferated even in the teeth of the Great Depression, the leading labs reached massive proportions in the subsequent three decades, as corporations plowed large chunks of their healthy postwar profits back into them and tapped new sources of funds available from government agencies such as NASA and the Department of Defense. With government footing roughly half the bill, R&D expenditures (in constant dollars) doubled between 1953 and 1960, and the ratio of such expenditures to GNP nearly doubled. By 1962, the

8   number of scientists and engineers employed in industrial research exceeded three hundred thousand.[22] At many corporations, the research labs served as showpieces, occupying elaborate campuses designed by award-winning architects and earning the highest laurels for their scientific accomplishments.[23]

In the heyday of corporate research labs a new generation of scholars, such as Kenneth Arrow and Richard Nelson, wrote papers explaining that the internalization of R&D within firms was the optimal way of organizing technological discovery because it solved information problems that otherwise inhibited the flow of technological knowledge. In the course of building and operating complex manufacturing operations and technological systems, enterprises accumulated distinctive firm-specific knowledge that could not transfer readily across corporate boundaries, and which companies generally did not wish to share in any event. Internalized R&D integrated these crucial proprietary assets in a common corporate entity while also tapping the more generalized knowledge of science, which was becoming essential for sustained improvement.[24]

Certainly the corporate labs could lay claim to a long legacy of accomplishments—radio (GE and RCA), celluloid movie film (Kodak), nylon (Du Pont), the transistor (AT&T), color television (RCA), and electronic computing (IBM)—as well as to a continuous stream of less dramatic improvements. But detailed studies of particular corporations, including several included in this book, show that information did not always flow smoothly within the large organizations that internalized research and development. The melding of newly acquired research capabilities with established functions in design, manufacturing, sales, and maintenance was never simple or easy. Jewett found a way to accomplish this at AT&T, where the paramount importance of the phone system (over which Bell enjoyed a monopoly) helped define common objectives across the firm. Managers at companies such as Du Pont and GE, which looked to develop a range of products for numerous competitive markets, faced a tougher challenge. They had to decide whether to keep research scientists in a centralized facility, where they enjoyed prestige and fellowship like that at Bell Labs, or farm them out to a particular product division or manufacturing facility. Close connections between labs and manufacturing plants were especially important in chemical process technologies and other settings where much of the critical learning occurred in the course of scaling up prototypes developed in the lab. The challenge was all the greater for a firm such as Du Pont, whose chemical plants were located in remote areas far distant from the corporate headquarters in Delaware.[25] More and more firms faced the problem of coordinating across an array of dispersed facilities after World War II, as companies sought to break up massive complexes such as Ford's

River Rouge facility outside Detroit, GE's corporate complex in Schenectady, and IBM's two corporate towns in rural New York.[26]

Even without the complicating factor of geographic remoteness, large companies struggled to align objectives of researchers with the more immediate concerns that occupied production and marketing. At IBM, a company discussed at greater length in Part III of this book, management grew frustrated when groups in the marketing side of the firm refused to incorporate into new products innovations generated at the labs. Marketers did not want to take risks that might alienate established accounts and interrupt the flow of annual commissions. Personnel in manufacturing, fearing downtime and other costs associated with learning, likewise balked when asked to make significant departures from routine that might lead to their failing to meet established standards. Often they dug in by redoubling their efforts with established techniques and procedures, achieving further economies that worked against more radical change. Such behavior repeatedly plagued mass producers such as Ford.[27]

In efforts to overcome such inertia, top management at many companies sought ways to signal clear priorities across the entire organization, often through grand symbolic gestures. David Sarnoff, who as president of RCA led one of the most successful corporate R&D programs for decades, flamboyantly committed his firm to targeted efforts aimed first at developing commercial television and later at adding color to it.[28] Ford executives sent a powerful message when they placed responsibility for building a new engine at a different plant than expected.[29] Tom Watson at IBM, in a move akin to Vail's promise that AT&T would achieve coast-to-coast telephone service in five years, simply declared to the assembled salesforce in 1949 that within a decade all IBM equipment would be built from electronics. Later he sent an edict that all machines would use transistors.[30] Managers at NCR and Polaroid resorted to similar techniques, not always successfully, when trying to drag those firms into the digital age.[31]

Moreover, although in-house R&D may have solved some important information problems that had inhibited technological change, it created others. Firms typically required employees to sign contracts in which they promised to assign all their patents to the company. Then, in order to reduce the negative effect that this requirement might have on their incentive to invent, they often offered employees monetary rewards for inventions leading to patents. Such bonuses, however, distorted employees' incentives. As one of AT&T's executives testified in a congressional hearing, this system encouraged employees to work "at counterpoints to their own associates," creating a situation in which "men would not work with each other, they would not confide with

10   each other, yet the problem which was before us was a problem which required team action."[32] Even worse, bonuses encouraged employees to put their own interests before those of the company by seeking multiple minor patents in cases in which a single broad one would better protect the company's intellectual property or by shirking on projects that would not result in patentable inventions. To counter these perverse incentives, companies such as AT&T and GE stopped paying bonuses for patents by the 1930s. Instead, they made salary and promotion decisions hinge on a number of different factors, of which patents obtained was only one. But this change in the structure of compensation meant that firms faced new difficulties both in measuring the output of their research employees and in creating credible incentives to encourage their productivity.[33]

These problems were compounded when firms hired scientists out of academia. The first generation of research directors, such as Jewett at Bell Labs, struggled continually to accommodate scientists who had grown accustomed to the freedoms of academic inquiry and were inculcated with the values of a "pure science" ideology. Willis Whitney of GE, who himself had left the academy reluctantly, let his best researchers publish more widely than his bosses preferred and extended them greater latitude in their choice of problems. The challenges were so stressful that Whitney suffered repeated bouts of exhaustion and eventually resigned.[34] At Du Pont, research managers attempting to build up capabilities in fields such as organic chemistry often met resistance from respected academics, who refused to send their best graduates into industry. Such resistance subsided during the interwar decades, as scientists working in corporate laboratories received Nobel Prizes and published papers of distinction, even while achieving notable commercial successes such as nylon. In many respects, however, such successes only complicated the tasks facing corporate employers. They raised unrealistic expectations among researchers about the freedom possible in corporate labs, at a time when these corporations needed to involve researchers more deeply in the follow-on activities necessary to sustain their commercial triumphs.[35]

Faced with these obstacles to bringing researchers into alignment with other parts of the organization, managers at many large companies resorted from time to time to a project-based approach. Taking an end run around established operating units, they joined personnel from the laboratories to teams that included their own independent sales and marketing capabilities. Sarnoff utilized the technique at RCA. IBM would later develop its personal computer in this way.[36] At least some companies, such as Corning, routinely pursued this more entrepreneurial approach. As Margaret Graham indicates in Chapter 2, researchers at Corning often worked with small groups of colleagues in its manufacturing facilities in highly targeted efforts to develop new products. Typically, these teams bypassed established marketing operations within the company by partnering with large first-users, such as AT&T in the case of fiber optics, or by working with fashion intermediaries, as in the case of Corning Ware cookery. Even Corning, however, drifted in the 1950s and 1960s toward tying research more narrowly to established product lines.

The Corning experience reflected broader trends. Across the economy, corporate research facilities foundered as firms encountered greater challenges in trying to manage their boundaries effectively. At firms such as AT&T, RCA, and Du Pont, researchers ensconced in attractive campuses grew ever more remote from commercial operations, while management held out hope they might repeat spectacular successes such as nylon or color television. Some companies, such as IBM, rewarded their most accomplished researchers with internal fellowships that freed them from all obligations. Biochemist Roy Vagelos, beginning a career that would ultimately take him to the helm of the pharmaceutical giant Merck, encountered a research environment more rarified (and comfortable) than that in any university.[37] Perhaps inevitably, firms with long records of success often exhibited a "not invented here syndrome," which denigrated techniques developed elsewhere. But companies that were willing to learn from other firms met resistance from government. A resurgence in federal antitrust policy, which had helped stimulate the internalization of research at the start of the century by blocking large companies from acquiring technologies through merger and pooling, now dashed attempts by firms such as Corning to enter alliances with corporate partners in innovation.[38]

By the late 1960s and early 1970s, a number of the firms that had pioneered in corporate research had lost the technical initiative, sometimes to smaller companies and sometimes to foreign competition. While AT&T became mired in an ill-fated effort to develop a picture phone, MCI exploited microwave technology to make serious inroads in the market for long distance. Du Pont poured vast sums into Corfam, while letting a former employee reap much larger returns from GoreTex. RCA cast its lot with the videodisc, only to see Japanese companies sweep the market for videotape recorders and ultimately displace RCA in the market for consumer electronics. In computing, IBM lost its leadership in high-performance machines to tiny Control Data in the mid-1960s and a decade later scrambled with only mixed success to catch up with Apple and other producers of personal computers. Despite their massive investments in research and unimpeachable accomplishments in science, these established giants came off as technically conservative organizations that could not keep up with the most innovative developments in their fields. By the early 1980s, large

12   firms seeking to recapture the initiative would look to redraw their corporate boundaries in ways that emulated and engaged these smaller companies.[39]

AMONG FIRMS

Redrawing corporate boundaries for the most part meant making them more porous. Large firms could regain some of the ground they had lost to smaller, more entrepreneurial enterprises by encouraging key employees to build connections with knowledgeable people outside the firm. In that way they could remain alert to developments that occurred elsewhere in the economy, position themselves to buy important new technologies or lure people with key expertise inside the firm, and also be well placed to partner with firms with the requisite know-how or financial resources. But building ties with outsiders could also pose problems for firms. Information necessarily flows two ways along these lines of connection, and firms risk losing control of proprietary technologies they deem vital for competitive advantage. They also risk losing valuable personnel.

Nowhere were the risks and rewards of such ties more apparent than in the region of northern California popularly known as Silicon Valley. Many of the electronics firms that populated this region in the 1970s shared a common ancestry. Their founders had either worked for the region's pioneering Fairchild Semiconductor Company or for firms founded by people who had worked for that enterprise. Fairchild itself was the product of a defection. A group of engineers known in Silicon Valley lore as the "traitorous eight" had left another firm, Shockley Semiconductor, to found their own enterprise with backing from the Fairchild Camera Company. Problems plagued the new firm, however, and the eight founders (and many of their employees) gradually left to start other companies. This pattern of defections continued with employees quitting their jobs to join new ventures, often in conjunction with colleagues from competing firms.[40]

At first firms fought these departures, instituting lawsuits against employees who formed or joined competing firms. Trade secrets cases are difficult to win, however, and under California law contracts that restrict the ability of employees to quit and join new firms are not enforceable. Firms in effect made a virtue of necessity and concentrated instead on maintaining good relations with departing employees, who often continued to hold ownership stakes in their former employers. These ongoing contacts provided a basis for the cooperative inter-firm behavior for which Silicon Valley is known, making it easier for companies to organize joint ventures, subcontract with each other, and even share information.[41]

Although Silicon Valley spawned several firms that grew large, and though some important enterprises headquartered elsewhere have maintained branches in the region, large firms generally found the high levels of employee mobility that characterize the region difficult to manage.[42] That may be one reason why the industry evolved so differently in Boston's "Route 128" region, another prime location for computer makers in the 1970s. Firms in the Boston area tended to be larger and more vertically integrated than those in Silicon Valley, and, as AnnaLee Saxenian has argued, less technologically dynamic as a result.[43] Massachusetts strictly enforces covenants not to compete, so it is much more difficult for employees in that state to leave their jobs and start new firms in related businesses. When employees of Massachusetts firms defect, therefore, they tend to move out of the region and especially to head for California, where judges refuse to enforce do-not-compete contracts signed in Massachusetts.[44]

Another important factor that helped to make Silicon Valley such a fertile technology region was that information flowed easily between local enterprises and Stanford University. Under the leadership of Frederick Terman, Stanford's dean of engineering and later provost, Stanford assisted local businesses in multiple ways: training their employees, consulting on research projects, developing technologies that could be commercialized by startups, and fostering networks among students and alumni.[45] Stephen Adams, in Chapter 5 in this volume, examines Terman's educational ideas, the role he played in making Stanford's engineering school a source of the region's dynamism, and the difficulties he faced when he tried to remake the whole university in the engineering school's image. After he retired from Stanford, Terman was much sought after to help build educational institutions that could serve as high-tech anchors in other regions of the country. As Adams discusses, the plans he drew up for a highly specialized university in New Jersey failed because the large-scale enterprises in that area, particularly pharmaceutical companies, ultimately were unwilling to put up the necessary funds. Similar efforts in other parts of the United States also failed.

Stuart Leslie and Robert Kargon have argued that one important reason for these failures was that large vertically integrated companies with strong in-house R&D labs rarely developed the culture of cooperation that enabled Valley firms to get so much from Stanford.[46] This is not to say that such firms did not form ties with researchers at academic institutions. Before the growth of government spending on higher education and research after World War II, large-scale enterprises were an important source of research funds for the nation's colleges and universities. That support continued in the postwar period (and more recently has again increased in relative importance), but it has

14   always had an episodic quality.[47] In some periods large firms have relied more or less exclusively on in-house expertise. In others, however, disruptive technological changes that occurred outside their bounds could send them scurrying to university researchers who possessed the requisite knowledge and skills. For example, when Merck found that recombinant DNA techniques made its existing capabilities in vaccine development obsolete, it developed research partnerships with academic scientists, as well as with some of the bio-tech start-ups that were forming around universities.[48]

Large companies sometimes shared technological information with each other, but these exchanges tended to occur in more contractually bounded, self-interested ways than in Silicon Valley. For example, in the early twentieth century automakers formed a patent pool in response to an attempt by the owner of a critical technology to hold them up. Their agreement mandated that producers cross-license all the patents they obtained to each other. That requirement soon came to seem costly for firms that invested more than the rest in technological discovery, and the pool broke down.[49] To give another example, large firms such as GE, Westinghouse, RCA, and AT&T acquired thickets of patents that had the potential to obstruct each other's core business activities. To resolve the stalemate they entered into cross-licensing agreements that had the added benefit of protecting them from external competition.[50]

Large firms were able to cooperate in a more sustained way under specific circumstances. For example, railroads had to be able to move their rolling stock over each other's track and couple their cars together. If one railroad adopted a technological improvement on its own, it could impede the movement of traffic. As a consequence, railroad managers were in continuous communication about technological innovations and participated actively in trade associations and engineering societies in which new devices were vetted.[51] Large firms also banded together when they faced a common danger or had mutual interests to promote, for example by forming trade associations to lobby government for protective tariffs or to oppose regulatory initiatives they thought would raise their costs or impede their activities.[52] In addition, they frequently joined forces in attempts to limit competition. Such efforts were more common and more overt before the passage of the Sherman Antitrust Act made price-fixing agreements illegal, but they continued long afterward. By collecting information about the output and sales of constituent firms, trade associations made it possible for firms informally to coordinate their pricing decisions in an environment in which formal coordination was illegal.[53] Of course, this kind of cooperation may have done more harm than good to the economy.

Other kinds of dangers could lead businesses to cooperate in a more socially beneficial way. As Joseph Pratt shows in Chapter 6, a series of devastating hurricanes in the 1960s literally "frightened" executives of the big oil companies who drilled off the Gulf Coast into collective action. The firms' geographic proximity to one another facilitated this response. Not only did these enterprises face a common danger, but their executives knew each other well and had long been sharing information in an uncoordinated fashion about the design of drilling platforms. The meetings that followed the big hurricanes of the 1960s were spontaneous responses to the disasters. No organization called them, but they provided the basis for a more coordinated effort under the auspices of the industry's trade association, the American Petroleum Institute, whose new Offshore Committee spearheaded the development of new design standards for offshore rigs with the aim of reducing the damage from future storms.

Although the oil companies that drilled in the Gulf of Mexico were more geographically clustered than is typically the case outside the natural-resource sector, they also benefited from a channel for the transmission of technological information that exists in many industries—common contractors and suppliers. In her study of the computer industry, for example, JoAnne Yates showed that in order to make sales, firms such as IBM had to help insurance companies solve their data-processing problems. They then passed the knowledge they acquired from these experiences on to other customers that bought computers.[54] Indeed, supplying this kind of information ultimately proved to be such a profitable business in its own right that in recent years firms such as IBM have increasingly focused on the provision of business services—that is, they have transformed themselves into consulting firms.

In today's economy consulting firms are one of the most important mechanisms for the transmission of information about state-of-the-art management practices among large firms, but the industry can be traced back at least to the systematic management movement of the late nineteenth and early twentieth centuries. In the decades following the Civil War there was an explosion of interest in developing techniques for managing the flow of production more efficiently, and increasing numbers of engineers went into the business of helping manufacturers streamline their operations.[55] The most famous was Frederick Winslow Taylor, who synthesized the movement's key ideas and successfully marketed them under the rubric of scientific management. Taylor's success attracted imitators, and the ranks of consulting firms steadily grew over the first third of the twentieth century. Although scientific management is most associated in the public mind with time and motion studies, most of the consultants' efforts went in other directions: reorganizing the flow of production to eliminate bottlenecks, instituting wage-incentive systems to increase labor productivity, deploying cost-accounting techniques to pinpoint inefficiencies, and improving communications throughout the enterprise.[56]

16   The biggest consulting firms today have a different history, however. As Christopher McKenna has shown, rather than evolving out of scientific management, they developed in response to the regulatory initiatives of the New Deal—particularly the Glass-Steagall and Securities and Exchange Acts. On the one hand, Glass-Steagall prohibited banks from performing the consulting services they had previously provided as part of the underwriting process. On the other, the Securities and Exchange Act specified that there had to be a due-diligence investigation by an independent party before a company could issue securities. Consulting firms emerged to provide these services in banks' stead.[57]

The consulting business mushroomed in the post–World War II era, when firms expanded both the scope of their services and the range of their clients. For example, in 1953 Arthur Andersen & Company oversaw the installation of a Univac computer at General Electric's Louisville, Kentucky, factory to handle the payroll and keep track of inventory—one of the earliest instances when a computer was used specifically for business purposes.[58] Around the same time, another leading firm, Cresap, McCormick and Paget, moved into the nonprofit sector, advising organizations as diverse as Yale University, the Republican Party, and the Metropolitan Museum of Art on how to improve their managerial controls.[59] During the postwar period, moreover, the consulting industry developed new links to universities, connecting business enterprises to the academic world in important but indirect ways. Business schools became important feeder institutions for consulting firms as well as for the companies they advised. In addition, as McKenna's chapter in this volume shows, information about cutting-edge managerial practices flowed in both directions between the consultancies and the schools' research-oriented faculty.

Although the institutionalization of these triangular relationships between businesses, consulting firms, and business schools provided the former with a way of securing information about advances in managerial practices without all the complications that would have been involved in learning directly from the pioneering firms, it also meant that information was increasingly subject to distorting cognitive biases. McKenna uses the series of reports and cases analyzing Honda's conquest of the American motorcycle market to show that as members of the academic-consulting complex built careers by promoting particular ideas about winning businesses' strategies, their analyses "accumulated like successive sedimentary layers of silt" that obstructed their ability to see what really happened. Just as entrepreneurs' commitment to their initial technological breakthroughs can make their businesses less innovative over time, consultants' (or academics') commitment to their own interpretations can blind them to what is truly innovative about a new successful business. And, of course, this blindness means that they will be less effective in keeping their clients informed about innovations outside their bounds—less effective, in other words, in helping them manage their boundaries creatively.[60]

FIRMS AND THE STATE

In the final section of the book we shift our attention to the influence of the federal government on the process of innovation. During the 1970s Schumpeter's view of the harmful consequences of government intervention came back into vogue, as critics blamed misguided Keynesian macroeconomic policies and burdensome regulation for the stagflation of that decade. Bolstered by complaints from both the left and the right that "captured" government agencies no longer operated in the interest of the public at large, support for deregulation soared.[61] In the 1990s, however, the resurgence of economic growth banished fears of stagnation, and scholars began to reassess this negative view of the state, with some going so far as to find cases in which state actions fostered economic development.[62] The chapters in this book continue this process of reassessment, acknowledging that the state has promoted economic development while also recognizing the costs and limits of government intervention.

The case against state involvement is perhaps strongest when applied to the topic of price regulation. According to standard neoclassical principles, when regulators set a price that is higher or lower than the one that would prevail in the market, they misallocate resources: prices that are too high induce producers to supply too much of a product relative to demand, and prices that are too low, too little.[63] In the 1970s the economist Alfred E. Kahn applied this simple theory to the case of the airline industry. In his view price regulation, in combination with government-imposed restrictions on entry, had kept airfares artificially high. As the head of the Civil Aeronautics Board in 1977 and 1978, Kahn worked to dismantle the controls on entry and pricing, turning what historian Thomas McCraw called a "non-competitive" industry into a "competitive" one.[64] Fares plummeted, benefiting consumers and increasing air travel. Of course, not everyone gained from the change. As Kahn had anticipated, some poorly managed airlines failed, and nearly all unions experienced losses in benefits for their members. What the deregulators failed adequately to appreciate, however, was the extent to which crowded airplanes and declining levels of service would undermine the quality of the airlines' product, making flying much less pleasant than it had been before.[65]

An exclusive focus on static comparisons, however, can overstate the relative gains from deregulation even within a simple neoclassical comparative static framework. Looking at New Deal farm regulation from a dynamic perspective, Sally Clarke provided a more nuanced view of price regulation. There

18   is no question that the minimum prices set by the Commodity Credit Corporation during the 1930s interfered with markets and misallocated resources in the short term. But taking a long-term perspective, Clarke found that New Deal regulation provided farmers with security against depressed prices and new credit instruments that they used to invest in technology and achieve new gains in productivity. Before the coming of New Deal regulation, farmers tried to protect themselves from the consequences of volatile prices for crops such as corn by conserving cash outlays and delaying purchases of expensive machinery such as tractors. As a result of New Deal price regulation and new federally sponsored credit policies, farmers had less to fear from low prices and could employ loans with lower interest rates and longer maturities to invest in farm equipment. Clarke found that in the Corn Belt farmers increased investments in tractors and raised levels of labor productivity despite the Great Depression. This trend persisted after the Depression. From the 1930s through the 1970s the agricultural sector maintained one of the highest rates of productivity growth among the various parts of the U.S. economy. Farmers' gains came not only from higher crop yields but also from the sustained increases in labor productivity that their large capital investments made possible.[66]

Another common target for critics of government regulation in the 1970s was antitrust policy. Big businesses were big, critics claimed, because they were more efficient—because they were able to exploit economies of scale and scope.[67] Penalizing them for their efficiency was not only wrongheaded but damaging to the economy as a whole. Moreover, antitrust policy was intellectually incoherent, based as it was on a set of mutually incompatible legal traditions and the confused economic reasoning of federal judges. The result, according to Robert Bork, was a policy "at war with itself," an economically irrational set of rules that "significantly impair[ed] both competition and the ability of the economy to produce goods and services efficiently."[68]

Again, recent literature offers a more complex interpretation of antitrust policy. Naomi Lamoreaux, for instance, found that there were problems with competing legal traditions in the years that followed the passage of the Sherman Antitrust Act in 1890. In particular, federal antitrust law initially conflicted with the states' well-established powers to charter and hence regulate corporations. These difficulties were largely resolved when the U.S. Supreme Court articulated the "rule of reason" in 1911, enabling the Court to break up Standard Oil, American Tobacco, and several other major consolidations. As is often the case, however, solving one problem led to others. In this case, large firms such as U.S. Steel soon learned to evade the tests the Court used to gauge anticompetitive behavior. The unfortunate consequence, Lamoreaux argued, was that regulators found it easier to prosecute aggressive firms in industries in which there was vigorous competition than dominant firms in industries controlled by tight oligopolies.[69]

The most obvious case of an aggressive firm targeted by federal authorities was IBM. In 1967, in the wake of a dramatic episode in which the computer giant "bet the company" by replacing virtually its entire product line with machines built from solid state components produced in a new in-house facility, the Department of Justice launched an investigation into IBM's business practices. The resultant suit stretched from January 1969 until January 1982, when William Baxter of the Reagan Administration's Justice Department negotiated a settlement that dismissed the case as "without merit." The judgment came down on the very same day government announced it would break up AT&T, launching one of the most dramatic episodes of boundary redrawing in the annals of American business. Together, these verdicts appeared to vindicate "Chicago school" critics such as Bork.[70] AT&T, a monopoly created by government, at last met its demise, opening telephony to competition of a sort deregulation had brought to the airlines. IBM, whose defense rested upon many tenets of the Chicago school, emerged as an exemplar of dynamic competition that had earned its large (but diminishing) market share.[71] When IBM's fortunes subsequently declined precipitously, it only confirmed for critics that the forces of competition were acting upon the firm all along, without need of government interference.

In Chapter 8, Steven Usselman refines this assessment and offers an alternative evaluation of the evolving history of antitrust across the twentieth century. Situating the 1969 case in a longer history of encounters between IBM and the Department of Justice stretching from the 1930s through recent actions against Microsoft, Usselman detects an underlying coherence in antitrust activity. Antitrust suits such as those against IBM provided a forum in which government, business, and an expanding community of lawyers and economists attempted to comprehend the changing nature of competition and the shifting dynamics of innovation. Some of the interventions pursued by government and by private litigants in antitrust suits addressed the very same concerns that occupied management at large firms such as IBM, who worried their large bureaucratic organizations might grow too removed from market inputs and suffer from disabling rigidities. While admitting that the forces of competition would often have exacted a price for such inefficiency even in the absence of government action, Usselman's portrait lends credence to other recent scholars who allow that antitrust retains some utility even in conditions of dynamic innovation.[72] At the very least, such work demonstrates that government has actively engaged—and will likely continue to engage—the process of boundary drawing that this book sees as a defining feature of the struggle to remain innovative.

Although price regulation and antitrust are perhaps the most conspicuous areas of government intervention in the economy, the state has also contributed to the process of innovation by funding scientific research. In the farm sector this role can be traced back to the first half of the nineteenth century, when the federal government sponsored experimentation with new seed varieties and cultivation techniques under the auspices of the U.S. Patent Office. One of the main responsibilities of the new Department of Agriculture, established in 1862, was to take over the Patent Office's programs and expand them. The Morrill Land Grant Act of the same year granted thirty thousand acres of federal land to the states for each senator and representative they had in Congress to finance the construction of agricultural and mechanical colleges. With the Hatch Act of 1887, Congress funded a system of experiment stations to conduct applied agricultural research, and in 1914 the Smith-Lever Act established the extension service, ensuring that farmers in each county would have an agent to help them make practical use of researchers' findings as well as report farmers' new problems to researchers.[73]

During the late nineteenth and early twentieth centuries farmers benefited to a great degree from government's support of agricultural research. Since its inception the federal system has played an instrumental role in the development of agricultural sciences such as agronomy and animal pathology while at the same time tailoring general knowledge to the specific problems faced by farmers' individual states and regions. It also provided critical support to private companies working on hybrid seeds or improving capital equipment such as tractors and mechanical pickers.[74] As Alan Olmstead and Paul Rhode have shown, the complex of agricultural institutions that the federal government funded made it possible for farmers to move out on the western prairies and plains during the second half of the nineteenth century. Without the new seed varieties that government-funded researchers supplied, there would have been much more significant crop failures in the harsh environmental conditions of the West. Indeed, in the absence of this stream of biological innovation, Olmstead and Rhode estimate, wheat yields would have been at least a third lower in 1909 than they actually were.[75] In the twentieth century the gains from agricultural research have been even greater. Zvi Griliches estimated the social rate of return on research on hybrid corn to be at least 700 percent over the period 1910 to 1955.[76]

Of course, the great increase in federal spending for research came in the aftermath of World War II. According to estimates compiled by Mowery and Rosenberg, the federal government accounted for slightly more than half of all R&D outlays in the U.S. economy in the early 1950s and nearly two-thirds during the 1960s.[77] Though these figures likely neglect the vast funds expended

by private industry on product development and process improvement, they
do not include support for innovation in the form of direct government purchases of high-tech products. Much of government's largess went to large firms
in the aerospace and armaments industries, but federal funds also helped lay
the foundations for innovation in highly entrepreneurial regions such as Silicon Valley. As Stuart Leslie has emphasized, "missing in virtually every account
of freewheeling entrepreneurs and visionary venture capitalists" is due recognition of "the military's role, intentional or otherwise, in creating and sustaining Silicon Valley." Defense contracts pumped billions annually into Valley
stalwarts such as Lockheed-Martin and Ford Aerospace, fueling rapid expansion of a vast engineering workforce that nurtured the region's innovative hob-
byist community. During the early 1960s, when Fairchild Semiconductor was
spinning off the Valley's first generation of semiconductor enterprises, the
Polaris and Minuteman missile programs for a brief time absorbed almost all
the integrated circuits that the "Fairchildren" produced, before the startups
found more lucrative commercial markets.[78]

Even the wave of innovation associated with the rise of networked computing and the dot.com boom, a phenomenon clearly driven primarily by private
capital and initiative, was seeded by a tiny military agency known as the Information Processing Techniques Office (IPTO). As Arthur Norberg and Judy
O'Neill have explained, IPTO was run by an astute group of scientists who
used government funds to provide seed money for highly risky undertakings in
new areas of computer science, such as time sharing, graphics, artificial intelligence, and, most famously, the ARPANET, which grew into the Internet.[79]
Key participants in these programs coalesced at Xerox's Palo Alto Research
Center before leaving to found firms such as Cisco and 3Com and contribute
significantly to Apple and Oracle.[80] Beyond these specific contributions, the
"centers of excellence" that IPTO funded in academia became many of today's
top-flight computer science departments. In this way IPTO created the institutional foundation for scholarly research in a wide range of additional computer fields.[81]

Beyond funding research and education, government has shaped innovation
through its efforts to address the harmful consequences that many technologies
have caused members of society. By the turn of the twentieth century, governments at all levels had already gone a long way toward protecting people and
property from the many dangers associated with railroading. Legislation required
carriers to fence tracks and install crossing gates, apply fire-preventing spark
arresters to their locomotives, and equip cars with an array of safety appliances.[82]
Some cities demanded that railroads eliminate irritating coal smoke by converting lines in tunnels and densely populated areas to electric traction. Such

22   measures broadened in the 1910s and 1920s into comprehensive efforts to cleanse urban spaces of smoke and other unsightly and unhealthy residues of the nation's massive commitment to industrial technology.[83] Interventions of this sort grew increasingly commonplace during the decades after World War II, as state and federal agencies asserted ever more influence over matters such as workplace safety and the environment.[84] Though businesses often complained that such activity imposed burdens and constraints that impeded their freedom to innovate, the new regulatory arrangements also spurred many technical developments. Catalytic converters developed by Corning, for instance, stimulated a return to good fortune at that firm while also enabling automakers to meet the Clean Air Act's stricter emissions standards and dramatically improve air quality in the Southland and other regions.[85]

Sally Clarke has explored the relationship between social costs and innovation in the automobile industry. Early autos were rudimentary, which is to say, defective, and when the nation's first car buyers purchased these crude devices, they assumed considerable risk of financial losses and physical injuries. As the market for automobiles grew in size, manufacturers voluntarily took some steps to improve the safety of their vehicles in order to secure consumers' repeat purchases. Their voluntary efforts were only a partial solution, however, because manufacturers faced a trade-off between introducing safety measures and pursuing other goals, such as reducing production costs. Through a regulatory web of private and public entities, managers were forced to pay more attention to the safety of their vehicles. Among private organizations, the Society of Automotive Engineers set industry standards, and the Underwriters' Laboratories, which protected insurance interests, rated key components that could be a fire or safety risk.[86] In the public sector, Judge Benjamin Cardozo's landmark product-liability ruling, *MacPherson* v. *Buick* (N.Y. 1916), put pressure on firms to improve their testing, as did state regulatory agencies. Government regulators were never able to eliminate the risk of injury completely, but the web of regulatory oversight did compel manufacturers to make their products safer.[87]

Whereas nearly all of the contributors to this volume, and indeed the majority of business scholars, have focused their attention on the creative process of innovation, it should be noted that the state has also shaped the social context in which different Americans acquired or failed to acquire corporations' innovative products. In other words, the state has been directly involved in the question, Who ultimately is the beneficiary of innovation in the economy? Many social and political historians have been critical of the resulting distribution. In the area of medical care, for instance, some historians have demonstrated that private corporations sought to limit government provision of health care. To the extent that they have succeeded, fewer working Americans acquired the insurance needed to benefit from improved medical technology.[88] In the important case of the GI Bill, Lizabeth Cohen has found that its benefits were not broadly distributed but were enjoyed more narrowly by white males from typically middle-class backgrounds. By implication, such men were more likely to benefit from the education subsidies in the GI Bill and thus be able to acquire better, higher-paying jobs.[89]

Many scholars have with good reason focused their attention on job discrimination, but credit discrimination also proved debilitating because credit had become so important to families during the twentieth century.[90] For instance, as home ownership increased from 49 percent of white households in 1900 to 67 percent in 1970 and from 24 percent of African American families in 1900 to 41 percent in 1970, the overall share of homes with mortgages doubled from 32 percent in 1900 to 61 percent in 1970.[91] Similarly, as the percentage of families owning at least one automobile rose from 54 percent in 1948 to 82 percent in 1970, the share of households using installment credit to purchase cars jumped from 35 percent in 1947 to 53 percent in 1970. Overall, consumers' installment debts rose more than tenfold from $40 billion (in inflation-adjusted 2005 dollars) in 1929 to $521 billion in 1969.[92]

Although credit became increasingly important, not all families enjoyed equal access to home mortgages or other types of financing. Discriminatory attitudes were reflected in the administration of some federal policies such as the practice of redlining mortgages. Beginning in the mid-1930s, the Home Owners' Loan Corporation (HOLC) coded neighborhoods with different risk ratings and included homeowners' race and ethnicity in its definition of risk; areas with large numbers of African American homes often were coded red, the riskiest category on the government scale. As other public and private lenders engaged in similar policies, they effectively limited African Americans' access to mortgages and diminished their living standards.[93] Sally Clarke's chapter in this book provides another example of how unequal access to credit resources limited Americans' ability to benefit from corporate innovation. Looking at one important consumer good, automobiles, she finds that the development of the market after World War II hinged on access to more lenient terms for installment credit. The Federal Reserve Board intervened in credit markets first to liberalize terms during the 1950s and then, during the 1970s, to root out unfair lending practices. The Fed's actions came as consumers, especially women who had experienced past discrimination, pressed Congress to pass the Equal Credit Opportunity Act.[94]

CONCLUSION

In taking steps to end discriminatory practices in the market for automobile loans, the federal government behaved in a very different fashion than Joseph

24   Schumpeter feared when he raised alarms about the future of capitalism during the 1930s. American history did not build toward a great domestic confrontation between capitalism and socialism. Instead, the nation's democratic politics fostered pursuit of far less grandiose objectives. Government policymakers typically showed greater concern with trying to make markets work more effectively than with bypassing them entirely. Measures such as credit reform looked to broaden participation in markets. Compulsory licensing of patents and other antitrust initiatives were means of restricting the reach of large firms and fostering more competitive markets. Unlike many Western European nations, the United States generally refused to anoint "national champions" who would assume responsibility for computing and other technologies deemed vital to national well-being.[95] Americans divided authority among multiple levels of government, and they subjected virtually all public agencies to perpetual legal review and tests of fairness.[96] When, during the 1970s and 1980s, they focused their attention on the shortcomings of regulation, the federal government dissolved publicly protected monopolies and cartels such as AT&T and the airlines. Americans looked to preserve the competitive business system, not particular businesses.

      Government policy sometimes stalled but sometimes assisted the Schumpeterian process of creative destruction. Regardless of size, businesses in the United States operated in a highly fluid environment, subject to inputs and pressures from many quarters. To confront the persistent challenge of remaining innovative, managers of large-scale enterprises needed to develop workable organizations and situate them effectively within shifting webs of institutions through which information circulated widely and rapidly. This complex undertaking demanded that they pay continual attention to organizational boundaries, both those among various units of their firm and those with outside entities such as customers, universities, and potential partners in industry, as well as government agencies. Managers searched for ways to read the external environment and signal changes in priority clearly to that outside world and across their own organizations. They had to impose sufficient control upon their innovative activities so that they might render the uncertainties inherent to innovation more manageable and reduce them to assessable risks, yet they had to avoid the temptation to settle into routines, become too insular, and grow captive to their own success. Above all, they had to acquire skills and experiences to weather crises. In an environment subject to the incessant pressures of both market competition and political democracy, no single approach or arrangement held sway for long.

      The chapters that follow offer rich insights into a diverse array of such efforts to meet the challenge of remaining innovative. They suggest why firms and inventors in the early decades of the century chose to form tight alliances, often through internalized programs of corporate research. They identify significant variations in the ways those firms linked researchers with other components of their operations and with outside partners, and they suggest why many of those prominent firms ultimately grew insular and encountered crises, often in the wake of extraordinary prosperity. The chapters document the increasing prominence of loose networks and distinctive regional clusters, often facilitated by intermediaries such as universities, consulting firms, and government agencies. They tell us how such arrangements altered supply chains, facilitated information flows, and helped manage the uncertainties of innovation, while also demonstrating that they were not universal panaceas, readily adoptable in any context or resolving the challenges of remaining innovative for all time. For in the final analysis, the historical studies assembled here point to no single, transcendent path that those looking to sustain innovation might best pursue. Rather, they suggest that the ability of firms to meet that challenge, individually and collectively, revolved to a considerable degree around their perpetual efforts to remake themselves.

NOTES

1. Joseph A. Schumpeter, *Capitalism, Socialism and Democracy*, 3rd ed. (New York: Harper & Row, 1950), 61. On Schumpeter's view of the impending threat of socialism, see Thomas K. McCraw, *Prophet of Innovation: Joseph Schumpeter and Creative Destruction* (Cambridge, Mass.: Harvard University Press, 2007), 347–74.

2. Schumpeter, *Capitalism, Socialism and Democracy*, 131–42.

3. Alfred D. Chandler Jr., *The Visible Hand: The Managerial Revolution in American Business* (Cambridge, Mass.: Harvard University Press, 1977), 8. Chandler noted that most managers accepted the role of the federal government in "maintaining aggregate demand" by the 1950s. Chandler, *The Visible Hand*, 496.

4. See Louis Galambos, "Technology, Political Economy, and Professionalization: Central Themes of the Organizational Synthesis," *Business History Review* 57 (Winter 1983): 471–93. This idea has been pushed furthest in the work of William Lazonick. See, for example, *Business Organization and the Myth of the Market Economy* (New York: Cambridge University Press, 1991).

5. The most insistent contrary voice was that of Philip Scranton. See *Proprietary Capitalism: The Textile Manufacture at Philadelphia* (New York: Cambridge University Press, 1983); *Figured Tapestry: Production, Markets, and Power in Philadelphia Textiles, 1885–1941* (New York: Cambridge University Press, 1989); and *Endless Novelty: Specialty Production and American Industrialization, 1865–1925* (Princeton: Princeton University Press, 1997). But see also Michael J. Piore and Charles F. Sabel, *The Second Industrial Divide: Possibilities for Prosperity* (New York: Basic Books, 1984); and Charles F. Sabel and Jonathan Zeitlin, eds., *World of Possibilities: Flexibility and Mass Production in Western Industrialization* (Cambridge, U.K.: Cambridge University Press, 1997).

6. See Richard N. Langlois, "The Vanishing Hand: The Changing Dynamics of Industrial Capitalism," *Industrial and Corporate Change* 12 (Apr. 2003): 351–85; and Naomi R. Lamoreaux, Daniel M. G. Raff, and Peter Temin, "Beyond Markets and Hierarchies: Toward a New Synthesis of American Business History," *American Historical Review* 108 (Apr. 2003): 404–33. See also "Symposium: Framing Business History," to which these authors as well as Charles F. Sabel and Jonathan Zeitlin contributed, in *Enterprise & Society* 5 (Sep. 2004): 353–403. All of the scholars involved in this revisionist effort have been deeply influenced by the evolutionary economics of Richard R. Nelson and Sidney G. Winter. See *An Evolutionary Theory of Economic Change* (Cambridge, Mass.: Harvard University Press, 1982).

7. The problems as well as the achievements of corporate R&D are highlighted in David A. Hounshell and John Kenly Smith Jr., *Science and Corporate Strategy: Du Pont R&D, 1902–1980* (New York: Cambridge University Press, 1988); Margaret B. W. Graham, *RCA and the VideoDisc: The Business of Research* (New York: Cambridge University Press, 1986); and Margaret B. W. Graham and Bettye H. Pruitt, *R&D for Industry: A Century of Technical Innovation at Alcoa* (New York: Cambridge University Press, 1990).

8. Susan Helper, for example, has argued that an important source of Japanese automakers' advantage over American automakers was their ability to encourage technological innovation throughout their supply chains. See "Strategy and Irreversibility in Supplier Relations: The Case of the U.S. Automobile Industry," *Business History Review* 65 (Winter 1991): 781–824. See as well Michael Schwartz's argument about the role of suppliers enabling Walter Chrysler to convert the two giant automakers into the "Big Three" during the 1920s and 1930s in "Markets, Networks, and the Rise of Chrysler in Old Detroit, 1920–1940," *Enterprise and Society* 1 (March 2000): 63–99.

9. JoAnne Yates has emphasized the role that consumers played in innovation in the insurance industry in *Structuring the Information Age: Life Insurance and Technology in the Twentieth Century* (Baltimore: Johns Hopkins University Press, 2005). Nancy F. Koehn traced the efforts of entrepreneurs to earn consumers' trust through building strong brands in *Brand New: How Entrepreneurs Earned Consumers' Trust from Wedgwood to Dell* (Boston: Harvard Business School Press, 2001). In *Imagining Consumers: Design and Innovation from Wedgwood to Corning* (Baltimore: Johns Hopkins University Press, 2000), Regina Lee Blaszczyk highlighted the important role that buyers for large retailers played in guiding innovation by collecting information about consumers' desires and conveying it to manufacturers. In contrast to these studies, Sally H. Clarke examined corporations' conflicted relations with consumers, wanting customers' loyalty but also imposing costs on consumers, in *Trust and Power: Consumers, the Modern Corporation, and the Making of the United States Automobile Market* (New York: Cambridge University Press, 2007).

10. For a good example, see Louis Galambos and Jeffrey L. Sturchio, "Pharmaceutical Firms and the Transition to Biotechnology: A Study in Strategic Innovation," *Business History Review* 72 (Summer 1998): 250–78.

11. See, for example, John K. Brown, *The Baldwin Locomotive Works, 1831–1915* (Baltimore: Johns Hopkins University Press, 1995).

12. Paul Israel, *Machine Shop to Industrial Laboratory: Telegraphy and the Changing Context of American Invention, 1830–1920* (Baltimore: Johns Hopkins University Press, 1992); and Paul Israel, *Edison: A Life of Invention* (New York: John Wiley & Sons, 1998).

13. W. Bernard Carlson, "The Coordination of Business Organization and Techno-logical Innovation Within the Firm: A Case Study of the Thomson-Houston Electric Company," in *Coordination and Information: Historical Perspectives on the Organization of Enterprise*, eds. Naomi L. Lamoreaux and Daniel M. G. Raff (Chicago: University of Chicago Press, 1995), 55–94; and W. Bernard Carlson, *Innovation as a Social Process: Elihu Thomson and the Rise of General Electric, 1870–1900* (New York: Cambridge University Press, 1991).

14. Thomas Parke Hughes, *American Genesis: A Century of Invention and Technological Enthusiasm* (New York: Viking, 1989).

15. Carlson, "The Coordination of Business Organization"; and Carlson, *Innovation as a Social Process*.

16. Stephen B. Adams and Orville B. Butler, *Manufacturing the Future: A History of Western Electric* (New York: Cambridge University Press, 1999), 28–38.

17. Lockwood did hire inventors from time to time but successfully opposed any sustained investment in in-house research. Lockwood is quoted in Naomi R. Lamoreaux and Kenneth L. Sokoloff, "Inventors, Firms, and the Market for Technology in the Late Nineteenth and Early Twentieth Centuries," in *Learning by Doing in Markets, Firms and Countries*, eds. Naomi R. Lamoreaux, Daniel M. G. Raff, and Peter Temin (Chicago: University of Chicago Press, 1999), 41–42. See also Louis Galambos, "Theodore N. Vail and the Role of Innovation in the Modern Bell System," *Business History Review* 66 (Spring 1992): 95–126; and, more generally, David Mowery, "The Boundaries of the U.S. Firm in R&D," in Lamoreaux and Raff, *Coordination and Information*, 147–76.

18. W. Bernard Carlson, "Innovation and the Modern Corporation: From Heroic Invention to Industrial Science," in *Science in the Twentieth Century*, eds. John Krige and Dominique Pestre (Australia: Harwood Academic Publishers, 1997), 203–26; Leonard S. Reich, *The Making of American Industrial Research: Science and Business at GE and Bell, 1876–1926* (New York: Cambridge University Press, 1985); Leonard S. Reich, "Lighting the Path to Profit: GE's Control of the Electric Lamp Industry, 1892–1941," *Business History Review* 66 (Summer 1992): 305–34; and George Wise, *Willis R. Whitney, General Electric, and the Origins of U.S. Industrial Research* (New York: Columbia University Press, 1985). Louis Galambos offered an early assessment of the origins of the industrial research laboratory, sketching the broad outlines in the economy. See "The American Economy and the Reorganization of the Sources of Knowledge," in *The Organization of Knowledge in Modern America, 1860–1920*, eds. Alexandra Olsen and John Voss (Baltimore: Johns Hopkins University Press, 1979), 269–82.

19. Carlson, "Innovation and the Modern Corporation"; Leonard S. Reich, "Research, Patents, and the Struggle to Control Radio: A Study of Big Business and the Uses of Industrial Research," *Business History Review* 51 (Summer 1977): 208–35, quote p. 231; Leonard S. Reich, "Industrial Research and the Pursuit of Corporate Security: The Early Years of Bell Labs," *Business History Review* 54 (Winter 1980): 504–29; Reich, *The Making of American Industrial Research*.

20. On this point, see Mowery, "The Boundaries of the U.S. Firm in R&D."

21. The figures in this paragraph are from David C. Mowery and Nathan Rosenberg, *Technology and the Pursuit of Economic Growth* (New York: Cambridge University Press, 1989), 64–71. See also David A. Hounshell, "Industrial Research and Manufacturing

28   Technology," in *Encyclopedia of the United States in the Twentieth Century*, ed. Stanley Kutler (New York: Charles Scribner's Sons, 1996), 831–57.

22. Richard R. Nelson and Gavin Wright, "The Rise and Fall of American Technological Leadership: The Postwar Era in Historical Perspective," *Journal of Economic Literature* 30 (Dec. 1992): 1931–64.

23. Hounshell, "Industrial Research and Manufacturing Technology," 846–49; and Scott G. Knowles and Stuart W. Leslie, "'Industrial Versailles': Eero Saarinen's Corporate Campuses for GM, IBM, and AT&T," *Isis* 92 (Mar. 2001), 1–33.

24. Richard Nelson, "The Simple Economics of Basic Scientific Research," *Journal of Political Economy* 67 (June 1959): 297–306; Kenneth J. Arrow, "Economic Welfare and the Allocation of Resources for Invention," in *The Rate and Direction of Inventive Activity*, Universities-National Bureau Committee for Economic Research (Princeton: Princeton University Press, 1962), 609–25; David J. Teece, "Technological Change and the Nature of the Firm," in Giovanni Dosi, Christopher Freeman, Richard Nelson, Gerard Silverberg, and Luc Soete, *Technical Change and Economic Theory* (London: Pinter, 1988), 256–81; and Mowery, "The Boundaries of the U.S. Firm in R&D."

25. Hounshell and Smith, *Science and Corporate Strategy*, esp. 11–189 and 331–364.

26. See David A. Hounshell, "Assets, Organizations, Strategies, and Traditions: Organizational Capabilities and Constraints in the Remaking of Ford Motor Company, 1946–1962," in Lamoreaux, Raff, and Temin, *Learning by Doing in Markets, Firms, and Countries*, 185–208; Ronald W. Schatz, *The Electrical Workers: A History of Labor at General Electric and Westinghouse, 1923-1960* (Urbana, Ill.: University of Illinois Press, 1983); and Emerson W. Pugh, *Building IBM: Shaping an Industry and Its Technology* (Cambridge, Mass.: MIT Press, 1995).

27. On IBM, see Steven W. Usselman, "Learning the Hard Way: IBM and the Sources of Innovation in Early Computing," in *Financing Innovation in the United States, 1870 to the Present*, eds. Naomi R. Lamoreaux and Kenneth L. Sokoloff (Cambridge, Mass.: MIT Press, 2007), 317–63. On Ford, see Hounshell, "Assets, Organizations, Strategies, and Traditions."

28. On Sarnoff, see Thomas K. McCraw, *American Business, 1920–2000: How It Worked* (Wheeling, Ill.: Harlan Davidson, 2000), 110–46. For a detailed analysis of innovation at RCA, see Margaret B. W. Graham, *RCA and the VideoDisc: The Business of Research* (New York: Cambridge University Press, 1986).

29. Hounshell, "Assets, Organizations, Strategies, and Traditions."

30. Usselman, "Learning the Hard Way."

31. Richard S. Rosenbloom, "Leadership, Capabilities, and Technological Change: The Transformation of NCR in the Electronic Era," and Mary Tripsas and Giovanni Gavetti, "Capabilities, Cognition, and Inertia: Evidence from Digital Imaging," both in *The SMS Blackwell Handbook of Organizational Capabilities: Emergence, Development, and Change*, ed. Constance E. Helfat (Malden, Mass.: Blackwell, 2003), 364–92 and 393–412.

32. Testimony of Dr. Frank Baldwin Jewett, U.S. Congress, House, *Hearings Before the Committee on Patents* (Washington, D.C.: Government Printing Office, 1936), part 1, 276–77.

33. Lamoreaux and Sokoloff, "Inventors, Firms, and the Market for Technology," 49–54.

34. Wise, *Willis R. Whitney*.

35. Hounshell and Smith, *Science and Corporate Strategy*, esp. 125–37, 365–83; and Hounshell, "Industrial Research and Manufacturing Technology," 844–46.

36. Graham, *RCA and the VideoDisc*; Paul Carroll, *Big Blues: The Unmaking of IBM* (New York: Crown, 1993), 3–44; and James Chposky and Ted Leonsis, *Blue Magic: The People, Power, and Politics Behind the IBM Personal Computer* (New York: Facts on File, 1988).

37. Leslie, "Industrial Versailles"; Hounshell and Smith, *Science and Corporate Strategy*, 360–83, 573–91; McCraw, *American Business*, 110–46; Charles J. Bashe, Lyle R. Johnson, John H. Palmer, and Emerson W. Pugh, *IBM's Early Computers* (Cambridge, Mass.: MIT Press, 1986), 456–58, 555–70; and P. Roy Vagelos with Louis Galambos, *Medicine, Science and Merck* (New York: Cambridge University Press, 2004).

38. On the role of antitrust in the emergence of corporate R&D, see Reich, *The Making of American Industrial Research*, 4–5 and 35–41; Hounshell and Smith, *Science and Corporate Strategy*, 6–7; Hounshell, "Industrial Research and Manufacturing Technology," 837–39; and Mowery, "The Boundaries of the U.S. Firm in R&D." On its subsequent role at particular firms, see chapters in this book by Graham, Lipartito, and Usselman.

39. Kenneth Lipartito, "The Social Meaning of Failure: Picturephone and the Information Age," *Technology and Culture* 44 (Jan. 2004): 50–81; Hounshell and Smith, *Science and Corporate Strategy*, 509–40; Graham, *RCA and the VideoDisc*; Paul E. Ceruzzi, *A History of Modern Computing* (Cambridge, Mass.: MIT Press, 1998); and Koehn, *Brand New*, 257–305.

40. See AnnaLee Saxenian, *Regional Advantage: Culture and Competition in Silicon Valley and Route 128* (Cambridge, Mass.: Harvard University Press, 1994); Christophe Lécuyer, *Making Silicon Valley: Innovation and the Growth of High-Tech, 1930–1970* (Cambridge, Mass.: MIT Press, 2006), 129–67; Emilio J. Castilla, Hokyu Hwang, Ellen Granovetter, and Mark Granovetter, "Social Networks in Silicon Valley," in *The Silicon Valley Edge: A Habitat for Innovation and Entrepreneurship*, eds. Chong-Moon Lee, William F. Miller, Marguerite Cong Hancock, and Henry S. Rowen et al. (Stanford, Calif.: Stanford University Press, 2000), 218–47; and Dimitris Assimakopoulos, Sean Everton, and Kiyoteru Tsutsui, "The Semiconductor Community in the Silicon Valley: A Network Analysis of the SEMI Genealogy Chart (1947–1986)," *International Journal of Technology Management* 25 (Nos. 1 and 2, 2003): 181–99.

41. Saxenian, *Regional Advantage*; Ronald J. Gilson, "The Legal Infrastructure of High Technology Industrial Districts: Silicon Valley, Route 128, and Covenants Not to Compete," *New York University Law Review* 74 (June 1999): 575–629; Alan Hyde, *Working in Silicon Valley: Economic and Legal Analysis of a High-Velocity Labor Market* (Armonk, N.Y.: M. E. Sharpe, 2003); and Shogo Hamasaki, "Patent Races with Information Spillovers: Covenants Not to Compete and the Silicon Valley Advantage" (unpublished paper, Faculty of Economics, University of California, Los Angeles, 2008).

42. Intel, for example, continues periodically to respond to employee departures with lawsuits. See Hyde, *Working in Silicon Valley*.

43. Saxenian, *Regional Advantage*.

44. Gilson, "Legal Infrastructure of High Technology Industrial Districts"; Paul Almeida and Bruce Kogut, "Localization of Knowledge and the Mobility of Engineers in Regional Networks," *Management Science* 45 (July 1999): 905–17.

45. See Stuart W. Leslie and Robert H. Kargon, "Selling Silicon Valley: Frederick Terman's Model for Regional Advantage," *Business History Review* 70 (Winter 1996): 435–72; and Saxenian, *Regional Advantage*.

46. Leslie and Kargon, "Selling Silicon Valley," 471.

47. See Mowery and Rosenberg, *Technology and the Pursuit of Economic Growth*.

48. Louis Galambos with Jane Eliot Sewell, *Networks of Innovation: Vaccine Development at Merck, Sharp & Dohme, and Multford, 1895–1995* (New York: Cambridge University Press, 1995). See also Rebecca Henderson, Luigi Orsenio, and Gary P. Pisano, "The Pharmaceutical Industry and the Revolution in Molecular Biology: Interactions Among Scientific, Institutional, and Organizational Change," in *Sources of Industrial Leadership: Studies of Seven Industries*, eds. David C. Mowery and Richard R. Nelson (New York: Cambridge University Press, 1999), 267–311; and Galambos and Sturchio, "Pharmaceutical Firms and the Transition to Biotechnology."

49. U.S. Temporary National Economic Committee (TNEC), *Verbatim Record of the Proceedings of the Temporary National Economic Committee*, vol. 1, no. 1 (Washington, D.C.: Government Printing Office, 1938), 134–35. See also John B. Rae, *American Automobile Manufacturers: The First Forty Years* (Philadelphia: Chilton, 1959), 75–81; and James J. Flink, *America Adopts the Automobile, 1895–1910* (Cambridge, Mass.: MIT Press, 1970), 289–92; 318–33.

50. Reich, "Research, Patents, and the Struggle to Control Radio"; Reich, *The Making of American Industrial Research*; Carlson, "Innovation and the Modern Corporation."

51. Steven W. Usselman, *Regulating Railroad Innovation: Business, Technology, and Politics in America, 1840–1920* (New York: Cambridge University Press, 2002). For a general treatment of the circumstances under which information sharing is likely to occur, see Robert Allen, "Collective Invention," *Journal of Economic Behavior and Organization* 4 (Jan. 1983): 1–24.

52. See, for examples, Louis Galambos, *Competition and Cooperation: The Emergence of a National Trade Association* (Baltimore: Johns Hopkins Press, 1966); Philip L. Merkel, "Going National: The Life Insurance Industry's Campaign for Federal Regulation After the Civil War," *Business History Review* 65 (Autumn 1991): 528–53; and Jerry Harrington, "The Midwest Agricultural Chemical Association: A Regional Study of an Industry on the Defensive," *Agricultural History* 70 (Spring 1996): 415–38.

53. For examples from the pre–Sherman Act period, see Robert H. Porter, "A Study of Cartel Stability: The Joint Executive Committee, 1880–1886," *Bell Journal of Economics* 14 (Autumn 1983): 301–14; Margaret Levenstein, "Mass Production Conquers the Pool: Firm Organization and the Nature of Competition in the Nineteenth Century," *Journal of Economic History* 55 (Sep. 1995): 575–611; and Naomi R. Lamoreaux, *The Great Merger Movement in American Business, 1895–1904* (New York: Cambridge University Press, 1985). For the post–Sherman Act period, see David Genesove and Wallace P. Mullin, "The Sugar Institute Learns to Organize Information Exchange," in Lamoreaux, Raff, and Temin, *Learning by Doing in Markets, Firms, and Countries*, 103–36; and Ellis W. Hawley, "Three Facets of Hooverian Associationalism: Lumber, Aviation, and Movies, 1921–1930," in *Regulation in Perspective: Historical Essays*, ed. Thomas K. McCraw (Cambridge, Mass.: Harvard University Press, 1981), 95–123.

54. Yates, *Structuring the Information Age*.

55. See Joseph A. Litterer, "Systematic Management: The Search for Order and Integration," *Business History Review* 35 (Winter 1961): 461–76; and JoAnne Yates, *Control Through Communication: The Rise of System in American Management* (Baltimore: Johns Hopkins University Press, 1989).

56. Daniel Nelson, "Industrial Engineering and the Industrial Enterprise, 1890–1940," in Lamoreaux and Raff, *Coordination and Information*, 35–50. See also Daniel Nelson, *Managers and Workers: Origins of the New Factory System in the United States, 1880–1920* (Madison, Wis.: University of Wisconsin Press, 1975); and Daniel Nelson, *Frederick W. Taylor and the Rise of Scientific Management* (Madison, Wis.: University of Wisconsin Press, 1980).

57. Christopher D. McKenna, *The World's Newest Profession: Management Consulting in the Twentieth Century* (New York: Cambridge University Press, 2006), 16–20.

58. McKenna, *The World's Newest Profession*, 72–73.

59. McKenna, *The World's Newest Profession*, 111–44.

60. This problem is now just beginning to receive attention from behavior theorists. For an agenda-setting article, see Andrew Watson, Terence Rodgers, and David Dudek, "The Human Nature of Management Consulting: Judgment and Expertise," *Managerial and Decision Economics* 19 (Nov.-Dec. 1998): 495–503.

61. The starting point for the discussion of capture theory on the Left was a little earlier. See Gabriel Kolko, *The Triumph of Conservatism* (New York: Free Press, 1963). On the Right, see George J. Stigler, "The Theory of Economic Regulation," *Bell Journal of Economics and Management Science* 2 (Spring 1971): 3–21. A historical assessment of the debates about regulation written during the 1980s can be found in Naomi R. Lamoreaux, "Regulatory Agencies," *Encyclopedia of American Political History* (New York: Charles Scribner's Sons, 1984), 1107–16.

62. For studies examining how the state promoted innovation in two very different sectors of the economy (computers and agriculture), see Arthur L. Norberg and Judy E. O'Neill, *Transforming Computer Technology: Information Processing for the Pentagon, 1962–1986* (Baltimore: Johns Hopkins University Press, 1996); and Sally H. Clarke, *Regulation and the Revolution in United States Farm Productivity* (New York: Cambridge University Press, 1994).

63. For a straightforward application of this view, see Albro Martin's critique of railroad regulation in the early twentieth century, *Enterprise Denied: The Origins of the Decline of American Railroads, 1897–1917* (New York: Columbia University Press, 1971). See also Richard H. K. Vietor, *Energy Policy in America Since 1945: A Study of Business-Government Relations* (New York: Cambridge University Press, 1984).

64. Thomas K. McCraw, *Prophets of Regulation: Charles Francis Adams, Louis D. Brandeis, James M. Landis, and Alfred E. Kahn* (Cambridge, Mass.: Harvard University Press, 1984), 222–99, quote 275.

65. Richard H. K. Vietor, *Contrived Competition: Regulation and Deregulation in America* (Cambridge, Mass.: Harvard University Press, 1994), 23–90.

66. Clarke, *Regulation and the Revolution in United States Farm Productivity*. On the diffusion of tractors, readers may also wish to consult Alan L. Olmstead and Paul W. Rhode, "Reshaping the Landscape: The Impact and Diffusion of the Tractor in American Agriculture, 1910–1960," *Journal of Economic History* 61 (Sep. 2001): 663–98;

32   and Todd Sorensen, Price Fishback, and Shawn Kantor, "The New Deal and the Diffusion of Tractors in the 1930s," unpublished paper, Faculty of Economics, University of California, Riverside, 2008. On the diffusion of tractors and the transformation of the South, a useful starting point is Gavin Wright's *Old South, New South: Revolutions in the Southern Economy Since the Civil War* (New York: Basic Books, 1986).

67. See Harold Demsetz, "Industry Structure, Market Rivalry, and Public Policy," *Journal of Law and Economics* 16 (Apr. 1973): 1–9; and, of course, Chandler, *The Visible Hand.*

68. Robert H. Bork, *The Antitrust Paradox: A Policy at War with Itself* (New York: Basic Books, 1978), 4–7, 15–49. See also Richard A. Posner, *Antitrust Law: An Economic Perspective* (Chicago: University of Chicago Press, 1976); and Alan Stone, *Economic Regulation and the Public Interest: The Federal Trade Commission in Theory and Practice* (Ithaca, N.Y.: Cornell University Press, 1977). On the rise of this "Chicago School" view of antitrust policy, see Tony A. Freyer, *Antitrust and Global Capitalism, 1930–2004* (New York: Cambridge University Press, 2006), 139; and Marc Allen Eisner, *Antitrust and the Triumph of Economics: Institutions, Enterprise, and Policy Change* (Chapel Hill, N.C.: University of North Carolina Press, 1991).

69. Lamoreaux, *The Great Merger Movement,* 159–86.

70. For an interesting perspective on shifting antitrust policy during this period, see Louis Galambos, "The Monopoly Enigma, the Reagan Administration's Antitrust Experiment, and the Global Economy," in *Constructing Corporate America: History, Politics, and Culture,* eds. Kenneth Lipartito and David Sicilia (New York: Oxford University Press, 2004), 149–67.

71. Peter Temin, with Louis Galambos, *The Fall of the Bell System: A Study in Prices and Politics* (New York: Cambridge University Press, 1987); and Franklin M. Fisher, John J. McGowan, and Joen E. Greenwood, *Folded, Spindled, and Mutilated: Economic Analysis and U.S. v. IBM* (Cambridge, Mass.: MIT Press, 1983).

72. New directions in antitrust scholarship can be found in Michael D. Whinston, "Tying, Foreclosure, and Exclusion," *American Economic Review* 80 (Sep. 1990): 837–59; Thomas M. Jorde and David J. Teece, eds., *Antitrust, Innovation, and Competitiveness* (New York: Oxford University Press, 1992); and Jerry Ellig, ed., *Dynamic Competition and Public Policy: Technology, Innovation, and Antitrust Issues* (New York: Cambridge University Press, 2001).

73. Clarke, *Regulation and the Revolution in United States Farm Productivity,* 28–33; Alan L. Olmstead and Paul W. Rhode, "The Red Queen and the Hard Reds: Productivity Growth in American Wheat, 1800–1940," *Journal of Economic History* 62 (Dec. 2002): 940.

74. See Deborah Kay Fitzgerald, *The Business of Breeding: Hybrid Corn in Illinois, 1890–1940* (Ithaca, N.Y.: Cornell University Press, 1990); Jack Ralph Kloppenburg Jr., *First the Seed: The Political Economy of Plant Biotechnology, 1492–2000* (New York: Cambridge University Press, 1988); and Clarke, *Regulation and the Revolution in United States Farm Productivity.*

75. Olmstead and Rhode, "The Red Queen and the Hard Reds."

76. Zvi Griliches, "Research Costs and Social Return: Hybrid Corn and Related Innovations," *Journal of Political Economy* 66 (Oct. 1958): 419–31. For a contrary view

arguing that Griliches overstated the gains to hybrid corn before 1937 and that new fertilizers were an important contributor to higher yields thereafter, see Richard C. Sutch, "Henry Agard Wallace, the Iowa Corn Yield Tests, and the Adoption of Hybrid Corn," NBER Working Paper No. 14141 (2008).

77. Mowery and Rosenberg, *Technology and the Pursuit of Economic Growth*, 129. On the origins of the military-industrial complex, see Gregory Hooks, *Forging the Military-Industrial Complex: World War II's Battle of the Potomac* (Urbana, Ill.: University of Illinois Press, 1991).

78. Stuart W. Leslie, "The Biggest 'Angel' of Them All: The Military and the Making of Silicon Valley," in *Understanding Silicon Valley: The Anatomy of an Entrepreneurial Region*, ed. Martin Kenney (Stanford, Calif.: Stanford University Press, 2000), 48–67, quote 49.

79. Norberg and O'Neill, *Transforming Computer Technology*.

80. Michael Hiltzik, *Dealers of Lightning: Xerox PARC & the Dawn of the Computer Age* (New York: HarperCollins, 1999).

81. On computer science departments, Norberg and O'Neill find that "[a]s late as 1959, there were virtually no formal programs in the field." But they also report that the academic landscape changed dramatically with IPTO funding: "At the beginning of the 1990s, more than 26 percent of the tenured members of those select science departments had received their Ph.D.s from one of the three main IPTO-sponsored institutions: MIT, Stanford, and CMU. This shows a remarkable penetration of IPTO in the computer education process. An even greater percentage, 42 percent, of the tenured faculty members in the ten top-ranked departments had received their Ph.D. degrees from one of the three schools, and more than 53 percent of the nontenured faculty members." Norberg and O'Neill, *Transforming Computer Technology*, esp. 289–90.

82. Usselman, *Regulating Railroad Innovation*, 273–326; and Barbara Young Welke, *Recasting American Liberty: Gender, Race, Law, and the Railroad Revolution, 1865–1920* (New York: Cambridge University Press, 2001).

83. David Stradling, *Smokestacks and Progressives: Environmentalists, Engineers, and Air Quality in America, 1881–1951* (Baltimore: Johns Hopkins University Press, 1999).

84. On the postwar suburban environment, see, for instance, Adam Rome, *The Bulldozer in the Countryside: Suburban Sprawl and the Rise of American Environmentalism* (New York: Cambridge University Press, 2001). An excellent account of the problem of social costs as applied to overfishing is Arthur F. McEvoy, *The Fisherman's Problem: Ecology and Law in the California Fisheries, 1850–1980* (New York: Cambridge University Press, 1986). On government intervention for workplace safety, see John Fabian Witt, *The Accidental Republic: Crippled Workingmen, Destitute Widows, and the Remaking of American Law* (Cambridge, Mass.: Harvard University Press, 2004); and Price Fishback and Shawn Everett Kantor, *Prelude to the Welfare State: The Origins of Workers' Compensation* (Chicago: University of Chicago Press, 2000).

85. See Margaret Graham on Corning in Chapter 2 in this volume; and Margaret B. W. Graham and Alec T. Shuldiner, *Corning and the Craft of Innovation* (New York: Oxford University Press, 2001), 350–58.

86. On the subject of private organizations influencing management, see Thomas J. Misa's study of the Society of Automotive Engineers in *A Nation of Steel: The Making*

of Modern America, 1865–1925 (Baltimore: Johns Hopkins University Press, 1995), 213–23, 229; and on the Underwriters' Laboratories, see Scott Gabriel Knowles, *Inventing Safety: Fire, Technology, and Trust in Modern America*, Ph.D. dissertation, Johns Hopkins University, 2003, 184–259.

87. Even with this web of regulatory oversight, Clarke found that firms never resolved the trade-off between safety and other business goals. As one upshot of this trade-off, firms helped develop the new market for product liability insurance. Clarke, *Trust and Power*, especially chapters 1, 2, and 4. Judge Benjamin Cardozo's landmark ruling was *MacPherson v. Buick*, 111 N.E. 1050 (N.Y. 1916). On *MacPherson*, see also Edward H. Levi, *An Introduction to Legal Reasoning* (Chicago: University of Chicago Press, 1949), 7–19; Steven P. Croley and Jon D. Hanson, "Rescuing the Revolution: The Revived Case for Enterprise Liability," *Michigan Law Review* 91 (Feb. 1993): 683–797; and Jonathan Lurie, "Lawyers, Judges, and Legal Change, 1852–1916: New York as a Case Study," *Working Papers from the Regional Economic History Research Center* 3 (1980): 31–56.

88. For background on the health care debate, see, for instance, Jennifer Klein, *For All These Rights: Business, Labor, and the Shaping of America's Public-Private Welfare State* (Princeton: Princeton University Press, 2003).

89. Lizabeth Cohen makes this argument in *A Consumers' Republic: The Politics of Mass Consumption in Postwar America* (New York: Knopf, 2003), 129–42, 156–60, and 166–73.

90. Lizabeth Cohen documented discrimination in access to credit in postwar America based on race and sex in *A Consumers' Republic*, 147–48, 170–72, 214, 221. See also Martha L. Olney, "When Your Word Is Not Enough: Race, Collateral, and Household Credit," *Journal of Economic History* 58 (June 1998): 408–31; and Louis Hyman, *Debtor Nation: Changing Credit Practices in 20th Century America* (Princeton: Princeton University Press, forthcoming), chapter 7.

91. Home ownership data is reported in Susan B. Carter et al., *Historical Statistics of the United States: Millennial Edition Online* (New York: Cambridge University Press, 2006), series Dc 761–762, http://hsus.cambridge.org/HSUSWeb/search/searchTable .do?id=Dc761-780 (accessed March 31, 2008). The percent of owner-occupied homes with mortgages is reported in U.S. Bureau of the Census, *Historical Statistics of the United States: From Colonial Times to the Present* (Washington, D.C.: U.S. Government Printing Office, 1976), series N 305, 651. On general trends in debt, see Andrew L. Yarrow, *Forgive Us Our Debts: The Intergenerational Dangers of Fiscal Irresponsibility* (New Haven, Conn.: Yale University Press, 2008). As Yarrow's title suggests, his account addresses the problem of too much debt for the nation and the burden it places on individual members.

92. For the use of installment credit with automobile purchases, see U.S. Bureau of the Census, *Historical Statistics of the United States*, Series Q 175 and Q 180, 717. It is worth noting that at some periods between 1947 and 1970, such as 1959 and 1960, the percentage of families using installment credit to buy cars rose to roughly 60 percent. On the increase in total installment debt, see Clarke, *Trust and Power*, 254. On installment credit before World War II, see Martha L. Olney, *Buy Now, Pay Later: Advertising, Credit, and Consumer Durables in the 1920s* (Chapel Hill: University of North Carolina Press, 1991), esp. 86–134.

93. Kenneth T. Jackson, *Crabgrass Frontier: The Suburbanization of the United States* (New York: Oxford University Press, 1985), 196–218; and Lizabeth Cohen, *Making a New Deal: Industrial Workers in Chicago, 1919–1939* (New York: Cambridge University Press, 1990), 273–77. On redlining, see also National Consumer Law Center, *Credit Discrimination*, 3rd ed. (Boston: National Consumer Law Center, 2002), 119–42. (Creditors who singled out certain areas for selling the most costly loans engaged in what became known as reverse redlining.) On credit discrimination experienced by African Americans, see also Olney, "When Your Word Is Not Enough."

94. Hyman also discussed the role of women who demanded protection against credit discrimination in *Debtor Nation*, chapter 7.

95. Steven W. Usselman, "IBM and Its Imitators: Organizational Capabilities and the Emergence of the International Computer Industry," *Business and Economic History* 22 (Winter 1993): 1–35, reprinted in *Industrial Research and Innovation in Business*, ed. David E. H. Edgerton (London: Edward Elgar, 1996), 452–86; and Steven W. Usselman, "Fostering a Capacity for Compromise: Business, Government, and the Stages of Innovation in American Computing," *Annals of the History of Computing* 18 (Summer 1996): 30–39.

96. Robert Kagan, *Adversarial Legalism: The American Way of Law* (Cambridge, Mass.: Harvard University Press, 2001), esp. 181–228. See also, Louis Galambos and Joseph Pratt, *The Rise of the Corporate Commonwealth: United States Business and Public Policy in the Twentieth Century* (New York: Basic Books, 1988).

PROLOGUE
REORGANIZING INNOVATION

Introduction to the Prologue

Sally H. Clarke, Naomi R. Lamoreaux,
and Steven W. Usselman

During the past two decades, executives at Du Pont, IBM, and other giant R&D-intensive firms have slashed their laboratory budgets, while CEOs throughout the economy have come under increasing pressure to demonstrate that dollars spent for research were contributing in a measurable way to their companies' bottom lines. According to the new conventional wisdom, large firms could best meet the challenge of remaining innovative by plugging into networks of entrepreneurs and scientists, by organizing collaborative ventures, and by acquiring high-tech startups rather than emphasizing in-house research.[1] As small firms once again became a prominent source of new technology, large firms had to learn to become "aggressive followers"—to use their technological know-how to keep abreast of ideas developed outside their labs, secure property rights to the most promising innovations, and commercialize them effectively.[2] In 1970 small firms obtained only about 5 percent of the patents issued anywhere in the world, but by the beginning of the twenty-first century the proportion was closer to a third.[3]

A hundred years earlier the trend ran in just the opposite direction. In "The Rise and Decline of the Independent Inventor," Naomi Lamoreaux and Kenneth Sokoloff analyze the circumstances under which inventive activity moved inside large firms. Using a novel body of data on patent assignments to track changes over time in patentees' careers, they show that by the early twentieth century inventors were finding it increasingly difficult to maintain their independence. Some took employment positions in the corporate research labs that large firms were building during this period. But this was only one of two major ways in which inventors responded to the changes wrought by the Second

40   Industrial Revolution. Many others formed ties with smaller independent firms in which they took a significant financial stake. A brand of Schumpeterian entrepreneurialism thus persisted, and thrived, alongside the emerging corporate giants.

This bifurcated pattern, the authors show, exhibited a distinct regional character. Inventors in the East, especially in the Middle Atlantic region, tended to ally with the large corporations that concentrated near the country's financial center in New York City, whereas those in the Midwest were more likely to be associated with firms in which they had a personal stake. The authors speculate that it was easier to organize startups in regions distant from New York, where capital tended to be pulled into the securities markets that raised funds for large-scale enterprises. The dual pathways seem also to have reflected trends in education. Young men with advanced degrees in science or engineering were increasingly likely to move to the Northeast and cast their lot with the large corporations located there.

Lamoreaux and Sokoloff argue against the conventional scholarly wisdom that in-house research laboratories were an inherently superior way of organizing innovative activity.[4] Although the labs solved some of the information and contracting problems that afflicted market trade in technological knowledge, they introduced new ones that corporate executives would struggle with over the next century (see Part I of this volume). What bringing R&D activities within the firm did accomplish, however, was to solve the problem of funding innovation in a context in which technologies were complex and expensive and financial markets had limited capabilities. In recent years, as the nature of cutting-edge technologies has shifted and venture capitalists and other new types of financial institutions have emerged, innovative activity has been reorganized yet again and large firms' in-house research labs have lost much of the ground they gained relative to small entrepreneurial enterprises.

NOTES

1. Robert Buderi, *Engines of Tomorrow: How the World's Best Companies Are Using Their Research Labs to Win the Future* (New York: Simon & Schuster, 2000).

2. Richard S. Rosenbloom and William J. Spencer, "Introduction: Technology's Vanishing Wellspring," in *Engines of Innovation: U.S. Industrial Research at the End of an Era*, eds. Richard S. Rosenbloom and William J. Spencer (Boston: Harvard Business School Press, 1996), 3–6. See also the other essays in the Rosenbloom and Spencer volume.

3. G. Steven McMillan and Diana Hicks, "Science and Corporate Strategy: A Bibliometric Update of Hounshell and Smith," *Technology Analysis & Strategic Manage-*

ment 13 (Dec. 2001): 497–505; and Diana Hicks, Tony Breitzman, Dominic Olivastro, and Kimberly Hamilton, "The Changing Composition of Innovative Activity in the U.S.—A Portrait Based on Patent Analysis," *Research Policy* 30 (Apr. 2001): 681–703.

4. On this point, see also their "Inventors, Firms, and the Market for Technology in the Late Nineteenth and Early Twentieth Centuries," in *Learning by Doing in Markets, Firms, and Countries*, eds. Naomi R. Lamoreaux, Daniel M. G. Raff, and Peter Temin (Chicago: University of Chicago Press, 1999), 19–60.

THE RISE AND DECLINE OF THE
INDEPENDENT INVENTOR

A Schumpeterian Story?

Naomi R. Lamoreaux and Kenneth L. Sokoloff

> The perfectly bureaucratized giant industrial unit
> not only ousts the small or medium-sized firm and
> "expropriates" its owners, but in the end it also ousts
> the entrepreneur and expropriates the bourgeoisie as
> a class which in the process stands to lose not only its
> income but also what is infinitely more important, its
> function. The true pacemakers of socialism were not
> the intellectuals or agitators who preached it but the
> Vanderbilts, Carnegies and Rockefellers.
>
> *Joseph A. Schumpeter*[1]

For Joseph Schumpeter, the heart of the capitalist system was the entrepreneur—
an extraordinary individual who had the foresight to see profit in new products
or production processes as well as the tenacity to overcome any obstacles that
stood in the way. Schumpeter believed that the rise of large firms in the early
twentieth century was making the entrepreneur obsolete. By investing in in-
house R&D laboratories staffed by teams of engineers and scientists, large firms
had routinized the process of innovation and "depersonalized and autonoma-
tized" technological change, so that advances were realized "as a matter of
course." In such an environment not only did "personality and will power," and
thus the entrepreneur, "count for less," but the greater efficiency of large-scale
enterprises was undermining the small- and medium-size firms that historically
had been spawning grounds for heroic innovators with radically new ideas
about how to do things. These developments, Schumpeter foretold, would have
profound consequences for the entire society. Because entrepreneurs were the
primary political supports for "private property and free contracting," their
eclipse would pave the way for socialist revolution.[2]

44   From the standpoint of the early twenty-first century, it is difficult to take
this vision of the decline of capitalism seriously. When Schumpeter was writing *Capitalism, Socialism and Democracy* on the eve of World War II, however, it seemed much more compelling. Gloom about the future was pervasive,
and rival systems, such as communism in the Soviet Union, were attracting
growing numbers of adherents throughout the West. Ironically, though, it was
not until the post–World War II boom that Schumpeter's ideas really took hold.
Large-scale, managerially directed enterprises dominated the economy as in
no other period of U.S. history, and the technological prowess exhibited by
General Electric, IBM, and other giant corporations spurred widespread
acceptance of Schumpeter's belief that large firms were undermining the basis
for individual entrepreneurship.

          The idea that technological discovery was most effectively pursued inside
large integrated enterprises became, if anything, even more dominant in the
scholarly literature during the 1980s, when the so-called "new economics of
information" supplied an alternative theoretical rationale. According to this
theory, problems of asymmetric information place severe limits on the
exchange of technological ideas in the market. Before firms will invest in a new
technology, they need to be able to assess its value. They need, for example, to
be able to estimate the extent to which an innovative process will lower production costs, or whether a novel product will likely appeal to consumers.
Inventors (or other sellers of new technology) fear that firms will steal their
ideas and so typically will not be willing to provide potential purchasers with
enough information to effectuate sales. Moving the process of technological
discovery inside the firm not only overcomes this source of market failure but
also, it is argued, yields other informational advantages. In particular, firms
with R&D labs will be better positioned to exploit ideas for innovation that
arise from the actual experience of producing or marketing goods. This kind
of knowledge is largely firm-specific and can be transmitted much more readily among personnel responsible for different functions within the organization than it can across organizational boundaries.[3]

          For proponents of this new informational theory, the rise of large firms was
an important advance without which the pace of innovation would likely have
stalled as the economy moved from the relatively simple technologies of the
First Industrial Revolution to the more complex technologies of the Second.
Schumpeter's position was somewhat more ambiguous. Although he believed
the economy gained from the spread of large-firm R&D, he also believed it suffered important losses. Certainly, the dynamics of technological change shifted
during the early twentieth century in ways that, at least on the surface, seem
to cast doubt on the idea that large-firm R&D was an unalloyed good. As
Figure 1.1 shows, patenting rates for U.S. residents increased dramatically across

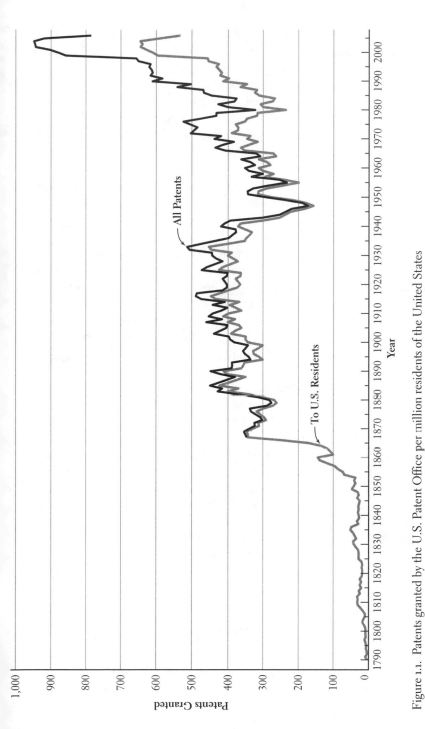

Figure 1.1. Patents granted by the U.S. Patent Office per million residents of the United States

SOURCES: Susan B. Carter et al., *Historical Statistics of the United States: Earliest Times to the Present, Millennial Edition* Vol. 1, 28–29, and Vol. 3, 426–28 (New York: Cambridge University Press, 2006); U.S. Patent Office, "U.S. Patent Activity: Calendar Years 1790 to the Present," http://www.uspto.gov/web/offices/ac/ido/oeip/taf/h_counts.pdf (accessed March 7, 2008); U.S. Bureau of the Census, "Population Estimates, 2000–2006," http://factfinder.census.gov/servlet/GCTTable?_bm=y&-geo_id=01000US&-_box_head_nbr=GCT-T1&-ds_name=PEP_2006_EST&-_lang=en&-format=US-9&-_sse=on (accessed March 7, 2008).

46   the nineteenth century, peaked during the early twentieth, and then stagnated
or even declined until very recently, when firms began to cut back on their
research budgets. In other words, patenting rates appear to have been inversely
correlated with the growth of in-house R&D, dropping in the twentieth century as more and more firms built their own research labs and then increasing
at the end of the century as firms began to shut these facilities down.[4]

In this chapter we explore the shift in the location of innovative activity
toward large firms in the early twentieth century with the aim of developing a
better understanding of what might have been gained and lost from this
change. Our approach is to use data on the award and assignment of patents
to analyze whether entrepreneurs were finding the economic environment less
hospitable. Of course, we recognize that there are objections to using patents
as a measure of innovation. Schumpeter himself explicitly distinguished the
concept of innovation from that of invention. What entrepreneurs did when
they innovated, according to Schumpeter, was to take a new idea (an invention) and make it work—that is, embody the idea in a productive enterprise
and generate profits. It was the latter achievement that was important to
Schumpeter, not the discovery of the invention.[5] We also recognize that patenting is an imperfect measure even of invention. Some valuable inventions are
never patented, and many inventions are patented that have little or no economic significance.

Nonetheless, we contend that data on trends in patenting are useful for testing Schumpeter's argument about the diminishing role of the entrepreneur in
technological change and, more generally, in capitalism. The essence of a
patent is the grant of an exclusive property right to a new technological idea. By
making property out of intangible knowledge—property that could be exploited
by the owner or sold or leased to someone else—the patent system created the
foundation of property rights upon which entrepreneurial innovation flourished.
Moreover, although Schumpeter may have been correct to distinguish the role
of the inventor from that of the entrepreneur as a matter of theory, in practice
inventors often behaved entrepreneurially. Indeed, as we will show, the flour-
ishing market for technology that the U.S. patent system made possible fostered
the emergence of a class of talented, highly entrepreneurial inventors who specialized in the production of new technological ideas for sale.

Before plunging into this demonstration, it is useful to provide a brief
description of our data. The starting point for our analysis is three random
cross-sectional samples (totaling about 6,600 patents) that we drew from the
Annual Reports of the Commissioner of Patents for the years 1870–71, 1890–91,
and 1910–11. For each patent in the samples we recorded a brief description of
the invention, the name and location of the patentee(s), and the names and

locations of any assignees who obtained rights to the invention before the patent was actually issued. We also linked this data on patents to other information, such as characteristics of the firms to which the patentees assigned their patent rights. Our second major data set is longitudinal and was obtained by selecting from the three cross-sectional samples all (561) inventors whose last names began with the letter "B."[6] We then collected information from *Patent Gazettes* and from the *Annual Reports of the Commissioner of Patents* for the (6,057) patents obtained by these patentees for the twenty-five years before and after they appeared in one of our samples, again linking this data to the same kinds of sources we used for the original cross-sections. For our third data set, we collected similar information for patents granted in selected years to "great inventors" born between 1820 and 1885. We defined great inventors as individuals whose technological discoveries were notable enough to earn them inclusion in the *Dictionary of American Biography*. In addition to specific information on a substantial subset of their patents, we collected biographical detail on these inventors from the dictionary entries, as well as the total number of patents each received over his or her career.[7]

In the analysis that follows we first describe the details of the patent system that were most important for supporting entrepreneurial innovation. After legislation in 1836 put the final elements of this system in place, both patenting and trade in patent rights boomed. The result, we show, was the emergence by the last third of the nineteenth century of a broad group of highly productive and highly entrepreneurial inventors. During the early twentieth century, however, it seems to have become increasingly difficult for talented inventors to pursue independent careers. We devote the rest of this chapter to examining the changes in the assignment patterns of these extraordinary individuals in order to understand why.

THE PATENT SYSTEM AND THE RISE OF INSTITUTIONS SUPPORTING A MARKET FOR TECHNOLOGY

The institutional foundation for the rise of the independent inventor in the nineteenth century was the U.S. patent system, created in accordance with the Commerce Clause of the Constitution. Although influenced by British law, the framers of the U.S. system self-consciously made a number of important innovations. Among them were dramatically lower fees, impersonal administrative procedures for handling applications, and the requirement that a patentee be the "first and true" inventor anywhere in the world. In addition, firms could not receive patents directly for ideas developed in their shops. Patentees had to be individual men or women, who had the option either of exploiting their inventions

48   themselves or selling (assigning) or leasing (licensing) them to other individu-
      als or to firms.[8] Taken together, these provisions made it possible for a broader
      group of inventors to obtain protection for their intellectual property than in
      Britain or elsewhere in Europe. Moreover, coupled (as they were) with effec-
      tive enforcement, these provisions meant that inventors could advantageously
      reveal information about their ideas to prospective buyers, even before they
      actually obtained the patents.[9]

      Although the main purpose of the patent system was to stimulate invention
      by granting creative individuals secure rights to their intellectual property,
      another important goal was to promote the diffusion of technological knowl-
      edge. The law required patentees to provide the Patent Office with detailed
      specifications of their inventions (including, when appropriate, working models),
      and the result was a central storehouse of information that was open to all. Any-
      one could journey to Washington and study the specifications of inventions
      already patented. In addition, more convenient means of tapping this rich
      source of information soon developed. The Patent Office itself began to publish
      on a regular basis lists of the patents it granted. By the middle of the century,
      moreover, a number of private journals had emerged to improve upon this offi-
      cial service. One of the most important was *Scientific American,* published by
      Munn and Company, the largest patent agency of the nineteenth century. Others
      included the *American Artisan,* published by the patent agency Brown, Coombs
      & Company; the *American Inventor,* by the American Patent Agency; and the
      Patent Right Gazette, by the United States Patent Right Association (which,
      despite its name, functioned as a general patent agency). Covering the full
      spectrum of technologies, these journals featured articles about important new
      inventions, printed complete lists of patents issued, and offered inventors advice
      about how to profit from their ideas. Many pages were devoted to classified
      advertisements by patent agents and lawyers soliciting clients, by inventors seek-
      ing partners with capital to invest, and by patentees hoping to sell or lease rights
      to their technologies. One of the primary aims of these journals, of course, was
      to drum up business for the patent agencies that published them.[10]

      During the early years of the century patent rights were awarded automati-
      cally to all inventors who registered their inventions and paid the necessary
      fees. This procedure effectively delegated to the courts responsibility for resolv-
      ing disputes about the originality, novelty, and appropriate scope of patent
      rights. New legislation in 1836 resolved the obvious problems this system cre-
      ated by requiring that each invention be scrutinized by technically trained
      examiners to ensure that it constituted an original advance in the state of the
      art and otherwise conformed to the law. Uncertainty about the value of patents
      decreased dramatically after the passage of this law, and trade in patent rights

boomed, attaining very quickly a volume of three to six times the number of patents issued.[11]

The bulk of this early commerce in patents (80 to 90 percent of transactions during the 1840s) involved efforts by inventors to make multiple, geographically delimited assignments of their rights to producers in different areas. Because markets were primarily local or regional, inventors with valuable intellectual property could use their ideas in their own manufacturing facilities and, at the same time, earn additional profits by assigning limited rights to producers elsewhere. Thomas Blanchard, inventor of the gunstocking lathe, a wood-bending machine, and a variety of other devices, exploited these possibilities to the hilt. For example, he used his lathe himself to make gunstocks for the Boston market and for export. He also leased the rights to use the invention to gun producers operating in other locations and to manufacturers making shoe lasts, tool handles, wheel spokes, and a variety of other goods in different places around the country.[12]

After the development of the railroad network and the emergence of national product markets, interest in purchasing geographically segmented patent rights declined, and manufacturers instead sought to acquire full national rights to important technologies. As a result, already by 1870 the proportion of assignments that were geographically specific had dropped to 23 percent of the total, and the proportion would fall to 5 percent by 1890. Moreover, in a legal environment in which many, if not most, inventions were protected by patents, and in which this kind of property right was vigorously enforced, maintaining one's competitive position often meant beating out rivals to secure exclusive rights to new devices. As firms eagerly snapped up new technological ideas, the proportion of patents that were assigned before they were even issued rose from 18 percent of the total in 1870–71 to 29 percent by 1890–91.[13]

As a consequence of the growth of national markets, the ways in which inventors could extract the returns from their ideas necessarily changed. Whereas Blanchard could make use of his invention himself and also sell off partial rights to others, patentees now typically faced a stark choice. They could either exploit their inventions directly by establishing enterprises capable of operating in national (or even international markets), or they could sell or license their rights to others better situated to develop and commercialize the technology on a large scale. Regardless, they could now benefit greatly from the assistance of specialized intermediaries who could help them find buyers or leasers for their national patent rights, or alternatively could help them secure the financial backing they needed to found their own enterprises (whether for the direct commercial exploitation of their inventions or for the generation and patenting of new technological knowledge to be licensed or sold).

Although the role of intermediary could be played by almost any kind of businessperson, patent agents and lawyers were particularly well suited for this task. Indeed, inventors who used the services of these professionals were able on average to sell their patents more quickly than anyone else (by the early 1870s, 47 percent of the assignments handled by patent agents or lawyers occurred before the issue of the patent, as opposed to 18 percent handled by other intermediaries, and only 9 percent handled by the parties themselves). Not only did patent agents and lawyers usually have considerable technical training, but in the course of their ordinary business they serviced both sides of the market for technology, helping inventors file patent applications and assisting buyers in evaluating inventions' technical merits. They were therefore in a good position to obtain information about new technologies coming on the market, as well as to know the kinds of technologies that particular buyers were interested in purchasing. Located in urban centers, these agents typically had connections with colleagues in other cities—sometimes formally through partnerships, sometimes informally through family ties, and sometimes simply through repeat interactions—and thus were able to weave into a national web the dense local networks that each of them created.[14]

As is evident from Table 1.1, both the propensity to assign patent rights and the location of concentrations of patent agents were positively associated across regions with rates of patenting per capita. New England, which had exhibited the highest patenting rates in the nation since the beginning of the nineteenth century, had both the largest proportion of patents assigned at issue and the most disproportionate concentration of patent agents relative to population. The Middle Atlantic and the East North Central regions were the next highest respectively in patenting, assignments rates, and the clustering of agents. The South ranked lowest on all three scales.[15]

This robust regional correspondence between the extent of the market for technology and patenting rates is exactly what one would expect to observe. On the one hand, intermediaries should, all other things being equal, concentrate in areas where rates of invention are already high. On the other, the presence of firms and institutions conducive to trade in technology should stimulate greater specialization in invention. In the first place, by increasing the net returns that inventors might expect from a given discovery, they should encourage individuals with a comparative advantage in invention to make appropriate investments in physical and human capital. Second, they should make it easier for inventors to raise capital to support their inventive activity. One might also expect that individuals already inclined to specialize in invention would move to cities or regions with more developed market institutions or greater demand for new technological knowledge. As we show in the next section, inventors did indeed respond to these various incentives in a manner befitting Schumpeterian entrepreneurs.

TABLE 1.1. Patents, assignments at issue, and patent attorneys, by region

	1870–1871 Subsample	1890–1891 Subsample	1910–1911 Subsample
New England			
Patents/population	775.8	772.0	534.3
% of patents assigned	26.5	40.8	50.0
Patent attorneys/population	—	2.7	2.0
Middle Atlantic			
Patents/population	563.4	607.0	488.6
% of patents assigned	20.6	29.1	36.1
Patent attorneys/population	—	2.2	2.0
East North Central			
Patents/population	312.3	429.9	442.3
% of patents assigned	14.7	27.9	32.3
Patent attorneys/population	—	1.1	1.1
West North Central			
Patents/population	146.5	248.7	272.0
% of patents assigned	9.0	21.8	17.5
Patent attorneys/population	—	0.3	0.7
South			
Patents/population	85.8	103.1	114.4
% of patents assigned	6.4	25.0	22.7
Patent attorneys/population	—	0.1	0.2
West			
Patents/population	366.7	381.6	458.4
% of patents assigned	0.0	25.4	21.4
Patent attorneys/population	—	0.5	1.1
All Patents, Including Foreign			
Patents/population	325.4	360.4	334.2
% of patents assigned	18.5	29.1	30.5
Patent attorneys/population	—	—	—

NOTES: Patents/population is the annual rate of patenting per million residents. The percentage of patents assigned includes only those assignments that occurred before the patent's date of issue. Patent attorneys/population is the proportion of attorneys registered with the patent office (excluding those in the District of Columbia) to the proportion of the U.S. population.

SOURCES: Regional counts of patents are built up from the state counts reported in U.S. Patent Office, *Annual Reports of the Commissioner of Patents* (Washington, D.C.: Government Printing Office, 1870, 1890, and 1910). New England includes the states of Connecticut, Maine, Massachusetts, New Hampshire, Rhode Island, and Vermont; the Middle Atlantic includes New Jersey, New York, and Pennsylvania; the East North Central includes Illinois, Indiana, Michigan, Ohio, and Wisconsin; the West North Central includes Iowa, Kansas, Minnesota, Missouri, Nebraska, North Dakota, and South Dakota; the South includes Alabama, Arkansas, Delaware, the District of Columbia, Florida, Georgia, Kentucky, Louisiana, Maryland, Mississippi, North Carolina, Oklahoma, South Carolina, Tennessee, Texas, Virginia, and West Virginia; and the West includes Arizona, California, Colorado, Idaho, Montana, Nevada, New Mexico, Oregon, Utah, Washington, and Wyoming. The percentage of patents assigned was calculated from our three cross-sectional samples (see text). The data on patent attorneys are from U.S. Patent Office, *Names and Addresses of Attorneys Practicing Before the United States Patent Office* (Washington, D.C.: Government Printing Office, 1883 and 1905). Population data are from Susan B. Carter et al., *Historical Statistics of the United States: Earliest Times to the Present, Millennial Edition* (New York: Cambridge University Press), Vol. 1, 180–379.

INDEPENDENT INVENTORS AS ENTREPRENEURS

Once one accepts the notion that inventions were a tradable good that, like other tradable goods, could be a source of profit, then it is not difficult to see how specialized inventors resembling Schumpeterian entrepreneurs emerged and thrived during the second half of the nineteenth century. The entrepreneurial bent of these specialized inventors manifested itself in a variety of ways. Perhaps the most significant was the extensive use they made of the market for technology—extracting the returns from their patented inventions by selling off the property rights to other individuals or firms. The resulting division of labor gave inventors the freedom to concentrate on what they did best—invent. It also allowed them to take better advantage of differences across firms in the ability or inclination to exploit the commercial potential of particular inventions.

That the evolution of this market for patented inventions did in fact expand opportunities for the technologically creative is suggested by the jump in the relative importance of specialized inventors between the first and last third of the nineteenth century (see Table 1.2). The proportion of patents in any given year that were awarded to individuals who received ten or more patents over their careers increased from below 5 percent during the early 1800s to 28.9 and 35.9 percent respectively in 1870–71 and 1890–91. (We will henceforth refer to patentees from our "B" sample who had ten or more patents as "productive"

TABLE 1.2. Distribution of patents by patentees' career total of patents, 1790–1911

	Number of Career Patents by Patentee (percentage)					
	1 Patent	2 Patents	3 Patents	4–5 Patents	6–9 Patents	10+ Patents
1790–1811	51.0	19.0	12.0	7.6	7.0	3.5
1812–29	57.5	17.4	7.1	7.6	5.5	4.9
1830–42	57.4	16.5	8.1	8.0	5.6	4.4
1870–71	21.1	12.5	9.9	15.8	11.8	28.9
1890–91	19.5	10.3	10.3	10.3	13.8	35.9
1910–11	33.2	14.3	8.2	9.8	9.4	25.0

NOTES and SOURCES: The figures from 1790 to 1842 are drawn from Kenneth L. Sokoloff and B. Zorina Khan, "The Democratization of Invention During Early Industrialization: Evidence from the United States, 1790–1846," *Journal of Economic History* 50 (June 1990): 363–78. The figures for the later years were computed from our longitudinal "B" data set (see text). Because our sampling ratios are higher for the earlier than the later cross-sections, these figures may slightly overstate the increase in specialization that occurred between the first three and last three time periods. However, because the method used to gather the career patenting totals for the early-nineteenth-century inventors was likely to be a bit more comprehensive than that used for the latter periods (especially 1910–11), there is a slight bias in the other direction as well.

TABLE 1.3. Assignment rates at issue of "B" patentees by career productivity

| | \multicolumn{4}{c}{Total Career Patents for Patentee} |
	1–2 Patents	3–5 Patents	6–9 Patents	10+ Patents
1870–71 Subsample				
Mean career patents	1.6	4.1	7.3	28.3
% assigned at issue	17.6	11.4	12.3	24.7
% assigned to companies	1.5	1.4	4.1	8.9
Number of patents	68	140	122	749
1890–91 Subsample				
Mean career patents	1.5	3.9	7.7	61.0
% assigned at issue	27.1	28.7	39.4	54.4
% assigned to companies	12.9	7.0	17.0	40.5
Number of patents	70	129	188	2060
1910–11 Subsample				
Mean career patents	1.5	4.0	7.1	107.4
% assigned at issue	15.0	22.0	42.5	62.4
% assigned to companies	3.8	7.7	30.8	56.2
Number of patents	133	155	120	1860

NOTES: These estimates were computed for the 545 "B" patents who were residents of the United States at some time during their careers. They include all of the patents these patentees received while living in the United States. For details on the "B" sample, see the text.

inventors to distinguish them from patents in our sample of "great" inventors.) Moreover, as indicated by their higher rates of assignment at issue by the late nineteenth century, inventors who obtained large numbers of patents were more likely to trade away the rights to their patents (see Table 1.3).[16] This association strengthened over time, so that, by the 1910–11 subsample, "B" inventors with 10 or more career patents assigned 62.4 percent of their patents by the date they were issued (56.2 percent to companies), whereas those with just 1 or 2 career patents assigned only 15.0 percent of their patents.

Because the inventors who were most successful at coming up with important new technological ideas were also likely to be those most inclined toward, and best able to mobilize resources for, continuing along that career path, it seems likely that these specialized, market-oriented patentees played an even more disproportionate role in generating significant (high-value) inventions. This view is consistent with the finding that the career patent totals for our sample of great inventors were far higher than they were for representative ("B") patentees and also that the great inventors were much more likely to assign away their inventions than were other patentees. For example, great inventors born between 1820 and 1839 received on average 35.1 patents over their careers, those born between 1840 and 1859 received on average 68.8 patents, and those born between 1860 and 1885 received 100.0. The respective rates of assignment for these cohorts were 29.3, 42.9, and 64.2 percent. By contrast, the average numbers of patents received by the different subsamples of our "B" inventors ranged between 7.8 and 15.6, and assignment rates were significantly lower. The evidence thus supports the view that by the last third of the nineteenth century there had emerged a class of highly productive and entrepreneurial inventors geared toward selling off the rights to their discoveries, and that this class was a crucial source of new technological knowledge.

Here skeptics might object that the high assignment rates we observe for productive and great inventors did not necessarily reflect their entrepreneurial orientation toward the market for technology but instead might simply indicate that they were employees of the firms to which they assigned their patents. Upon closer examination, however, we can reject this possibility. During the middle of the nineteenth century, as we have already discussed, most assignments involved patentees selling geographically delimited rights to firms or individuals in different jurisdictions. Such geographic assignments were virtually always arms-length transactions. Moreover, in earlier work we traced the occupations of "B" patentees wherever possible in city directories. We found that, even as the prevalence of geographic assignments declined over time with the emergence of national product markets, it remained quite unusual for productive inventors to be employees of the firms that obtained their patent rights. To the contrary, the great majority of these patentees either had no long-term attachment to their assignees or, as was increasingly the case by the early twentieth century, were principals in the firms to which they assigned their patents.[17]

The entrepreneurial opportunities that the ability to sell off patent rights afforded specialized inventors were not merely potential. Productive inventors seem actively to have pursued them by assigning their inventions to multiple individuals and firms. This pattern is evident from Table 1.4, which examines the assignment behavior of the 545 patentees from our "B" sample who were residents of the United States during at least some period of their careers. Of these patentees, 168 (or 30.8 percent) received ten or more patents over the fifty years we followed them (accounting for 80.6 percent of the total 5,794 patents awarded to the 545 inventors), and 51 of these 168 transferred their patents at issue to four or more different assignees. These 51 patentees (9.4 percent of the total number of patentees) received 2,034 patents (or more than 35 percent of total patents). In comparison, the 36 patentees with ten or more patents (6.6 percent of the sample) who dealt with only one assignee over their careers received only 727 patents (12.5 percent of the total).[18]

Even stronger evidence for such entrepreneurial behavior comes from the
sample of great inventors. Although our estimates (in Table 1.5) of their contractual mobility are based only on a 20 percent sample of their patents and
thus are downwardly biased, we still find that 26 percent of these renowned
inventors had four or more distinct assignees over their careers, accounting for

TABLE 1.4. Contractual mobility of "B" patentees:
Distributions of patents and patentees by inventors' total career patents

	Total Career Patents for Patentee				
	1–2 Patents	3–5 Patents	6–9 Patents	10+ Patents	n (col. %)
Panel A: Distribution of Patents					
No Assignees					
Row %	23.8	25.3	15.2	35.8	875
Col. %	76.8	52.1	30.9	6.7	(15.1)
1 Assignee					
Row %	6.1	15.0	9.2	69.8	1042
Col. %	23.3	36.8	22.3	15.6	(18.0)
2–3 Different Assignees					
Row %	0.0	2.4	8.1	89.6	1781
Col. %	0.0	9.9	33.5	34.2	(15.8)
4+ Different Assignees					
Row %	0.0	0.2	2.7	97.0	2096
Col. %	0.0	1.2	13.3	43.6	(36.2)
n	271	424	430	4669	5794
(row %)	(4.7)	(7.3)	(7.4)	(80.6)	(100.0)
Panel B: Distribution of Patentees					
No Assignees					
Row %	59.9	22.5	7.5	10.1	267
Col. %	78.8	53.6	32.3	16.1	(49.9)
1 Assignee					
Row %	32.3	30.1	10.5	27.1	133
Col. %	21.2	35.7	22.6	21.4	(24.4)
2–3 Different Assignees					
Row %	0.0	12.8	24.4	62.8	86
Col. %	0.0	9.8	33.9	32.1	(15.8)
4+ Different Assignees					
Row %	0.0	1.7	11.9	86.4	59
Col. %	0.0	0.9	11.3	30.4	(10.8)
n	203	112	62	168	545
(row %)	(37.3)	(20.6)	(11.4)	(30.8)	(100.0)

NOTE: On the "B" sample, see Table 1.3 and the text. Only assignments at issue are included in the figures.

roughly 43 percent of all the patents in the sample. Beginning with the cohort born between 1840 and 1859, the difference between patentees with multiple assignees and those who assigned to only one individual or firm became especially pronounced. More than 53 percent of the patents obtained by members of that cohort went to inventors with four or more assignees, and only 14 percent went to those with only one assignee.[19]

Entrepreneurs are generally thought to be geographically mobile—willing to relocate to take advantage of better opportunities. Hence another way of examining the entrepreneurial qualities of these specialized and productive inventors is to track their movements. The great-inventor sample allows us to compare the residences of patentees at the times they received their various patents with each other and also with their place of birth, and it is evident from the data that these talented inventors were far more geographically mobile than the general population. For example, the fraction of patents awarded to U.S.-born great inventors who resided outside their state of birth was nearly double the migration estimates (36 percent and 30 percent) that Joseph Ferrie has compiled for U.S.-born adults over the thirty-year periods from 1850 to 1880 and 1880 to 1910 respectively.[20] Perhaps more directly comparable (and impressive),

TABLE 1.5. Contractual mobility of great inventors:
Distributions of patents and patentees by birth cohort

| | Birth Cohort | | | |
	1820–1839	1840–1859	1860–1885	All Cohorts
Panel A: Distribution of Patents (col. %)				
No assignees	17.2	6.1	2.3	7.3
1 assignee	16.4	14.0	34.1	21.1
2–3 different assignees	49.5	26.6	16.7	28.4
4+ different assignees	17.0	53.4	46.9	43.2
n	548	1117	815	2480
Panel B: Distribution of Inventors (col. %)				
No assignees	20.0	8.1	4.2	11.5
1 assignee	22.9	24.3	37.5	27.1
2–3 different assignees	42.8	35.1	25.0	35.4
4+ different assignees	14.3	32.4	33.3	26.0
n	35	37	24	96

NOTES: The distributions are computed for the ninety-six great inventors for whom we sampled ten or more patents. Because we have sampled only one-fifth of the patents on average and are confining our analysis to assignments at issue, we are underestimating the full extent of contractual mobility. For a description of the sample, see the text.

about 55 percent of these patents went to great inventors who obtained their last patent in a different state from the one where they had obtained their first. For great inventors whose patenting careers (defined as the period between their first and last patents) lasted between twenty and thirty-five years, the figure was fully two-thirds, and for those with careers over thirty-five years, nearly three-quarters. Moreover, patentees who had the most career patents were on average the most mobile, and hence seemingly the most entrepreneurial, even after adjusting for the duration of their careers.

Not only were the great inventors highly mobile, but their locational decisions followed systematic patterns (see Table 1.6). Patenting by great inventors remained very concentrated in the Northeast throughout the period under study. For the cohort born between 1820 and 1839, this pattern could be attributed to the disproportionate numbers of great inventors born in New England and the Middle Atlantic, but later cohorts evinced a powerful tendency to relocate to these regions and particularly to the Middle Atlantic states. For example, members of the 1840 to 1859 birth cohort who were born in the Middle Atlantic accounted for nearly 18 percent of total patents, whereas those who resided in that region received 47.3 percent. By the third cohort (those born between 1860 and 1885), great inventors who were residents of the Middle Atlantic obtained nearly two-thirds of all patents, even though those born in the region generated less than 10 percent of the total. This extreme concentration of great-inventor patents in the Middle Atlantic contrasts starkly with the contemporaneous trend toward regional convergence in overall patenting rates, suggesting that conditions in that region were especially favorable to the most talented inventors. Inventors and the enterprises committed to technological innovation may have found it increasingly important over time to have ready access to the legal services and financial intermediation provided by the region's dense cluster of patent agents, lawyers, and financial institutions.

Elmer Sperry is an excellent example of a great inventor who assiduously pursued entrepreneurial opportunities.[21] Born into an upstate New York family of modest means, Sperry attended public schools and then went on to college at Cortland Normal. While at Cortland, Sperry decided he wanted to be an inventor and set about learning as much as possible about electricity. Acting on the suggestion of one of the professors he sought out at nearby Cornell University, he designed a generator capable of supplying a constant current regardless of the load on its circuits and then began systematically to scour the local business community in search of a financial backer. The Cortland Wagon Company, whose executives included both inventors and investors interested in supporting new technological developments, took Sperry in, providing him with the advice and services of a patent lawyer, as well as money to live on and

TABLE 1.6. Great-inventor patents by region of birth and region of residence

Region of Birth	Region of Residence at Date of Patent					
	New England	Middle Atlantic	Midwest	Other U.S.	Foreign	Total
Panel A: 1820–1839 Birth Cohort						
New England						
Number	306	83	25	3	0	417
Row %	73.4	19.9	6.0	0.7	0.0	100.0
Col. %	80.7	21.9	24.8	8.6	0.0	46.5
Middle Atlantic						
Number	35	220	29	4	0	288
Row %	12.2	76.4	10.1	1.4	0.0	100.0
Col. %	9.2	58.1	28.7	11.4	0.0	32.1
Midwest						
Number	0	11	21	9	3	41
Row %	0.0	26.8	51.2	27.0	7.3	100.0
Col. %	0.0	2.9	20.8	25.7	100.0	4.6
Other U.S.						
Number	2	2	0	7	0	11
Row %	18.2	18.2	0.0	63.6	0.0	100.0
Col. %	0.5	0.5	0	20.0	0.0	4.6
Foreign						
Number	36	63	26	15	0	140
Row %	25.7	45.0	18.6	10.7	0.0	100.0
Col. %	9.5	16.6	25.7	42.9	0.0	15.6
Total						
Number	379	379	101	35	3	897
Row %	42.3	42.3	11.3	3.9	0.3	100.0
Panel B: 1840–1859 Birth Cohort						
New England						
Number	51	94	9	4	25	183
Row %	27.9	51.4	4.9	2.2	13.7	100.0
Col. %	17.9	15.1	2.9	6.4	78.1	13.9
Middle Atlantic						
Number	30	149	51	2	4	236
Row %	12.7	63.1	21.6	0.9	1.7	100.0
Col. %	10.5	23.9	16.3	3.2	12.5	17.9
Midwest						
Number	3	207	167	7	2	386
Row %	0.8	53.6	43.3	1.8	0.5	100.0
Col. %	1.1	33.2	53.4	11.1	6.3	29.3

TABLE 1.6. *(continued)*

| Region of Birth | Region of Residence at Date of Patent | | | | | |
	New England	Middle Atlantic	Midwest	Other U.S.	Foreign	Total
Panel B: 1840–1859 Birth Cohort (continued)						
Other U.S.						
Number	1	9	8	37	0	55
Row %	1.8	16.4	14.6	67.3	0.00	100.0
Col. %	0.4	1.4	2.6	58.7	0.00	4.2
Foreign						
Number	200	164	78	13	1	456
Row %	43.9	36.0	17.1	2.9	0.2	100.0
Col. %	70.2	26.3	24.9	20.6	3.1	34.7
Total						
Number	285	623	313	63	32	1316
Row %	21.7	47.3	23.8	4.8	2.4	100.0
Panel C: 1860–1885 Birth Cohort						
New England						
Number	58	156	1	0	0	215
Row %	27.0	72.6	0.5	0.0	0.0	100.0
Col. %	43.3	25.8	0.8	0.0	0.0	23.5
Middle Atlantic						
Number	42	286	40	5	0	373
Row %	11.3	76.7	10.7	1.3	0.0	100.0
Col. %	31.3	47.4	33.1	8.8	0.0	40.7
Midwest						
Number	2	43	78	25	0	148
Row %	1.4	29.1	52.7	16.9	0.0	100.0
Col. %	1.5	7.1	64.5	43.9	0.0	16.2
Other U.S.						
Number	0	7	0	9	0	16
Row %	0.0	43.8	0.0	56.3	0.0	100.0
Col. %	0.0	1.2	0.0	15.8	0.0	1.8
Foreign						
Number	32	112	2	18	0	164
Row %	19.5	68.3	1.2	11.0	0.0	100.0
Col. %	23.9	18.5	1.7	31.6	0.0	17.9
Total						
Number	134	604	121	57	0	916
Row %	14.6	65.9	13.2	6.2	0.0	100.0

NOTE: For a description of the great inventors sample, see the text. For definitions of the regions, see Table 1.1.

60   a shop in which to work. The year was 1880, and in this sheltered environment, Sperry not only perfected his dynamo but over the next two years developed a complete system of arc lighting to go with it. The company had a branch in Chicago, and there the company's officers, with additional backing from local Chicago investors, organized The Sperry Electric Light, Motor, and Car Brake Company in 1883, with Sperry (who owned a big chunk of the company's stock) serving as electrician, inventor, and superintendent of the mechanical department.

Although the company launched Sperry's career as an inventor, it was never a financial success. For Sperry, it proved to be a constant source of anxiety that absorbed all his attention and left him little time and energy for creative pursuits. Indeed, this period was the low point of his career in terms of generating new technological ideas. The nineteen patents he applied for during his five years with the company amounted collectively to half his *annual* average over a career as an inventor that stretched from 1880 to 1930. Sperry emerged from this experience determined to devote his energies to invention and never again to become so deeply involved in the internal affairs of a company. But he was also determined to profit from his discoveries. To this end, he sold off many of his inventions to companies that were well placed to put them to productive use. Others he commercialized himself, founding with the help of a wide assortment of financial backers a variety of companies that bore his name, such as the Sperry Electric Mining Machine Company, the Sperry Streetcar Electric Railway Company, and the Sperry Gyroscope Company. Although Sperry often played an active role in these companies in their early stages, he always downgraded his position to technical consultant as quickly as possible and went on to something else.

Like other great inventors, Sperry demonstrated a willingness throughout his career to uproot himself and his family whenever new opportunities beckoned. He moved from Cortland to Chicago to commercialize his arc lighting system, departed for Cleveland when a group of wealthy investors offered financial backing for his work on electric streetcars, and finally headed back East to Brooklyn to work on electrolytic methods for producing tin plate for American Can. Apparently opportunities in the Middle Atlantic were sufficiently rich that he was able to spend the last thirty-plus years of his long, productive career in New York.

GROWING DIFFICULTIES FACING INDEPENDENT INVENTORS

If Sperry had been born twenty years later, his career might have been entirely different, for the economic environment changed in ways that seem to have made it more difficult for technologically creative people to embark on careers as independent inventors. After the turn of the century patenting rates were essentially stagnant in the country as a whole (see Figure 1.1). Moreover, they were declining dramatically in the two northeastern regions where inventive activity had long been concentrated. Whereas New Englanders had obtained 775.8 patents per million residents per year in 1870–71, their rate had dropped to 534.3 by 1910–11. In the Middle Atlantic, grants declined from 563.4 to 488.6 patents per million residents per year over the same period (see Table 1.1).

Certainly, there is evidence that barriers to entry were rising for inventors. The most obvious change was the apparent advantage that accrued to inventors who had obtained formal training in science and engineering. Figure 1.2 documents the shift for successive cohorts of our sample of great inventors. Up through the cohort of inventors born between 1820 and 1845, individuals without an advanced formal education accounted for most major inventions. Indeed, roughly 75 to 80 percent of the patents awarded to great inventors in these cohorts went to those with only primary or secondary schooling. There was an abrupt change, however, in the educational backgrounds of the great inventors born after 1845. Whereas only 10 percent of the patents awarded to those born between

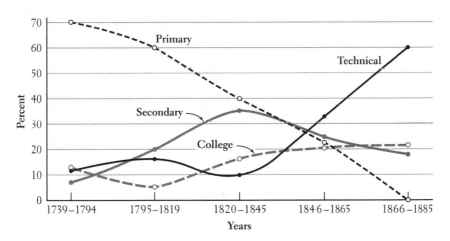

FIGURE 1.2. Distribution of great-inventor patents by formal schooling and birth cohort (percent)

NOTES AND SOURCE: Great inventors were classified as having a primary education if they did not attend school beyond age twelve (or did not attend school at all). The secondary-school category includes those who were identified as spending one or more years in an academy or who attended school after the age of twelve but did not attend a college or seminary. Those who spent any time at all in college were counted either in the college category or (if they attended a school with an engineering orientation or followed a course of study in medicine or a natural science) in the technical category. For more discussion of the data set, see B. Zorina Khan and Kenneth L. Sokoloff, "Institutions and Technological Innovation During Early Economic Growth: Evidence from Great Inventors in the United States, 1790–1930," in *Institutions, Development, and Economic Growth*, eds. Theo S. Eicher and Cecilia Garcia-Peñalosa (Cambridge, Mass.: MIT Press, 2006), 123–58.

1820 and 1845 went to individuals with college or graduate training in a technical field (engineering, natural science, or medicine), the figure jumped to around 60 percent for the cohort born between 1865 and 1885 (with still another 20 percent awarded to those who studied other subjects in college).[22] To be an important inventor in the new science-based technologies of the Second Industrial Revolution, it seems, required more formal schooling than had been the case for the mechanical technologies of the First Industrial Revolution.

There is also evidence, albeit necessarily indirect, that the amount of capital required for successful invention had risen as well. One of the most difficult hurdles that would-be inventors had to surmount was to raise sufficient capital to finance their technological explorations. The career patterns of patentees in our "B" sample suggest that it was becoming more difficult to do this over time. We classify patents granted within five years of the first patent as belonging to the early stage of an inventor's career, those granted five to fifteen years after the first patent as belonging to the intermediate stage, and those thereafter as late. As Table 1.7 shows, the patents awarded to specialized inventors in the 1870–71 subsample were spread fairly evenly over their careers. Inventors obtained about a third of their patents in the early stage of their careers, a third in the intermediate stage, and a third later on. By contrast, the patents of productive members of the 1910–11 subsample were concentrated in the later stages of their careers. Inventors in this group obtained only about 15 percent of their patents in the first stage of their careers and nearly 60 percent in the last.[23] This shift in career patterns is consistent with the idea that the capital required for effective invention had increased over time and that, as a result, inventors could not gain access to the resources they needed to be productive at their vocation until they had first proved their worth in some way.

Although the rise in the amount of human and physical capital required for effective invention might be enough in and of itself to explain the increasing

TABLE 1.7. Number of patents obtained by productive patentees in the "B" subsamples at different stages of their careers

	Stage of Career		
	Early	Intermediate	Late
1870–71 subsample	243	255	251
1890–91 subsample	323	651	1086
1910–11 subsample	305	479	1076

NOTES: Only patents awarded to patentees with ten or more career patents are included in this table. We classify patents granted within five years of the first patent as belonging to the early stage of an inventor's career, those granted five to fifteen years after the first patent as intermediate stage, and those thereafter as late. For a description of the "B" sample, see Table 1.3 and the text.

difficulty of pursuing a career as an independent inventor, Schumpeter's theory   63
offers an alternative possibility: that the in-house R&D laboratories that large
firms began to build during this period were so productive at technological discovery, particularly in the science-based industries associated with the Second
Industrial Revolution, that independent inventors could not survive in competition. According to this view, technologically creative individuals had no
choice, if they wanted to engage in inventive activity, but to take positions in
large firms' R&D facilities. It is likely, moreover, that this alternative career
path was only open to those with educational credentials showing that they had
the requisite scientific training or to those who already had a track record that
demonstrated their inventive talent.

Consistent with this Schumpeterian explanation is evidence that inventors
were increasingly forming some kind of long-term attachment with a firm.
Table 1.8 shows that, after rising between the 1820–39 and 1840–59 birth
cohorts, the contractual mobility of great inventors with ten or more patents in
our sample was beginning to decline. Although the fraction of great inventors
in the 1860–85 cohort with four or more different assignees remained essentially the same, their share of total patents had slipped, and an increasing proportion had only one assignee over their entire careers. The contractual
mobility of "B" inventors declined even more strikingly by the early twentieth
century. Whereas the proportion of patentees who assigned their patents at
issue to four or more assignees had risen from 7.5 to 15.1 percent between the
1870–71 and 1890–91 subsamples, it fell to 9.4 percent by the 1910–11 group.
Intriguingly, the decline in contractual mobility that occurred during this last
interval was virtually all a result of developments in New England and the Middle Atlantic. Contractual mobility actually increased in the East North Central between the 1890–91 and 1910–11 subsamples, though its level was still
below that of New England and only slightly greater than that of the Middle
Atlantic. We will come back to this regional difference later on.

To explore the possibility that inventors were increasingly taking employment positions with large firms, we tracked changes over time in the kinds of
assignees to which patentees were transferring their inventions. Our analysis
was complicated by the difficulty of classifying the numerous assignees in our
samples in a way that enabled us to distinguish firms with the characteristics
Schumpeter had in mind. The solution we adopted was to sort patents by
whether or not they were assigned at issue and to what kind of assignee—for
example, whether they were assigned to an individual, a company that bore
the same name as the patentee, a national company, or some other type. We
defined a national company as one for which financial information was
reported in the *Commercial and Financial Chronicle* or in *Poor's* or *Moody's
Manual of Industrial Securities* (indicating that the company was important

TABLE 1.8. Changes in the contractual mobility of "B" patentees, by region

Number of Assignees at Issue	Region of Patentee				
	New England (col. %)	Middle Atlantic (col. %)	East North Central (col. %)	Other U.S. (col. %)	Entire U.S. (col. %)
No Assignees					
1870–71 subsample	51.2	64.8	60.5	57.1	59.2
1890–91 subsample	25.7	42.4	36.7	41.7	37.3
1910–11 subsample	34.8	48.4	55.0	60.4	52.1
1 Assignee					
1870–71 subsample	39.0	18.5	26.3	14.3	25.9
1890–91 subsample	28.6	12.1	38.3	33.3	26.5
1910–11 subsample	17.4	29.0	13.8	27.1	21.6
2–3 Different Assignees					
1870–71 subsample	2.4	9.3	5.3	21.4	7.5
1890–91 subsample	25.7	25.8	16.7	12.5	21.1
1910–11 subsample	34.8	12.9	18.8	10.4	16.9
4+ Different Assignees					
1870–71 subsample	7.3	7.4	7.9	7.1	7.5
1890–91 subsample	20.0	19.7	8.3	12.5	15.1
1910–11 subsample	13.0	9.7	12.5	2.1	9.4
Number of Patentees					
1870–71 subsample	41	54	38	14	147
1890–91 subsample	35	66	60	24	185
1910–11 subsample	23	62	80	48	213

NOTES: Each inventor is represented in the table by one patent that was randomly selected from the patentee's record. On the "B" sample, see Table 1.3 and the text. For definitions of the regions, see Table 1.1.

enough to tap the national capital markets) or, alternatively, that was listed in an early 1920s National Research Council directory of companies with research laboratories.[24] We think of assignees of this type as being closest to the bureaucratic enterprises to which Schumpeter attributed the decline of the entrepreneur, and assume that the patentees who assigned to them were more likely to be employees than principals. Companies with the same name as the patentee, by contrast, were enterprises that were likely run by inventors, their relatives, or other close personal associates and thus most closely resemble Schumpeter's concept of an entrepreneurial enterprise—that is, a company set up to exploit the profit potential of a particular innovation.

Table 1.9 presents cross-tabulations for each of our "B" subsamples in which we break down the distribution of patents by the total number of career patents

TABLE 1.9. Changes in the distribution of patents in the "B" sample
by type of assignment at issue and number of career patents

Type of Assignment at Issue	Total Number of Career Patents			
	1–2 Patents (col. %)	3–5 Patents (col. %)	6–9 Patents (col. %)	10+ Patents (col. %)
Not Assigned				
1870–71 subsample	82.4	88.6	87.7	75.3
1890–91 subsample	72.9	70.5	60.6	45.6
1910–11 subsample	85.0	78.1	57.5	39.3
Assigned Share to Individual				
1870–71 subsample	10.3	3.6	4.1	5.5
1890–91 subsample	10.0	11.6	12.8	3.9
1910–11 subsample	7.5	6.5	5.8	2.8
Assigned in Full to Individual				
1870–71 subsample	2.9	5.0	2.5	8.8
1890–91 subsample	2.9	8.5	6.4	9.6
1910–11 subsample	1.5	3.2	1.7	3.2
Assigned to Company with Same Name				
1870–71 subsample	0.0	0.0	0.0	1.7
1890–91 subsample	0.0	1.6	3.7	6.1
1910–11 subsample	0.0	0.0	5.8	24.6
Assigned to Large National Company				
1870–71 subsample	0.0	0.0	0.8	1.2
1890–91 subsample	1.4	0.0	0.5	9.9
1910–11 subsample	0.0	1.9	0.0	14.8
Assigned to Other Local Company				
1870–71 subsample	1.5	0.7	2.5	4.5
1890–91 subsample	10.0	3.9	5.3	15.9
1910–11 subsample	1.5	3.9	15.8	8.4
Assigned to Other Non-Local Company				
1870–71 subsample	2.9	2.1	2.5	2.9
1890–91 subsample	4.3	3.9	10.6	9.0
1910–11 subsample	3.9	6.5	13.3	7.0
Number of Patents				
1870–71 subsample	68	140	122	749
1890–91 subsample	80	129	188	2060
1910–11 subsample	133	155	120	1860

NOTES AND SOURCES: On the "B" sample, see Table 1.3 and the text. The definitions of the types of assignments at issue are as follows: "Not Assigned"—the inventor did not assign the patent at issue; "Assigned Share to Individual"—the inventor made a partial assignment at issue to an individual (often a partner); "Assigned in Full to Individual"—the inventor transferred all rights to the patent to another individual; "Assigned to Company with the Same Name"—the inventor assigned the patent to a company that bore his or her surname (indicating that the inventor was likely to be a principal in the firm); "Assigned to Large Integrated Company"—the inventor assigned to a company for which financial information was reported in the *Commercial and Financial Chronicle* or in *Poor's* or *Moody's Manual of Industrial Securities* (indicating that the company was important enough to tap national capital markets) or, alternatively, that was listed in the National Research Council's *Research Laboratories in Industrial Establishments of the United States, 1920–21* (New York: R. R. Bowker, 1920); "Assigned to Other Local Company"—the assignment was to a not-already-classified company that was located in the same city as the patentee; "Assigned to Other Non-Local Company"—the assignment was to a company that did not fall into any of the preceding categories.

66   the patentee received (our measure of specialization at invention) and by the type of assignment. In all three subsamples patentees who were more specialized at invention had markedly higher assignment rates. Although inventors in all categories assigned some of their patents to individuals, those who were more specialized or productive over their careers were increasingly likely to assign to companies. By the 1910–11 subsample, inventors with ten or more career patents assigned at issue nearly two-thirds of all the patents they were awarded, and nearly two-thirds of these assignments went to companies with the same name as the inventor or that we classified as national enterprises. Indeed, productive inventors accounted for virtually all of the patents assigned to companies of these types.[25]

These results, however, are only partially consistent with Schumpeter's view of the growing importance of large-firm R&D. Although specialized inventors transferred more and more of their patents to large national companies, over time they assigned an even greater proportion of their patents to companies that bore their name—that is, to companies in which they were more likely to be principals than employees. Inventors with ten or more career patents in the 1910–11 subsample assigned 24.6 percent of their patents to such companies (accounting for fully 40.5 percent of their assignments). Although these findings suggest that by the early twentieth century inventors needed to attach themselves to firms in order to remain specialized and productive at invention, clearly working for a large-scale enterprise was not the only available option.

Examination of the data on great inventors yields essentially the same findings, though the relative importance of assignments to large national versus family-name companies is reversed. Table 1.10 reports the disposition at issue of the patents awarded to inventors in each of the three birth cohorts. Members of the 1860–85 cohort would have been roughly similar in age to the "B" patentees in the 1910–11 subsample, so it is this comparison that is most instructive. Inventors in both the 1860–85 cohort of great inventors and the 1910–11 "B" subsample assigned about two-thirds of their patents at issue, with more than 60 percent of their assigned patents going either to large national companies or to companies bearing the inventor's family name. The similarity of the results for these two very different data sets provides powerful support for the idea that productive inventors were finding it increasingly difficult to maintain their independence. However, it also underscores the idea that they had other options besides working for large firms.

To get a better sense of what might be driving inventors' occupational choices, we break down the patents awarded to productive "B" inventors by the stage of the inventor's career as well as by the type of assignment. As Table 1.11 indicates, productive patentees seem to have had much more difficulty finding

TABLE 1.10. Change over birth cohorts in the types
of assignments at issue made by great inventors

	Birth Cohort			
Type of Assignment at Issue	1820–1839 (col. %)	1840–1859 (col. %)	1860–1885 (col. %)	All Cohorts (col. %)
Not assigned	69.5	56.5	35.3	54.1
Assigned to individual	9.6	5.6	3.5	6.1
Assigned to company with the same name	7.1	11.3	14.3	11.0
Assigned to large national company	1.5	11.8	27.0	13.3
Assigned to other company	12.3	14.8	19.9	15.6
Number of patents	908	1351	921	3180

NOTES: The definitions of the types of assignments at issue are the same as in Table 1.9, except "Assigned to individual" means that the inventor made a full or partial assignment to another individual and "Assigned to other company" includes both local and non-local companies not in the preceding categories. For a description of the great inventors sample, see the text.

assignees for their patents when they were first starting out than during the later phases of their careers. Although over time the overall assignment rate increased to such an extent that patentees from the 1910–11 subsample had higher assignment rates in the early stage of their careers than those from the 1870–71 subsample who were in the last stage of their careers, even for the 1910–11 group there were still dramatic differences in assignment rates across career stages.

These differences were particularly evident in the case of assignments to companies that bore the inventors' surname. Here again there was a general time trend. The propensity to assign to family-name companies increased quite steeply over time; as late as the 1890–91 group even inventors in the last stage of their careers assigned only 8.4 percent of their patents (amounting to only 13.2 percent of their assignments) to family-name companies. Nonetheless, inventors in the 1910–11 subsample who were in the early stage of their careers still assigned only a trivial fraction of their patents to family-name companies, whereas those who were in the final stage transferred fully 35.4 percent of their patents (amounting to more than half of their assignments) to such companies. We interpret these trends as indicating, first, that over time patentees increasingly were choosing to exploit their technological discoveries by forming their own firms, but second, that this option was generally not available to them until they had already established themselves as inventors.

The difference in the proportion of patents assigned to large national companies in the early and late stages of inventors' careers was not nearly as great as

TABLE 1.11. Changes in the distribution of patents
awarded to "B" inventors with ten or more patents, by type
of assignment at issue and stage of the patentee's career

Type of Assignment at Issue	Stage of Career		
	Early (col. %)	Intermediate (col. %)	Late (col. %)
Not Assigned			
1870–71 subsample	81.9	75.3	68.9
1890–91 subsample	62.0	52.7	36.6
1910–11 subsample	45.6	50.3	32.1
Assigned Share to Individual			
1870–71 subsample	6.2	6.7	3.6
1890–91 subsample	4.0	5.4	3.0
1910–11 subsample	6.9	4.0	0.9
Assigned in Full to Individual			
1870–71 subsample	4.1	11.4	10.8
1890–91 subsample	12.1	11.1	8.0
1910–11 subsample	7.2	3.1	1.9
Assigned to Company with Same Name			
1870–71 subsample	0.4	0.0	4.8
1890–91 subsample	2.2	4.2	8.4
1910–11 subsample	1.3	17.1	35.4
Assigned to Large National Company			
1870–71 subsample	0.0	0.0	0.0
1890–91 subsample	7.1	6.3	12.9
1910–11 subsample	12.1	7.3	19.2
Assigned to Other Local Company			
1870–71 subsample	6.6	3.1	7.6
1890–91 subsample	8.4	15.1	18.6
1910–11 subsample	17.1	11.7	4.1
Assigned to Other Company			
1870–71 subsample	0.8	3.5	4.4
1890–91 subsample	4.3	5.4	12.6
1910–11 subsample	9.8	6.5	6.3

NOTE: See Tables 1.3, 1.7, 1.9, and the text.

the difference for patents assigned to family-name companies. Inventors in the
1910–11 subsample assigned 12.1 percent of their patents (22.2 percent of their
assignments) to large national companies in the early stages of their careers
and 19.2 percent of their patents (28.3 percent of their assignments) in the later
stages. It may be that the relatively high proportion of these assignments early
on reflects a propensity on the part of these enterprises to hire university-trained
researchers for whom the college degree could serve as a signal of the quality,
making a previous record of successful invention less necessary to secure a

job.[26] Circumstantial evidence for this idea comes from the data set on great inventors. As we have seen, there was a substantial increase over time in the proportion of great inventors who attended college in engineering or the natural sciences. Those who had such backgrounds attained higher rates of patenting at younger ages than those without similar educational achievements.[27] Moreover, by the 1860–85 birth cohort great inventors who had college training in science and engineering had much higher assignment rates than other members of the group and were also far more likely to assign their patents to large national companies (Table 1.12).

TABLE 1.12. Changes in the distribution of great-inventor patents by type of assignment at issue and educational background

	Educational Background		
Type of Assignment at Issue	Primary or Secondary Education (col. %)	College in Non-Scientific Subject (col. %)	College in Engineering, Natural Science, or Medicine (col. %)
Not Assigned			
1820–39 birth cohort	64.9	82.0	82.9
1840–59 birth cohort	49.9	63.1	64.9
1860–85 birth cohort	49.1	44.0	24.5
Assigned to Individual			
1820–39 birth cohort	11.2	3.1	7.2
1840–59 birth cohort	5.8	8.6	2.9
1860–85 birth cohort	4.7	7.9	0.9
Assigned to Company with Same Name			
1820–39 birth cohort	8.4	3.9	2.7
1840–59 birth cohort	11.5	1.4	19.0
1860–85 birth cohort	3.5	25.0	14.8
Assigned to Large National Company			
1820–39 birth cohort	1.8	0.8	0.9
1840–59 birth cohort	15.3	11.1	4.9
1860–85 birth cohort	11.2	12.5	41.4
Assigned to Other Company			
1820–39 birth cohort	13.8	10.2	6.3
1840–59 birth cohort	17.5	15.8	8.3
1860–85 birth cohort	31.5	10.7	18.4
Number of Patents			
1820–39 birth cohort	669	128	111
1840–59 birth cohort	724	279	348
1860–85 birth cohort	232	216	473

NOTE: See Figure 1.2, Table 1.10, and the text.

These findings on patentees' career patterns lend credence to the idea that the technological advances of the Second Industrial Revolution had made it more difficult for inventors to maintain their independent status. To remain productive at invention, technologically creative individuals increasingly had to form some kind of long-term attachment with a firm, the main choices being either to take an employment position with a large-scale enterprise that was devoting significant resources to in-house R&D or to find financial backing for an entrepreneurial startup. Intriguingly, in choosing between these options inventors displayed a pronounced geographical bias. Although great inventors born between 1860 and 1885 assigned a large fraction of their inventions to national companies in all of the three main patenting regions, only in the East

TABLE 1.13. Changes in the distribution of great-inventor patents by type of assignment at issue and region

	Region		
Type of Assignment at Issue	New England (col. %)	Middle Atlantic (col. %)	East North Central (col. %)
Not Assigned			
1820–39 birth cohort	58.8	76.8	81.2
1840–59 birth cohort	47.7	60.9	52.4
1860–85 birth cohort	35.1	33.6	32.2
Assigned to Individual			
1820–39 birth cohort	9.0	12.9	2.0
1840–59 birth cohort	2.4	7.6	6.1
1860–85 birth cohort	5.2	2.8	2.5
Assigned to Company with Same Name			
1820–39 birth cohort	11.6	4.5	2.0
1840–59 birth cohort	19.2	5.4	17.9
1860–85 birth cohort	6.0	13.3	25.6
Assigned to Large National Company			
1820–39 birth cohort	2.4	0.5	3.0
1840–59 birth cohort	25.8	9.1	7.0
1860–85 birth cohort	22.4	30.5	26.5
Assigned to Other Company			
1820–39 birth cohort	18.2	5.3	11.9
1840–59 birth cohort	4.9	17.0	16.6
1860–85 birth cohort	31.3	19.9	13.2
Number of Patents			
1820–39 birth cohort	379	379	101
1840–59 birth cohort	287	647	313
1860–85 birth cohort	134	604	121

NOTE: See Tables 1.1 and 1.10 and the text.

TABLE 1.14. Changes in the distribution of patents in the

"B" sample by type of assignment at issue and region

	Region		
Type of Assignment at Issue	New England (col. %)	Middle Atlantic (col. %)	East North Central (col. %)
Not Assigned			
1870–71 subsample	76.1	75.6	83.0
1890–91 subsample	24.7	58.1	51.3
1910–11 subsample	35.0	38.1	44.6
Assigned Share to Individual			
1870–71 subsample	3.7	5.5	8.3
1890–91 subsample	3.8	5.3	4.8
1910–11 subsample	3.7	2.0	3.1
Assigned in Full to Individual			
1870–71 subsample	10.6	8.3	2.3
1890–91 subsample	7.8	4.5	18.3
1910–11 subsample	5.2	3.2	2.1
Assigned to Company with Same Name			
1870–71 subsample	0.6	2.3	0.5
1890–91 subsample	3.4	5.0	6.8
1910–11 subsample	23.0	2.7	31.4
Assigned to Large National Company			
1870–71 subsample	0.0	0.0	0.0
1890–91 subsample	15.5	9.4	3.8
1910–11 subsample	23.0	22.1	4.1
Assigned to Other Local Company			
1870–71 subsample	7.5	3.9	1.0
1890–91 subsample	30.8	9.5	10.6
1910–11 subsample	3.7	8.2	8.4
Assigned to Other Company			
1870–71 subsample	1.6	4.4	0.0
1890–91 subsample	14.1	8.2	4.4
1910–11 subsample	6.5	23.8	6.4
Number of Patents			
1870–71 subsample	322	434	218
1890–91 subsample	555	947	707
1910–11 subsample	383	601	1050

NOTE: See Tables 1.1, 1.3, and 1.9, and the text.

North Central did they assign an equivalent proportion to family-name companies (Table 1.13). The distinctiveness of the East North Central region is also apparent in the "B" data set (Table 1.14). By 1910–11, inventors in the Middle Atlantic were assigning 22.1 percent of their patents to large national companies and only 2.7 percent to family-name companies, whereas the figures for

72   inventors in the East North Central were just the reverse—only 4.1 percent of patents went to large national companies and 31.4 percent to companies that bore the inventor's name. New England was an intermediate case, with inventors assigning equivalent numbers of patents to both family-name and large national companies.

The relatively high proportion of assignments to large national companies in the Middle Atlantic is not surprising, given the tendency of such firms to locate near New York City, which was rapidly becoming the financial center of the nation. Indeed, the concentration of large firms near New York may account as well for the migration of great inventors to that region that we observed in Table 1.6. What is more puzzling is the disproportionately high fraction of assignments to family-name companies that occurred in the East North Central. One obvious hypothesis is that the well-developed financial markets of the Middle Atlantic efficiently funneled savings into the large-scale enterprises that were headquartered there, making it difficult for other kinds of enterprises to get financing. The Midwestern economy was not yet fully integrated into the national financial system.[28] It may well be that capital markets in this region, because they were more confined to local investment projects, offered entrepreneurs greater opportunities to raise funds. Other research that we have done on Cleveland, an East North Central city that was a hotbed of high-tech startups during the Second Industrial Revolution, highlights the existence of a vibrant local venture-capital sector—a sector that may have flourished (like Silicon Valley in the late twentieth century) in part because of its distance from New York.[29]

WAS SCHUMPETER RIGHT?

Although more work needs to be done, clearly there was some truth to Schumpeter's observation that the advance of technology had transformed the organization of invention and innovation by the early twentieth century—that the higher cost of conducting R&D was making it increasingly difficult for talented inventors to realize their creative potential on their own. This change shows up in our finding that the contractual mobility of specialized inventors was declining (that is, there was a sharp reduction in the number of different individuals and firms to which they assigned their patents) and that inventors were increasingly assigning their patents to firms with which they seem to have formed some kind of long-term attachment. It also shows up in the changing career patterns of specialized inventors. By the 1910–11 subsample of "B" patentees, even the most productive inventors were obtaining on average many fewer patents in the early stages of their careers than they were later on, after they had formed such attachments.

Inventors could procure the funds and complementary inputs they needed to pursue their vocation by taking employment positions in the R&D labs of large national firms. That many important inventors were indeed choosing this career option during the early twentieth century is readily apparent from our great-inventors data set. These highly talented individuals migrated in large numbers to the Middle Atlantic, where they increasingly assigned their patents to the large national companies that were concentrated in that region. Great inventors who had studied science or engineering in college were particularly likely to assign their patents to this type of firm. A big advantage of this career option for inventors with scientific training was that their university credentials enabled them to secure desirable positions early in their careers. A big disadvantage, of course, was their loss of independence, which, Schumpeter suggested, may have affected their creativity.

Schumpeter's account stops here, but our own analysis suggests that there is much more to the story—that talented individuals had other ways of gaining access to the resources needed for effective invention. In particular, they could become principals in firms formed for the purpose of exploiting their inventions. This alternative seems to have been most readily available in East North Central cities such as Cleveland, where there were local pools of venture capital. We find that productive inventors in the East North Central region were much more likely than those in other regions to assign their patents to enterprises that bore their name. This pattern, along with the recent emergence in areas such as Silicon Valley of similar pockets of entrepreneurship by the technologically creative, suggests that Schumpeter's pessimism about the future role for entrepreneurs in technological innovation was more than a bit extreme. Nonetheless, because inventors who chose this route first had to prove themselves in order to attract capital, they faced substantial hurdles to success—hurdles that, along with the greater amount of education that effectiveness at invention required, may help to explain the decline in patenting rates that occurred in the twentieth century.

The simultaneous emergence of these two alternative career paths, in combination with the geographic pattern of inventors' choice of one or the other career, casts doubt on the idea that large firms' R&D labs were an inherently superior environment for innovation. Rather, these trends are more consistent with the idea that the main advantage that large firms possessed was their superior access to the nation's main financial markets. Inventors seem to have found it a lot easier to organize their own firms in the Midwest, where local capital markets were not yet fully integrated into national ones. As the degree of financial integration increased over the course of the twentieth century, however, inventors would find this alternative more difficult to pursue, even in the Midwest. Not until changes in the regulatory environment in the late twentieth

74   century once again made it possible for local pools of venture capital to emerge in outlying areas, this time mainly in the West, was founding their own enterprises a viable strategy for significant numbers of inventors operating at the technological cutting edge.[30] In the interim, the economy's ability to meet the challenge of remaining innovative would mainly depend on developments within large firms — on the effectiveness with which the managers of in-house R&D labs learned to organize the creation and exploitation of new technological knowledge.[31]

NOTES

We would like to express our appreciation to our research assistants Marigee Bacolod, Young-Nahn Baek, Dalit Baranoff, Lisa Boehmer, Nancy Cole, Homan Dayani, Yael Elad, Gina Franco, Svjetlana Gacinovic, Brian Houghton, Charles Kaljian, Anna Maria Lagiss, Huagang Li, Catherine Truong Ly, David Madero Suarez, John Majewski, Yolanda McDonough Summerhill, Brian Rivera, Ludmila Skulkina, Dhanoos Sutthiphisal, Matthew Wiswall, and Tamara Zavaliyenko. We are also indebted to Zorina Khan for her collaboration with us in collecting the great-inventor data sets and for her invaluable advice, as well as to Marjorie Ciarlante and Carolyn Cooper for their help in accessing the records of the Patent Office at the National Archives. We have also benefited from the comments of Robert Allen, Ashish Arora, Sally Clarke, Iain Cockburn, Wes Cohen, Paul David, Stanley Engerman, Joseph Ferrie, Catherine Fisk, Louis Galambos, Avner Greif, Timothy Guinnane, Stephen Haber, Rebecca Henderson, David Hounshell, Thomas P. Hughes, Margaret Jacob, Adam Jaffe, Martin Kenney, Josh Lerner, Margaret Levenstein, Joel Mokyr, Ariel Pakes, Daniel Raff, Jean-Laurent Rosenthal, Bhaven Sampat, Suzanne Scotchmer, John Kenly Smith Jr., Scott Stern, William Summerhill, Peter Temin, Ross Thomson, Manuel Trajtenberg, Steven Usselman, Michael Waldman, John Wallis, Norton Wise, Gavin Wright, and Mary Yeager, as well as two anonymous referees and participants in seminar presentations at the Washington Area Seminar in Economic History, Harvard University, Northwestern University, Oxford University, Stanford University, and Yale University; at meetings of the Economic History Association and the NBER's Entrepreneurship Group; and at the Johns Hopkins University Conference, "Organizing for Innovation." We gratefully acknowledge the financial support we have received for this research from the National Science Foundation, as well as from All Souls College at Oxford University, the Russell Sage Foundation, the Social Science Research Council, the Collins Endowment, and the Academic Senate at the University of California, Los Angeles.

     1. Joseph A. Schumpeter, *Capitalism, Socialism and Democracy*, 3rd ed. (New York: Harper & Row, 1950), 134.

     2. Schumpeter, *Capitalism, Socialism and Democracy*, 131–63.

     3. See Kenneth J. Arrow, "Economic Welfare and the Allocation of Resources for Invention," in *The Rate and Direction of Economic Activity*, Universities–National Bureau Committee for Economic Research (Princeton: Princeton University Press,

1962), 609–25; David C. Mowery, "The Relationship Between Intrafirm and Contractual Forms of Industrial Research in American Manufacturing, 1900–1940," *Explorations in Economic History* 20 (Oct. 1983): 351–74; David C. Mowery, "The Boundaries of the U.S. Firm in R&D," in *Coordination and Information: Historical Perspectives on the Organization of Enterprise*, eds. Naomi R. Lamoreaux and Daniel M. G. Raff (Chicago: University of Chicago Press, 1995), 147–76; and David J. Teece, "Technological Change and the Nature of the Firm," in *Technical Change and Economic Theory*, eds. Giovanni Dosi, Christopher Freeman, Richard Nelson, Gerald Silverberg, and Luc Soete (London: Pinter, 1998), 256–81.

In recent years, this view of the inherent superiority of large-firm R&D has come under attack from scholars who show that it both overestimates the problems associated with the market exchange of technological ideas and underestimates the difficulty of managing technological information within the firm. See Ashish Arora, Andrea Fosfuri, and Alfonso Gambardella, *Markets for Technology: The Economics of Innovation and Corporate Strategy* (Cambridge, Mass.: MIT Press, 2001); Joshua Gans and Scott Stern, "The Product Market and the Market for 'Ideas': Commercialization Strategies for Technology Entrepreneurs," *Research Policy* 32 (Feb. 2003): 333–50; and Naomi R. Lamoreaux and Kenneth L. Sokoloff, "Inventors, Firms, and the Market for Technology in the Late Nineteenth and Early Twentieth Centuries," in *Learning by Doing in Markets, Firms, and Countries*, eds. Naomi R. Lamoreaux, Daniel M. G. Raff, and Peter Temin (Chicago: University of Chicago Press, 1999), 19–60.

4. For data on the timing of large firms' investments in R&D laboratories, see David C. Mowery, "Industrial Research and Firm Size, Survival, and Growth in American Manufacturing, 1921-1946: An Assessment," *Journal of Economic History* 43 (Dec. 1983): 953–80; and David C. Mowery and Nathan Rosenberg, *Technology and the Pursuit of Economic Growth* (New York: Cambridge University Press, 1989), 61-74. On the shift away from in-house R&D laboratories, see the essays in Richard S. Rosenbloom and William J. Spencer, eds., *Engines of Innovation: U.S. Industrial Research at the End of an Era* (Boston: Harvard Business School Press, 1996).

5. See Joseph A. Schumpeter, "The Creative Response in Economic History," *Journal of Economic History* 7 (Nov. 1947): 152.

6. Because of the way that the patent data was recorded, we had to choose a letter of the alphabet. We picked "B" because it generated the largest sample.

7. The data set was compiled in collaboration with B. Zorina Khan. The approach to the collection of information was basically the same as that previously followed by Khan and Sokoloff for the late eighteenth and early nineteenth century, except that we systematically collected information on assignments at issue and opted to sample all patents awarded through 1846 and then those in every fifth year from 1850 on, rather than obtain detailed information on every patent the inventor was granted. See B. Zorina Khan and Kenneth L. Sokoloff, "'Schemes of Practical Utility': Entrepreneurship and Innovation Among 'Great Inventors' During Early American Industrialization, 1790–1865," *Journal of Economic History* 53 (June 1993): 289–307. We also refer in this article to earlier work based on samples taken from the manuscript records of patent assignments. To be legally binding, a contract for the sale or transfer of a patent right had to be recorded with the Patent Office. These records are now stored in Record Group 241 (Records of the Patent and Trademark Office), National Archives II, Washington, D.C.

76   8. Of course, employers could require inventors to assign patents obtained on the job to them, but these contracts were relatively rare until at least the 1920s and were narrowly construed by the courts. See Lamoreaux and Sokoloff, "Inventors, Firms, and the Market for Technology"; and Catherine L. Fisk, "Removing the 'Fuel of Interest' from the 'Fire of Genius': Law and the Employee-Inventor, 1830–1930," *University of Chicago Law Review* 65 (Fall 1998): 1127–98.

9. For a comparison of the U.S. and British patent systems, see B. Zorina Khan and Kenneth L. Sokoloff, "Two Paths to Industrial Development and Technological Change," in *Technological Revolutions in Europe: Historical Perspectives*, eds. Maxine Berg and Kristine Bruland (Cheltenham, U.K.: Edward Elgar, 1998), 292–313.

10. See Lamoreaux and Sokoloff, "Inventors, Firms, and the Market for Technology"; and Naomi R. Lamoreaux and Kenneth L. Sokoloff, "Intermediaries in the U.S. Market for Technology, 1870–1920," in *Finance, Intermediaries, and Economic Development*, eds. Stanley L. Engerman, Philip T. Hoffman, Jean-Laurent Rosenthal, and Kenneth L. Sokoloff (New York: Cambridge University Press, 2003), 209–46.

11. B. Zorina Khan, "Property Rights and Patent Litigation in Early Nineteenth-Century America," *Journal of Economic History* 55 (Mar. 1995): 58–97; Naomi R. Lamoreaux and Kenneth L. Sokoloff, "The Market for Technology and the Organization of Invention in U.S. History," in *Entrepreneurship, Innovation, and the Growth Mechanism of the Free-Market Economies*, eds. Eytan Sheshinski, Robert J. Strom, and William J. Baumol (Princeton: Princeton University Press, 2007), 213–43; and Lamoreaux and Sokoloff, "Intermediaries in the U.S. Market for Technology."

12. Carolyn C. Cooper, *Shaping Invention: Thomas Blanchard's Machinery and Patent Management in Nineteenth-Century America* (New York: Columbia University Press, 1991).

13. Lamoreaux and Sokoloff, "Inventors, Firms, and the Market for Technology," 26. These figures are based on a sample of assignment contracts taken from the manuscript records in Record Group 241, National Archives II.

14. Lamoreaux and Sokoloff, "Intermediaries in the U.S. Market for Technology."

15. For definitions of these regions, see Table 1.1.

16. In previous work, based on the manuscript assignment records stored at the National Archives, we have shown that the more productive inventors were also more likely to use specialized intermediaries. See Lamoreaux and Sokoloff, "Intermediaries in the U.S. Market for Technology."

17. Lamoreaux and Sokoloff, "Inventors, Firms, and the Market for Technology."

18. Because the possibility of having more assignees increases with the number of patents, this way of describing the patterns in the data slightly overstates the strength of the relationship we want to highlight, but the qualitative result is robust to other approaches. It is also important to emphasize that our data include only assignments at issue. If we were able to track assignments that occurred subsequent to issue, our estimates of the number of different assignees would undoubtedly increase.

19. Note that contractual mobility was beginning to decline again for the cohort born between 1860 and 1885. We discuss this change in the next section.

20. It is also striking that the foreign born were significantly overrepresented, relative to the general population, among the great inventors. Those born abroad composed

22.9 percent (19.7, 33.0, and 11.4 percent of the respective three birth cohorts) of the great inventors born between 1820 and 1885 and received 24.9 percent (15.6, 34.7, and 17.9 percent for the respective cohorts) of the great inventors' patents. The foreign-born great inventors were more likely to locate in New England than their U.S.-born counterparts. Ferrie's estimate is from Joseph P. Ferrie, "Longitudinal Data for the Study of American Geographic, Occupational, and Financial Mobility, 1850–1990" (unpublished paper, Faculty of Economics, Northwestern University, 2003).

21. The following discussion is based on Thomas Parke Hughes, *Elmer Sperry: Inventor and Engineer* (Baltimore: Johns Hopkins University Press, 1971).

22. Those who received some schooling at institutions of higher learning are strikingly overrepresented among great inventors relative to the general population, in which the proportions of cohorts graduating from secondary school and college were under 10 percent and 3 percent respectively as late as 1900. See Thomas D. Snyder, *120 Years of American Education: A Statistical Portrait* (Washington, D.C.: U.S. Department of Education, 1993), Figures 11 and 17. See also B. Zorina Khan and Kenneth L. Sokoloff, "Institutions and Technological Innovation During Early Economic Growth: Evidence from the Great Inventors of the United States, 1790–1930," in *Institutions, Development, and Economic Growth*, eds. Theo S. Eicher and Cecilia Garcia-Peñalosa (Cambridge, Mass.: MIT Press, 2006), 123–58.

23. These figures may somewhat understate the trend because we are likely to have undersampled patents for both the first career stage of the first subsample and the last stage of the last subsample.

24. National Research Council, *Research Laboratories in Industrial Establishments of the United States, 1920–21* (New York: R. R. Bowker, 1921).

25. Not surprisingly, the trend toward assignments to large national and family-name companies was especially pronounced for patents in the manufacturing and energy and communications sectors, presumably the most capital-intensive on average in the economy. Using descriptions of the patents, we classified each of the patents in the three cross-sectional samples (1870–71, 1890–91, and 1910–11) into six sectors: agriculture and food processing; construction; energy and communications; manufacturing; transportation; and miscellaneous. Although the differences across sectors in 1870–71 were small and insignificant, for the latter two cross-sections patents in energy and communications and in manufacturing were much more likely to be assigned at issue than patents in other sectors and also more likely to be assigned to companies than to individuals. In addition, they were much more likely (energy and communications patents especially) to be assigned to large integrated companies, and somewhat more likely to be assigned to companies with the same name as the inventor.

26. On this point, see Naomi R. Lamoreaux and Kenneth L. Sokoloff, "Market Trade in Patents and the Rise of a Class of Specialized Inventors in the 19th-Century United States," *American Economic Review* 91 (May 2001): 39–44.

27. See Khan and Sokoloff, "Institutions and Technological Innovation."

28. Lance E. Davis, "The Investment Market, 1870–1914: The Evolution of a National Market," *Journal of Economic History* 25 (Sep. 1965): 355–99.

29. Naomi R. Lamoreaux, Margaret Levenstein, and Kenneth L. Sokoloff, "Financing Invention During the Second Industrial Revolution: Cleveland, Ohio, 1870–1920,"

78   in *Financing Innovation in the United States: 1870 to the Present*, eds. Naomi R. Lamoreaux and Kenneth L. Sokoloff (Cambridge, Mass.: MIT Press, 2007), 39–84.

30. See Samuel Kortum and Josh Lerner, "Assessing the Contribution of Venture Capital to Innovation," *RAND Journal of Economics* 31 (Winter 2000): 674–92.

31. For important examples, see the chapters in this volume by Margaret Graham (Chapter 2), Kenneth Lipartito (Chapter 4), and Steven Usselman (Chapter 8).

PART I
WITHIN FIRMS

Introduction to Part I

Sally H. Clarke, Naomi R. Lamoreaux,
and Steven W. Usselman

The three chapters in Part I dig beneath the broad statistical patterns documented by Lamoreaux and Sokoloff in Chapter 1 and delve deeply into
research and development activities at two notable firms: Corning Glass and
AT&T. Both of these companies reorganized their approach to innovation at
the dawn of the twentieth century, hiring scientists with advanced degrees and
developing significant internal R&D capabilities, and both sustained remarkable records of innovation for many decades thereafter. As these chapters
demonstrate, however, their ability to sustain innovation did not flow simply
from the initial choice to bring new expertise within the firm. Success resulted
from the ability of managers to reassess and reconfigure both the external and
the internal boundaries of their organizations on a continual basis. Neither
firm, moreover, pulled off the balancing act without crisis.

Though publicly traded and occupying a spot near the middle of the Fortune 500 for most of the century, Corning in many respects fits the pattern of
the entrepreneurial firms Lamoreaux and Sokoloff associate with the upper Midwest. Located in rural New York, it has long been run by a family possessing
considerable firsthand knowledge of its core technology. As historian Margaret
Graham shows in "Corning as Creative Responder: A Schumpeterian Interpretation of Disruptive Innovation," the company shunned an opportunity to
become an integrated mass producer of industrial glass and concentrated instead
on developing high-value specialty products from novel composites. This risky
strategy hinged upon the ability to link laboratory scientists with personnel in
other parts of the firm, particularly craft workers and machinery designers in
Corning's manufacturing facilities. The company's project-based approach also
frequently involved partnerships with groups outside the firm, including key

82   industrial customers. Managing such alliances required great sensitivity. Top management learned the value of sending appropriate signals to all parties. Through conspicuous technical accomplishments such as the grinding of a mirror for the famous telescope at Mt. Palomar, for instance, they demonstrated technical prowess externally while also conveying the value of a team-based approach to their own employees.

In the years following World War II, Corning adjusted its approach to innovation, at least partly in response to changes in its external environment. Impeded by new antitrust policies from forging the inter-firm alliances essential to its innovative strategy, Corning tilted toward integrated mass production of commodities for the booming American market. The departure from previous practice generated healthy returns for a while but eventually drew Corning to the brink of the sort of crisis that confronted automakers and other mass producers in the late 1960s and early 1970s. The firm recovered by reverting to a project-based strategy. Consciously drawing upon its own legacy from the 1930s, management again opened space for entrepreneurial individuals within the firm to pursue high-risk ventures. Working in partnership with large industrial customers, Corning achieved two stunning successes when it developed materials essential for catalytic converters and optical cable.

One of Corning's partners in the latter venture was AT&T, the focus of the remaining two chapters in Part I. These chapters examine the history of Bell Laboratories, perhaps the most lauded of all corporate research facilities. They probe the origins and ongoing evolution of internalized research and development at a classic Middle Atlantic corporation, one devoted above all to operating a sprawling, complex system for long-distance telephony and generating predictable returns for its investors.

In the first of these chapters, "Probability Theory and the Challenge of Sustaining Innovation: Traffic Management at the Bell System, 1900–1929," accounting historian Paul Miranti takes us back to the turn of the century, when AT&T was struggling to build out its long-distance network. The company had lost its patent monopoly and faced intense competition from numerous rivals. It embarked on a strategy of constructing urban exchanges that would serve dense aggregations of the most affluent customers. The approach aimed to generate significant returns on capital, thus satisfying the Wall Street financiers who kept a close watch on the company while also preserving funds to work on the sticky technical problems of long-distance telephony. The strategy hinged on AT&T's ability to identify the most attractive segments of its many urban markets. Miranti shows how the firm acquired such knowledge and gradually utilized it to coordinate activities across the entire company, including the scheduling of production at its manufacturing subsidiary, Western Electric, and the choices of hardware innovations pursued by its engineering designers.

A key step in this process occurred when AT&T's management, headed by
Theodore N. Vail, redrew the internal boundaries of the firm. Vail separated
responsibility for marketing and traffic management from the engineering
department that had long dominated the Bell companies. At first, the new commercial department relied on market surveys. Soon, however, it learned to
bypass the costly surveys and project potential demand using readily available
Census data on rental properties. Mathematicians within the department
devised advanced statistical techniques that enabled them to anticipate demand
on the basis of probabilities calculated from this data. Soon they incorporated
further refinements into the formulas, such as a mix of rate plans designed to
shape consumer demand. The array of statistical data, in addition to helping
AT&T coordinate production schedules and eliminate waste, enabled the firm
to assess potential returns from new innovations. Such calculations, for
instance, led Bell to invest in more trunk lines and automatic switching technology. The new form of information thus effectively reduced the uncertainties involved in building out the telephone system and enabled AT&T to
proceed on the basis of calculable risks.

The statistical techniques pioneered by the commercial department grew
so central to AT&T's success that management eventually placed the statistical group within Bell Laboratories, the new central research arm that the firm
formally established in the early 1920s. As Kenneth Lipartito shows in "Rethink-
ing the Invention Factory: Bell Laboratories in Perspective," the second of our
two chapters on AT&T, the firm was in no sense relegating statistics to a
peripheral or subordinate role by locating it within the Labs. Contrary to what
their later reputation as a bastion of "pure science" might suggest, AT&T
intended these research facilities to play a vital role in the management of the
phone system. Under the astute leadership of research director Frank Jewett,
the Labs retained a critical place in the line of authority. They functioned as
a mediator between installations in the field and the manufacturing operations
at Western Electric. All orders for equipment flowed through the Labs, where
personnel evaluated the particulars of technological practice and interpreted
them in the more universal language of scientific principles and technical standards, much as the commercial department expressed the complexities of consumer demand in statistical probabilities. Thus positioned, the Labs became
the arbitrators of technical practice, and innovation, across the entire corporation. Lipartito emphasizes that this function involved not only an ongoing effort
to manage the risks and uncertainties associated with potential innovation but
also an effort to portray the corporation as one devoted to relentless pursuit of
technical excellence. As at Corning, management of research required astute
manipulation of symbols intended to signal to insiders and outsiders alike that
the company valued innovation.

84 Like Corning, AT&T lost its way amidst the postwar affluence. The Labs continued to generate impressive technical accomplishments, such as the transistors used to further automate telephone switching. Such successes, however, gradually led management to venerate science and isolate it from other elements of the firm. Whereas Jewett saw science as a servant to problems dictated by commercial ends, and thus took care to situate researchers squarely in the line of authority, his successors viewed science as a force dictating the shape of the phone system and the enterprise, and perversely allowed it to grow more insulated. As Corning abandoned its history by coupling its innovative activities too closely to the short-term objectives of integrated mass production, AT&T misread its past by allowing its researchers to grow too remote from commercial objectives. New emphases on decentralization and systems engineering failed to right the balance and restructure the boundaries of the firm in ways more conducive to innovation. Meanwhile, competitors found ways to link innovations more directly to the needs of consumers. In the end, Lipartito concludes, the experience of AT&T belies Schumpeter's notion that science-based innovation could easily be routinized within the large firm.

2

CORNING AS CREATIVE RESPONDER

A Schumpeterian Interpretation
of Disruptive Innovation

Margaret B. W. Graham

Sooner or later, every historian of innovation, or of the innovative company, must deal with Joseph Schumpeter, the Austrian economist who is now best known for his prescient phrase "creative destruction." Nathan Rosenberg has called Schumpeter the only true "radical economist" for his discrediting of the neoclassical paradigm. For our purposes, Schumpeter merits special attention as an economist who believed that the mechanisms and actors of creative destruction could only be understood *ex post* by understanding their sociology and studying their history.[1]

Schumpeter distinguished between two types of innovators, whether entre-preneurial individuals or firms. One was the "adaptive responder" coping with economic change, either gradual or sudden, in economically predictable (or what Schumpeter termed "objectively rational") ways, while the other was the "creative responder,"[2] who initiated economic change in unpredictable (that is, "subjectively rational") ways.[3] It was the creative responder, Schumpeter theo-rized, who was the chief dynamic element in economic life, taking deliberate decisions that led to disruptive outcomes, setting up the renewed cycle of cre-ative destruction, and forcing other firms to make abrupt, often unforeseen tran-sitions.[4] Here I offer an account of one such episode of creative response, at Corning Glass Works, later Corning Incorporated.[5]

Corning belongs to a relatively select group of recidivists, creative respon-ders who engage in repeated acts of creative destruction, generation after gen-eration, decade after decade. Seen from this point of view, Corning and other creative responders form a special category of successful enterprises that do not fit the most common model of U.S. industrial success—the Chandlerian multi-divisional company that takes growth as its primary objective, relying on inte-gration and scale in production and distribution to achieve a dominant position

86   in its industry. Corning chose instead a more open and participative route to
profitability, involving technology sharing and permeable boundaries supported
by a set of inter-company relationships that were first called associations and
later became known as alliances.[6] The innovative culture that Corning established was interrupted for several decades in the mid-twentieth century. At that
time, an activist U.S. Justice Department interpreted Corning's strategy of knowledge generation and knowledge sharing as a predatory effort to collude with
other members of what it termed the Glass Trust to monopolize glass technology in machinery and container production. When the antimonopoly pressures
subsided in the 1980s, especially for companies that aimed to be competitive in
international markets, Corning eventually resumed, and improved upon, the
patterns of innovation that it had developed in its earlier era.

Corning's patterns of behavior in managing innovation closely resembled
what Schumpeter identified as signs of creative response. These included
acquiring and nurturing "superior" people as defined by the company's particular needs for expertise, participating in fields of activity that attracted and
motivated such superior employees' best efforts, and taking the risk of allowing these individuals degrees of freedom to pursue projects of their own choosing in their own way.

Protecting the autonomy of individual agents both inside and outside the
company to take entrepreneurial risk was a major factor in Corning's ability to
survive terrible economic transitions by responding creatively. This freedom to
take potentially profitable risk, rare in most organizations, was established early
at Corning and maintained in an organizational context that was deliberately
managed for innovation. In this chapter I propose a historical explanation for
Corning's successful transition as a creative responder through one especially
difficult economic period. I argue that the central patterns and intentional strategies that evolved at Corning before World War II were enduring resources that
enabled Corning to respond creatively in the 1970s, a nightmare era in U.S.
industrial life and one that caused many innovative companies to abandon innovation altogether, or at least to fall out of the ranks of creative responders.

Corning did not emerge from the 1970s transition in a better place because
it enjoyed uniquely favorable conditions. To the contrary, it fared worse than
many other companies. Its main customers for television tube glass, its major
growth product, disappeared; windshield safety glass, in which it had invested
ten years and many millions of dollars, proved to be a dud; its headquarters survived a devastating flood caused by Hurricane Agnes; its stock price dropped
from noteworthy highs at forty times earnings to unique lows, and it had the
first management layoffs ever ("the guns of August"). Yet during a period that
subdued or defeated many formerly innovative North American companies

Corning introduced two highly successful new products for important existing
markets, meanwhile developing to prototype stage its proprietary version of
optical fiber that helped transform the market for telecommunications. Corning's commercialization of optical fiber accelerated the adoption of fiber optics
in the worldwide telecommunications industry by about twenty years.[7]

BUILDING THE INNOVATIVE CULTURE

As economists have observed, even simple "dynamic capabilities" are not easily developed.[8] But the capacity for creative response is a complicated combination of dynamic capabilities, evolving over time along multiple historical
pathways. Corning's capacity for repeated creative response in the twentieth
century rested on a foundation that took half a century to build, was articulated
as an explicit strategy before World War I, and came to fruition in the interwar
period. Much depended on the company's early circumstances, decades of
experimentation, and success and failure in technologies and markets, leading
eventually to the formation of both an identity based on innovation and a strategic pattern of entrepreneurial risk-taking.

Corning started as a family company and remained one even after it was
publicly listed in 1945.[9] It was founded by the Houghtons—entrepreneurial
glassmakers who became not only industrialists but professional politicians and
diplomats, involving social skills that contributed much to the way they conducted their business.[10] Unlike the professional managers—agents—who lead
and staff most large managerial hierarchies, families are possessed of long memories and enduring interests. The Houghtons were first motivated to pursue
innovation by their shared memory of early disasters, the most vivid of which
was a loss of their company for more than a year in 1871–72. The family guarded
against future loss by investing heavily in the capacity to innovate—a combination of acquired and accumulated expertise; new organizational arrangements, including early corporate R&D laboratories of several kinds; and
leadership responsibilities for fostering innovation that became the particular
duty of each new CEO.

The location in the town of Corning, arrived at after two relocations, was
also a big factor in the Houghtons' decision to choose an innovation strategy
rather than the more common pattern for process industries at the time—
growth and consolidation. Located in remote upstate New York, the town has
remained a constant in the company's history and a factor in its distinctive
identity. Invited by a group of decorative glassmakers and leading citizen-
investors to locate their glassworks in Corning in 1868, the Houghtons took the
town's name for their company. At first the family believed local conditions

88   were ideal for a glassworks—both in supplies of raw materials and fuel and also in proximity to transportation. But what looked like a chance to follow the economic trend of the times and move to large-scale glassmaking turned out to be a disappointment. Although Corning had good connections by rail, the haul across the mountains from the huge markets opening to the West was still costly, and local material supplies of sand and coal were better suited to specialty versions of lead glass than to cheaper forms of lime glass used for high-volume containers and windows.[11] The establishment of thriving glass industries in Pittsburgh, Wheeling, and Toledo based on cheaper raw materials forced Corning to adopt a new strategy. One strategic advantage Corning had over the westerners was its exceptional access to good sources of specialty expertise. It was close to the city of Rochester, which had companies (such as Bausch & Lomb, American Optical, and Eastman Kodak) specializing in optical goods, and it was even closer to Cornell University in Ithaca, which pioneered in the use of microscopy and spectrometry as research techniques. It also maintained a strong association with the Carnegie Institute in Washington, D.C., as well as with the Mellon Institute in Pittsburgh.

Shut out of direct competition with the Western glass companies, Corning turned to specialty glass, which allowed it to capitalize on its secret store of glass compositions. At first its products were hard-to-produce items such as thermometer glass and glass tubing, specially designed railroad semaphores, and consistently colored red kerosene signal lanterns. To support its shift to ever-more-demanding specialty products Corning turned first to consultants out of universities and public and private laboratories. Then it formed one of the first corporate research laboratories in the United States. In 1908 the company hired as "chief chemist" Eugene Sullivan, a Leipzig-educated Ph.D. chemist who gathered around him a small group of Ph.D.s in physical chemistry, physics, and optics.[12] The members of Corning's interdisciplinary Chemical Department referred to themselves collectively as glass technologists even before the first university glass technology department opened at the University of Sheffield in England in 1912. From the business standpoint Corning's products had an odd defect: they were too durable to need frequent replacement. A constant stream of new products had to be found. The objectives of the early Chemical Department, therefore, were to study the physical properties of the company's many recorded glass compositions, to devise better production methods for existing products, and to search for new products.

Corning learned to anticipate product opportunities in changing social needs such as railroad and highway safety problems, or in major technical movements leading to new high-growth industries such as the electrochemical revolution, which required glasses with different properties for transmission line insulators, molded envelopes for lightbulbs or vacuum tubes, or chemically neutral test tubes for chemical research.[13] Early successes came when large customers with specialty needs—Thomas Edison with his lightbulb problem, or the Association of Railway Engineers wanting better standards for a range of signal colors— sought out Corning because of its reputation for problem solving and multi-faceted expertise.

By the time other glass companies, along with other process companies such as the iron and steel producers, were forming large industrial combinations, Corning was in a position to formulate a clear alternative strategy to combination, one based not on size and scale advantages but on superior knowledge. Large by the standards of glass companies prior to combination, but small by process-company standards in general, Corning benefited from close relationships with much larger customers such as General Electric, which could be persuaded to trade some of its vast financial resources and superior management know-how for Corning's research and craft capabilities. Corning had little choice if it wanted to avoid competing head on with these large companies; it had to adopt a different strategy.

It was Dr. Arthur Day, distinguished vulcanologist at the Carnegie Institute, who first proposed an explicit knowledge strategy for Corning in 1911. Knowing that Day had already served as a respected advisor to companies such as General Electric, Corning's new president Alanson Houghton asked for his help in managing what he and his brother regarded as their "growth problem," the very real possibility that the investments they were already making in advanced knowledge and expertise might require more resources to exploit than their company could generate internally, and might bring them into conflict with their customers. Against the wishes of their father, who had vigorously opposed investing in research, for fear of disclosing closely guarded compositions to the wider scientific community, the Houghton brothers Alanson and Arthur had promoted extramural efforts to engage in equipment invention. When they took over the company, they were free to pursue their own inclinations.[14] Arthur Day proposed that there was indeed an alternative to the strategy of combination and restraint of trade that was causing public outrage against the steel, transportation, oil, and electrical industries of the time.[15] This alternative strategy was to continue on a grander scale and in a more formal way what they had been doing already: investing in superior knowledge and parlaying that knowledge into advanced products and processes that giant, well-endowed companies formed from the merger of smaller enterprises were unlikely to come up with by themselves.

Having proffered this advice, Day agreed to help the Houghtons implement it. First he became a consultant, later a Corning vice president in charge of

90   manufacturing, and then for many years a vice president at large, while continuing to pursue his own scientific research and administrative activities in
Washington. Day thus became the first of many "superior people," in Schumpeter's term, that Corning went out of the way to accommodate so as to benefit from their ingenuity and expertise.

The knowledge Day considered to be Corning's best asset was the specialist knowledge of glass technology generated by combining craft and science,
that is, experimentation on the factory floor supported by laboratory research
both outside the company and in its own Chemical Department. Day advocated adding in-house process research to Corning's knowledge agenda as a
way of capitalizing on its existing advantage, for the Houghtons had already
recognized great value in the craft expertise resident in their skilled workforce.[16]
The Houghtons took Day's advice and initiated process equipment research,
at first by contracting with the Carnegie Institution. Like most company owners they were leery at first of antagonizing their workforces by doing process
research in-house. Their hope was that investment in process would take them
into completely unoccupied territory—the "uncluttered path," as they liked to
call it, not above, but *beyond* competition.[17]

To be sustained, the knowledge strategy soon required direct investment in
in-house research, regardless of the risks of making it visible to the workforce.
In 1916, Corning broadened its corporate research establishment, formerly only
the Chemical Department, by adding a process laboratory called the Mechanical Department. The Mechanical Department was headed by David Gray, a
distinguished engineer from MIT who shared the philosophy of combining
craft and science. Another "superior" person attracted to Corning by the
chance to put his philosophy into practice, Gray envisioned a department that
had full responsibility for the design and manufacture of all machinery pertaining to specialty glass. It would monitor what was going on in other companies, keep track of patents, conduct experimentation, anticipate the
requirements of other Corning departments, implement the best ideas of Corning workmen, and keep good records of all work done, especially invention.

The use of the ideas of skilled workmen in particular led to the invention
of several important machines that proved to be considerably better than any
that could be licensed abroad, and they found a market in the broader glass
industry. The company was able to attract craftsman inventors such as Austrian
tubing expert Ed Wellech, who perfected the Vello tubing machine. It also
filed the first of several patents on the ribbon machine—jointly invented by
David Gray and glassblower William Woods—which transformed the making
of glass lightbulb envelopes from a semiskilled blowing and molding process
to a fully mechanized operation achieving one thousand bulbs per hour by

1939. When most large manufacturing companies were striving to stabilize their processes, not change them, process innovation became a continuous habit at Corning.

Over time Gray's philosophy of respect for all levels of expertise also became engrained in the Corning culture. Forms of knowledge and the people who held them were accorded respect based on their contribution, rather than their provenance or the status of their contributor. Glass chemistry was too complex to be turned into an exact science, and some of the best solutions to problems came from those whose credentials were experience-based, not degree-based. Whether a laboratory was staffed by Ph.D. scientists, MIT-trained engineers, or craftsmen who had worked their way up on the factory floor, it was considered an important part of the whole. In the era between the wars, Corning's chemical research and process laboratories received equal budgets as a matter of policy.[18]

If workers were to be major sources of invention, they had to be treated well, and Corning's workforce policies became gradually more progressive. Long after most manufacturing companies had segregated their knowledge in specialized departments run by trained engineers, excluding craft for credential, Corning would continue to encourage and promote local experts whose knowledge was acquired on the job. Such men were sent to share their best ideas with other Corning plants, and their contributions received the same attention as knowledge possessed by holders of advanced degrees. This spirit of balance and equal treatment between different forms of expertise continued in the Pilot Plants, where television tube development—black and white and then color— was conducted in the 1940s and 1950s. In the 1960s, when the new Sullivan Park campus opened for business, the process-development section was the first to enjoy the new facilities, and, unusually by that time, it was still staffed partly by engineers and partly by craftsmen and employees with floor experience.

Having such an integrated research establishment gave the company the absorptive capacity to take advantage of technologies far beyond what it could have produced with its own resources.[19] Corning positioned itself to leverage parallel developments from much larger companies in the United States and Europe, as well as from private research establishments such as the Carnegie Institution and the Mellon Institute, sources Corning tapped regularly by hiring their employees.

In addition to attracting superior people, Corning's knowledge strategy benefited from participating in activities that posed difficult technical problems and generated public excitement. Just how far the value of such activities could go beyond the immediate financial return they generated was demonstrated by the huge two-hundred-inch mirror that Corning designed and manufactured in the 1930s for the Mount Palomar telescope. In the depths of the Depression this telescope mirror, made of Corning's proprietary material Pyrex, brought Corning considerable public attention. Astronomy was widely followed by the popular science media, and Corning's achievement was all the more news-worthy because the much larger and better-known General Electric Company had failed in an earlier attempt to produce the mirror.

The two-hundred-inch mirror project came Corning's way because the company had developed a capability in telescope mirrors as a kind of technical "loss leader." Not an especially profitable market, it offered a way of keeping the skilled workforce gainfully employed when regular product demand was low. It was also highly visible in the American scientific community, helping to attract to Corning leading scientists and any future business they might generate. The company had already produced numerous small telescope mirrors at cost in the 1930s when the badly-behind-schedule project for the new observatory on Mount Palomar presented itself in 1932.

General Electric had originally persuaded the Palomar funders to give it an open-ended contract to produce the Palomar telescope mirror using a radical new process of fusing powdered quartz. Tired of supporting a still-experimental process, and wary of all further claims that such a large mirror could be done at a reasonable cost and in a reasonable time period, the funders were persuaded by Arthur Day to transfer the project to Corning.

Corning's successful attempt to produce the huge mirror is usually credited to Corning physicist George McCauley, who started the telescope mirror line of projects and managed the Palomar project. McCaulay's superb organizing skills and ability to work across scientific boundaries were indeed crucial to completing the mirror on time and on budget. But McCaulay was not just being modest when he later stressed other factors in the Corning culture that were unusual for the time—the interdisciplinary cooperation and absence of hierarchy of expertise that contributed to making Corning both inventive and an able innovator.[20]

As frequently happened at Corning, the most important invention to come out of the Palomar mirror episode was not one used for the immediate project. This was vacuum deposition, a technique that stands as one of Corning's critical inventions because it eventually became crucial to Corning's success with fiber optics. Almost unnoticed at the time, the invention resulted from an independent research project that Frank Hyde, an organic chemist from Harvard, was inspired by the mirror project to undertake. Corning had hired Hyde to do exploratory research in organic chemistry at the time when the two-hundred-inch mirror project transferred from GE to Corning. While "monkeying around" with organo-silicons, Hyde became intrigued by GE's attempt to use

powdered quartz to make the huge disk. Using silicon tetrachloride, Hyde perfected a flame hydrolysis technique known as vapor deposition to produce a
pure glass-like substance. Corning filed for a patent on this technique in 1934,
and it was issued in 1942.

At the time vapor deposition appeared to have little utility for Corning. Not
until the 1950s when Corning received a small military contract to make radar
delay lines did the company use the technique.[21] Nevertheless, it was the gradual application of this proprietary technique to various small projects that made
it possible for Corning to develop its own proprietary process for making optical fiber in the late 1960s. By developing technologies in this way, not keeping
them "on the shelf" in codified formulas in lab books but actively using them
for small projects, Corning created multiple technical pathways that could be
pursued, combined, and recombined. Small individual projects might yield
little profit in themselves, but they could often be funded by other companies
or the government, and they offered much learning and experience that
remained with Corning. When sudden unforeseen customer needs or opportunities for knowledge sharing through alliances came along, the learning was
resident and ready to be activated in the Corning technical community.

ASSOCIATIONS—PRECURSORS TO ALLIANCES AND NETWORKS

One of the chief drawbacks to Corning's emphasis on knowledge accumulation as key to its strategy, and perhaps a more potent risk than exposing its glass
composition secrets, was the risk of creating too many usable ideas, more ideas
than it could afford to fund. Corning's solution to this problem went beyond
the organizational practice of the time—it formed associations, inter-company
relationships that would later be called alliances. So, for instance, when Frank
Hyde, who was exploring the boundary areas between glass and plastics, discovered new glass fiber binders, Corning exploited this technology by forming
a fiber-glass alliance with Owens Illinois, a 50-50 association called Owens-
Corning. And when the prolific Hyde later produced an economic version of
silicone, Corning formed another alliance with Dow Chemical to form Dow-
Corning. These and several other similar arrangements set the standard for
alliances as Corning would typically form and manage them. Corning's ability to make these relationships uncommonly effective reflected not so much
the skills of the glassmaker or industrialist as the other Houghton family expertise, politics and diplomacy. To give these alliances the maximum chance to be
creative, Corning tried where feasible to form a 50-50 joint venture with its
counterparty, always putting the Corning name second. The policy reflected
Corning management's belief in creative autonomy as a necessary condition

94   for innovation, whether inside its own operations or in its alliances. Corning reinforced the aims of the 50-50 governance structure by giving the new associations access to its own laboratories through regular monthly meetings with the heads of the companies and their research operations. Corning's associations formed the core of a broader group of interconnected companies that might now be called a knowledge network, a set of enterprises linked by a flow of information and expertise and committed to sharing their knowledge with Corning, and, through Corning, with other participants.[22]

Another way Corning leveraged its knowledge strategy without expanding beyond the level it could fund and control was to invent and patent process inventions for licensing to a wider network beyond the core of allied companies, to a large group of domestic and overseas licensees linked through non-compete agreements.[23] For many years regulators treated practices of technology sharing as acceptable as long as the outcomes for consumers were positive, or the products involved had important uses for the military. Corning kept its glass compositions to itself, but the company licensed larger glass companies and glassmakers such as General Electric to use its process inventions for glass machinery and for volume-oriented operations such as containers and light-bulbs. Although the mass production companies used its inventions as labor-saving devices designed to replace workers, Corning used the equipment to maintain a stable workforce. When demand for products such as lightbulb envelopes and vacuum tube housings was growing far more rapidly than Corning's ability to add and train workers, machines such as its ribbon machine kept Corning from exceeding the supply of semiskilled workers available in its local area as well as from having to import less-skilled immigrants that might undermine Corning's innovative culture.

Government Shock and Business Transition

In the face of the changing political circumstances in the 1930s, Corning's extended knowledge strategy of providing technology through its network of allies and licensees to much of the glass industry ran into determined opposition. Franklin D. Roosevelt's choice of Thurman Arnold to lead the Antitrust Division of the Justice Department led to a different view of technical monopoly than previous administrations had held, and Corning became one of the chief targets in the New Deal government's prosecutions of what it ominously termed the "Glass Trust." One memorable diagram titled "Major Inter-Company Relations in the Glass Container Industry" depicted Corning as the spider in a vast web of conspiracy using technology to control competition. Much larger companies were shown as mere nodes on the web, whereas the smaller Corning and its associations, Owens Illinois and Hartford Empire—the latter jointly owned by General Electric—were depicted as controlling a network of connections crisscrossing and binding the entire U.S. glass industry.[24] On the eve of the U.S. entry into World War II, while serving as a dollar-a-year man working for the War Production Board in Washington, Corning's young CEO, Amory Houghton, left his government post to lead a campaign of legal appeals against the company's and his own personal conviction for monopolistic behavior, ultimately ending in the Supreme Court.

Even before the matter was settled Corning had to abandon one of the strongest planks of its knowledge strategy, technology sharing through non-compete agreements with its wide network of licensees at home and abroad. In a terse letter to its Japanese licensee, Asahi Glass Company, in 1941, Corning waived its previous contractual covenant denying Asahi access to the U.S. market, allowing the Japanese company forthwith to import any goods it wished into the United States whether or not they contained Corning technology or competed with Corning goods.[25] As Thurman Arnold had intended when he began vigorous prosecution of antitrust legislation that had previously received mainly lip service, the experience and cost of defending the company's intellectual property strategy in court, and even the seeming arbitrary nature of the target companies and the penalties they received, caused Corning to adopt a more cautious approach both to patenting and to technology sharing for several decades.[26]

Going It "Alone"

The end of World War II thus found Corning with more serious regulatory limits on its knowledge sharing, and a reluctant acceptance of growth as a now unavoidable consequence of innovation. Though the Houghton family still controlled a large part of the business, the company issued stock not only to settle estate taxes but also to counter a public perception that it feared had damaged the company's reputation as the antitrust cases wound their way through the courts. The demand for high-tech companies, as Corning was now recognized to be, placed a whole new set of pressures on Corning's management. On the one hand much greater resources were available, but on the other hand, the resources had more strings attached.

Still, the culture of innovation that Corning had fostered before the war stood the company in good stead as the entire U.S. manufacturing establishment faced heavy demand on two fronts simultaneously. The U.S. military need for specialty glasses in such major programs as radar and its huge use of vacuum tubes for radio communication continued, but they also paved the way for many new and expanded peacetime businesses oriented toward mass consumer culture. With its background in electrical glasses, Corning was one of the few companies that could move quickly to supply the glasses that were needed for the explosive postwar entertainment product, television. For some time glass was the limiting factor in television production, and it continued to be a much-resented fact among television producers that the glassmakers were in the most profitable part of the television business. Because of their specialist knowledge, including hard-to-imitate craft-knowledge and knowledge embedded in high-cost capital equipment, they were able to keep it that way.

On two successive occasions Corning was challenged to "bet the company" on television glass—first when it supported the many black and white television producers with related but different cathode ray tube (CRT) designs, and second when it went out on a limb to help RCA produce color CRTs while the rest of the industry boycotted color television and tried to undermine RCA's pioneering efforts. At one point Corning could boast of having 125 percent of the market in color television glass—this number taking into account the inevitable breakage that occurred in shipping.[27]

In the circumstances the company found itself swept along by a universal tide of mass production. The needs of the high-volume business that television glass was becoming were quite different from their previous specialty glass businesses and caught them off-guard. It seemed that all the company's efforts needed to focus in this one area. By dwarfing more experimental efforts, television sucked all Corning's talent into minor improvements and all the investment money into funding further versions of the same products, and diminished the store of small developing techniques that could be teed up quickly if an existing product failed to materialize.

Meanwhile Corning Ware, the ovenware made from Pyroceram, a glass ceramic material also used for missile nose cones, became another high-volume success for a new Corning material. This new generation of versatile cookware—from icebox to oven to table—never broke and hardly ever wore out. It was a hard act that had to be followed—because it created a huge position in consumer distribution channels that would eventually have to be filled with another large consumer product.

A succession of "dual-use" (suitable for use in defense and civilian products alike) Corning innovations such as Pyroceram and television glass took the company further and further from its old niche strategy. From "uncluttered paths" Corning found itself on broad, if temporarily thinly traveled, highways where there was a very good chance of being overtaken by larger and well-endowed companies in search of new postwar products. The company thus joined the ranks of research-based companies that were expected to produce blockbusters, or "home runs," as they were referred to in the company.

At the same time the unintended result of Corning's postwar developments was a drift in the style of its activities, away from the exciting and technically motivating areas of leading-edge technologies into mass production. The high-volume consumer products drained their broadly diversified and interdiscipli-nary R&D organization, already depleted of talent and ideas both by wartime and by the need to support vast postwar manufacturing programs. Company leaders, still Houghton family members, accepted their personal responsibility to promote innovation, but construed this duty as a need to look for inventions that were predictably large from the start. The industries they aimed to supply were high-volume ones such as the car industry or the big consumer retail out-lets for tableware, generally less demanding customers technologically than their prewar customers.

To accompany its new self-contained, growth-oriented strategy, Corning rebuilt its central research capability in the 1950s. The emerging technology of the day most closely related to Corning's existing businesses was solid state electronics. Corning had produced simple electronic components related to its radar business, but it lacked a serious electronics research capability. In the early 1950s Corning sought out Edward U. Condon, an eminent physicist with unparalleled experience in the academy, industrial research, and government, to rebuild and reorganize Corning's research capability. His mandate included designing a fundamental research program, attracting the best people in elec-tronics, and gaining access for Corning to the leading-edge areas of solid state and nuclear technology. Condon also certainly qualified as one of Corning's "superior people," but even though his achievements and contacts in solid state and nuclear physics were gold standard, the plan proved to be unachievable in the political climate of the time.[28] As a high-profile advocate of international sharing of scientific knowledge, Condon became the target of Richard Nixon and his anticommunist allies. Despite repeated gestures of support by Presi-dent Truman, Condon lost his security clearance, and with it the ability to lead a company research program at a time when much depended on government connections. Condon did succeed in rebuilding Corning's research staffs, men-toring a young optics expert, William Armistead, to take his place when he resigned prematurely in 1954. Although he remained on Corning's payroll in much the same way Day had in an earlier time, Condon's ability to act on Corning's behalf concerning semiconductor research was compromised, and Corning failed to develop the kind of program that it needed to support its aspi-rations in solid state electronics.

The federal government's efforts to control access to leading-edge technical arenas during the Cold War led Corning to maintain a cautious distance, approaching diversification beyond glass with care. To protect its intellectual prop-erty from forced licensing or additional government march-in rights associated

98   with government-funded research contracting, it chose to keep much of its technology in the form of trade secret. In areas in which Corning had a clear advantage, as in massive glass structures, it developed new technology at its own expense and then offered it to the government. Meanwhile its efforts to participate in electronics and nuclear reactors—the leading scientific areas in which the brightest technical people of their generation were concentrated— encountered difficulty, threatening its independence and requiring resources that were too large for Corning alone to supply.[29]

Meanwhile from the late 1950s to the late 1960s Corning's traditional customer relationships, a perennial source of leading-edge ideas that had been sustained for more than a decade after the war, began to fray. All parties were under increasing cost pressures, and sought both greater productivity from their internal operations and greater efficiency from their suppliers. The G.E. relationship with Corning, thrown into the spotlight by the case against Hartford Empire and the Glass Trust, dissolved in mutual recrimination when G.E. researchers were discovered to have filed patents on research they had originally observed in Corning's laboratories. Meanwhile Westinghouse, another customer of long standing, threatened to integrate backward into glassmaking, going so far as to erect and equip a glass plant in Alabama before Corning agreed to negotiate a better price for its lightbulb enclosures. Finally, in the late 1960s, RCA, one of Corning's biggest and most lucrative customers, set up its own glassmaking facility to produce its highest-volume (therefore most profitable for Corning) CRTs. After supporting RCA through its difficult period of pioneering in color television, Corning had counted on a longer-term relationship. But all television producers were coming under pressure from foreign television manufacturers, and loyalty no longer mattered to the new generation of management at RCA.[30]

In the early 1960s Corning found itself in an unfamiliar situation, facing a new kind of "growth problem." No longer worried about controlling growth, it was under stockholder pressure to grow faster and larger, and to demonstrate that the successes of the postwar period were no mere accidents. Although it had many inventive ideas in early stages, tremendous cash flow, and an elevated stock price, it lacked the immediate late-stage investment opportunities to satisfy the urgent demand for growth articulated by analysts.

Had things gone according to plan with solid state electronics, Corning would not only have had a lucrative television glass business, it would also have developed into a significant player in electronic components. But things had not gone according to plan. Having suffered crucial years of delay in the rapidly developing business of solid state electronics, Corning opted to gain rapid entry into the semiconductor business by using its highly valued stock to acquire Signetics, a spinoff of the Fairchild Semiconductor company.[31] This uncharacteristic move to acquire rather than merge or develop internally turned out badly, partly because, despite early estimates of manifold potential synergies, Signetics lacked the intellectual property it needed, and consumed excessive amounts of management time and resources.[32] Like so many acquisitions made for stock and under time pressure, Signetics was neither best of breed nor did it possess a culture compatible with the innovative culture that Corning took pains to foster in its own technical community.[33]

Internally Corning concentrated its innovation forces on one huge follow-on project to television glass, "malleable glass" for automobile windows. Like television glass before it, this project, driven by senior management, consumed much of the capacity of the technical center and pilot facilities alike. This concentration of effort did not seem especially risky to Corning's managers. Like many other companies that had enjoyed success with science-based innovation after World War II, Corning's leaders believed in a research-based model of large-scale innovation, and had developed faith in their own ability to innovate on demand.[34] As they conceived of the process, all the key elements of disruptive innovation were in place. The research base was strong; senior management was driving the windshield glass project as it should; the federal government had mandated major safety improvements in cars; and the windshield glass had already been used in one production model of a vehicle, the AMC Javelin. Although the highly integrated auto companies had their own glass-production facilities, Corning was confident that the auto company capabilities in specialty glass were no match for Corning's.

Meanwhile the smaller projects waiting in the wings ready for development, but far from commercialization, languished because they received neither management attention nor significant development resources. Like the radar delay line project, they proceeded mainly where they could gain outside funding, but could not compete with the needs of the promising high-volume projects that were calculated to be bigger opportunities.[35]

THE 1970S—MAJOR ECONOMIC TRANSITION

The consumer electronics industry went into steep recession in the 1970s, the Middle East oil boycott combining with the latter stages of the Vietnam War to cause huge disruption in the industrial economy. The U.S. government responded with price controls, producing a period of stagflation, high interest rates, and capital scarcity. As Corning entered this period it sustained several shocks, a few completely unforeseeable and some of its own making.

In fact, the first half of the decade was a litany of Corning disasters. Market demand for glass for television tubes, after hitting an air pocket when RCA integrated backward into TV glass, leveled off again when Japanese competitors moved the center of television manufacturing operations, and therefore the main markets for all consumer electronics suppliers, to Japan. With this, the product that had been funding all of Corning's innovation efforts suddenly ceased to throw off cash. At the same time, Corning's main new product effort, safety windshields made of "malleable glass," also ran into difficulty. After ten years of effort and many millions of dollars in investment, the project had to be abandoned in the face of a superior float-glass windshield product made by the English company Pilkington Glass, Ltd. Float glass should have been anticipated as a threat and wasn't. In 1972 even Mother Nature seemed to endorse the company's humiliation when a hurricane caused severe flooding in the Chemung Valley, sweeping away part of Corning's Main Plant and headquarters and threatening to put a stop to the two new product programs that were Corning's immediate hope of staying afloat. For the first time, in 1975, Corning laid off many management-level people, its stock price plummeted from the heady ranks of the Nifty Fifty, and in the same year it sold Signetics, the semiconductor company acquisition that it had expected to achieve its diversification into electronic components.

Creative Response — Version Two

Faced with the prospect of leveling sales for television glass, and no big new product coming, Corning took a serious look at the measures other companies were adopting. It adopted stringent new financial controls, hired consultants to review its cost structure, and emphasized productivity and process controls. It set up a powerful new manufacturing engineering force that was empowered to require certain practices to be followed and brooked little opposition. What it steadfastly resisted, however, was allowing these measures to destroy its culture of innovation.[36] Although Corning's leaders, especially CEO Amory Houghton and president Tom MacAvoy, were stunned at the many adverse turns of events, they recognized an obligation to continue assessing their own situation independent of the prevailing wisdom purveyed by consultants. Corning's cautious approach to adopting the new tools and techniques of professional management — Net Present Value Analysis, Overhead Value Analysis, and various other approaches to increased productivity — was unusual. Its leaders continued to make room for risk-taking, and they found a number of risky projects to pursue.[37]

Looking at opposite conditions from a decade earlier, when there was too much money chasing too little opportunity, Corning turned in the 1970s to filling its "revenue gap" with smaller, shorter-term projects. With no home runs in sight, it reluctantly resorted to investing in singles, doubles, and even bunts.[38] Two immediate opportunities presented themselves. The first, Celcor, rose phoenix-

like from the ashes of the failed windshield glass; the second, Corelle, adapted a   101
laminated material originally intended for roofing tiles for use as a follow-on
tableware product to Corning Ware. The other innovation launched during the
1970s was a much longer-term and therefore more controversial investment that
turned out to be the sustaining business for the next generation: fiber optics.

Celcor and Corelle: Small Urgent Needs Become Big Products

When Corning's new president Tom MacAvoy audited Corning's list of new
product commitments and investments in 1971 he found a dismaying prospect.
Corning's plan to go into production on its inventive and demanding windshield product for the auto industry had one serious flaw: the auto companies
had no intention of buying it. Convinced that the social need for improved
automobile safety would not be met by seatbelts alone, and naively certain that
the government mandate would guarantee the market, Corning had developed
its laminated safety glass, using a leading-edge proprietary technique called
fusion glass. MacAvoy found that, however impressive the technological
achievements involved, what Corning termed *malleable glass* had a serious
downside from the auto companies' point of view that Corning had not taken
into account. It required massive retooling and reinvestment on the part of the
car companies. Meanwhile Corning's erstwhile licensee and long-time coop-
erator Pilkington Glass had found a way to adapt its own very successful process
innovation, float glass, for use in safety windshields. Although the results would
not match Corning's malleable glass in performance, the float glass promised
to be much less costly for the car companies to adopt. In an earlier era, when
Corning was in close touch with glass producers both internationally and
domestically, it would have been well aware of Pilkington's capabilities, but
in recent years the companies had exchanged only superficial information,
glossy annual research reports of "achievements" containing little in the way
of substance.[39]

Meanwhile MacAvoy learned on a visit to General Motors that the car companies were having difficulty with a different regulatory requirement, the Clean
Air Act for reduced auto emissions. Standards for this act had to be met in the
1974 model year. Though the auto companies' own laboratories had been working on ways of reducing emissions through devices attached to the exhaust system, these were proving to be expensive and hard to produce. Priority would
go to an internally developed device if it could be completed on time, but GM
was looking for a backup. A related material accompanied by a pocket of relevant experience existed at Corning: one of Corning's little projects had been a

form of ceramic material for use in automotive heat exchangers. It seemed just possible that Corning might supply this material as a more effective approach to improved emissions control.

There were at least two obvious problems with the emissions control opportunity. The timetable was extremely short, and the opportunity appeared to be time limited. In the mid-1970s a tremendous effort was going into developing entirely new forms of car engines such as the rotary engine. Any add-on device designed to control emissions would almost certainly be superceded in the space of about five years. Given the major technical uncertainties, and the brief time in which to recoup the investment, no reasonable financial analysis based on the prevailing technique of calculating the present value of future earnings would support such a project. The project's primary, almost intuitive, appeal was that it was the sort of risk that Corning had taken before—stepping in with a cheaper, more suitable material when bigger, better-endowed companies had faltered or failed.

MacAvoy returned to Corning's Technical Center bearing one of the more urgent stretch needs ever expressed by a customer. The allure of the project was the chance to maintain the relationship with the auto companies and access to their research organizations that was about to be lost with the demise of the windshield. Although it was not the only potential supplier, nor even the most obvious one, Corning took the chance.

When Corning decided to pursue the catalytic converter opportunity the company had certain advantages. First and most important was speed. The centralized nature of its technical organization allowed it to redeploy technical personnel very quickly. Second, it had already been doing small projects related to ceramics for automobiles—especially heat exchangers—and it had some familiarity with the potential problems. Third, it was accustomed to taking a bake-off approach to technical initiatives, trying out as many different ideas as it could come up with and then seeing which ones might work. In the case of the catalytic converter substrate, Celcor, two individuals, one in materials and the other in process equipment, worked together to develop the ceramic material and the intricate extrusion process that Corning ultimately chose to meet the auto companies' deadline and also their volume requirements. The interdependency of the material composition and the equipment were key here—based on deep Corning expertise and impossible to imitate.

The production process for Celcor, put together and integrated by an interdisciplinary team of engineers and researchers, was demanding, and the problems of achieving acceptable yields took several unprofitable years to resolve, but Corning managed to produce a product that was crucial to the auto companies at a price they would accept at the time they needed it for the new model year.[40] At first, General Motors, which had invited Corning into the project to start with, decided not to buy Celcor, using its own in-house design instead. But Chrysler and Ford did adopt the Corning approach, and after a couple of years of using its own version, featuring expensive platinum-coated beads in a flat metal pan, General Motors also switched to the Corning version.

The Celcor project turned out to be far more critical for Corning than simply filling a temporary revenue gap. It reaffirmed a business model that had been crucial to Corning's earlier success but had fallen out of fashion when growth and scale became the primary objective. It also launched Corning on an enduring business that proved to have follow-on opportunities in several directions—diesel engines for one, and filtration devices for chemical companies for another. More important, Celcor provided the opportunity for Corning to perfect its approach to high-volume process control techniques for extremely demanding materials. With Celcor, Corning developed vital new ways of working that transferred to other projects. First, the integrative project team effort that was so difficult to pull off for Celcor served as a model for later projects—especially for the fiber optics program, on which volume production started in the late 1970s. Second, the exacting production control methods that were necessary to achieve yields in the extruded ceramic product were also crucial later for the optical fiber project, and the same team worked on both of them. Third, it was on Celcor that Corning learned to "turn up the clock speed" on its competitors, increasing both the performance and the production difficulty of the product simultaneously, even when customers had not requested improved performance, and even when the price could not be increased.

With continuous improvement of Celcor, Corning took advantage of the path dependence of its customers. The presence of this form of catalytic converter, improving over time with Corning and others' efforts in performance and cost, made the switch to other forms of engine seem less vital to the auto industry. What was to have been a five-year product turned into an established automotive technology in the United States. Celcor substrates for catalytic converters qualified as a sustaining innovation in Christensen's terms much as Pilkington's float glass did in the matter of safety glass windshields. Its availability helped to head off the prospective new engine technology that would have been more disruptive to the auto industry even though it had been long anticipated.

The other innovation that provided a plank for Corning's life raft in the 1970s was Corelle, a lightweight laminated ceramic dishware that ended up selling billions of pieces well into the 1990s. Corelle, which turned out to be the follow-on product to Corning Ware, was made possible by the intense mechanical ingenuity of one of Corning's most prized maverick inventors, Jim Giffen. Giffen's career had begun at the end of World War II. It was his ingenious and completely intuitive approach to glass molding that solved the problem of making square television tubes, a problem Corning's scientists had declared to be insoluble. Giffen had been working on a machine to produce laminated ceramic material as a potential roofing material, a product aimed at another high-volume market that Corning had tried unsuccessfully to penetrate, the architectural materials industry. Unlike Celcor, aspects of Corelle had been under development since the mid-1950s, and the glass formulations and product specifications had been decided in the 1960s. But the high-speed vacuum-forming process previously used to mold various pieces of Corning Ware, called the hub machine, was very difficult to make operational for laminates.

Corelle posed the same kind of interdependency problems between material and machine that Celcor had had to solve. The high-speed continuous molding process had to be applied to a laminated product that derived its incredible strength from two different formulations of glass in three different layers. There were so many production inefficiencies in the process initially that Corning was losing money on every piece of Corelle it sold. It was only when the much improved process-engineering group that headed Corning's productivity efforts in the 1970s was able to move Giffen and his crew aside that the process became rationalized and eventually profitable.

Once Corning finally was producing dishware at the planned level, Corelle took very little time to reach profitability. From four hundred thousand pieces sold in the last quarter of 1970, Corelle sales leaped to almost forty million pieces in the next full year, and doubled the year after that. Sales reached into the hundreds of millions of dollars annually in the late 1970s, and over a ten-year period Corning sold literally billions of pieces of the lightweight dishware. The unanticipated reason for its outstanding success was a big change in market channels during the 1970s. Corning had relied on department stores as its main distribution channel for Corning Ware, but in the late 1960s and throughout the 1970s the big discount stores—the big box stores—began to exert a powerful effect on consumer buying. Corelle was a perfect product for these channels. The ability to make high volumes of product for low prices made Corelle a natural for these discount chains.

Neither Celcor nor Corelle were conceived of as disruptive innovations. In fact they were sustaining for their customers. By contributing to the delay of new internal combustion technologies (though another factor in the 1970s was the high cost of capital that reduced spending for new technologies), Celcor altered the pace of change in the automobile industry. Both products turned out to be much higher-volume products and for much longer than anticipated. Although both Celcor and Corelle could be construed as adaptive responses to Corning's

situation that also had unanticipated disruptive effects, together they paved the
way for a major episode of creative response for Corning—optical fiber.

Optical Fiber: Intentional Disrupter

The clearest example of creative response Corning undertook in the 1970s, Corning's optical fiber project, involved taking large-scale risk in the face of an unknown technology and a completely uncertain market. This was recognized to be a "big megilla" from the start, and the first unusual thing about it was that it was supported by Corning's top management so soon after its previous top-management-supported project—the safety windshield—had ended in abject failure.

Corning's optical fiber project history is far too complicated to be summarized within the scope of this chapter, and it has been recounted in detail several times elsewhere.[41] Here it is only possible to sketch out the ways in which it followed previous Corning patterns; drew on accumulated Corning technologies, expertise, and relationships; and, even so, involved serious risk-taking behavior on the part of many committed individuals inside and outside the company.

The fiber optics project began with a request for samples from a new leading-edge customer, the British Post Office, soon to be British Telecom. This customer was neither in a jam nor in a hurry: it was seeking the purest glass available, for purposes of experimentation. One Corning researcher at Sullivan Park, Robert Maurer, an MIT physicist who had earlier spent the most boring part of his career manufacturing radar delay lines in the laboratories using the vacuum deposition process, recognized an important advantage for Corning. Corning potentially had a combination of process and material no one else had—a glass-like substance purer than the purest glass anyone could make that could be doped to achieve desired qualities. This line of invention could be available only to an organization that, like Corning, had worked with extremely high-temperature furnaces. For the time being that put Corning in a class by itself. Although other large and distinguished laboratories, far richer and better equipped—were potentially interested in the use of optical fiber for telecommunications, and certainly had the resources to develop all needed capabilities, they were focused on other technologies. In the first instance then, Maurer defined a stretch problem for Corning to solve—technically so difficult to do, both in knowledge and know-how, that Corning could have a unique and probably patentable advantage if its projects were rigorously documented.

As already noted, the little fiber research project, starting up in the late 1960s, could hardly have emerged at a less auspicious time for Corning. The company was subjecting all small projects to new levels of financial analysis, and the fiber project came nowhere near clearing the bar. Although all interested customers expected to need something like optical fiber in twenty years, most of them expected to make it themselves. The only domestic customer for Corning, AT&T, had negotiated rights to everything Corning invented in the optical realm, leaving no chance that Corning would have solo control of the technology. Corning's technical leadership, still clinging to its home-run mind-set, initially rejected fiber optics as a strategic nonstarter.

But a team of Corning researchers, who in the spirit of creative autonomy were free to pursue ideas they cared about when their expertise wasn't urgently required for other more strategic projects, managed by using proprietary materials and processes developed for other projects, with the problem-solving skills of some key technicians, to break through a critical light-attenuation barrier. Less than 20 decibels of light loss was theoretically the target, and they demonstrated 16, later dropping the number to 2. The result was an announcement that, as the head of Corning research quipped, was a twentieth-century version of the "shot heard round the world." Major communications laboratories realized immediately that their timing was off, and that fiber optics could feasibly be used for long-distance communication in the foreseeable future.

Following the old pre–World War II pattern of leveraging pockets of proprietary expertise to work with much larger organizations, the optical fiber research team used the existing cross-licensing relationship with AT&T to gain access to Bell Labs and the laser work going on there. By the time the Bell Labs team realized two years later that Corning was making real headway with fiber, it was too late to slow them down. Though the doors to Bell Labs were barred from then on, Corning researchers could now find help and support from other quarters—especially the largest international cable producers.

Many of the important contributors to Corning's fiber project had prior interests in the technology behind it. Among these individuals, Chuck Lucy had managed the Navy's contract with Corning to produce radar delay lines. Lucy, who had not endured Corning's period of antitrust litigation, contrived to find financial support for the research project outside the company by creating a "joint development alliance," a new form of alliance reminiscent of the prewar network that had landed Corning in antitrust trouble.

The environment was becoming more favorable to U.S. companies that could roll out their technologies on an international stage, and Lucy convinced each of five different cable producers from Europe and Japan to put up $100,000 per year for five years. This amount of money was sufficient to support early phases of development, making the team exempt from Corning's financial hurdles. It was also enough to protect project personnel through Corning's worst period of layoffs and downsizing in 1975. Having to show con-

tinued progress in return for the financial support accelerated the project's timing considerably. By the mid-1970s Corning had filed for many key process
patents that would give it a commanding patent position in the production of
fiber optics, and twelve of those would eventually be granted. Defending these
patents by prosecuting infringers was both costly and risky, but an aggressive
campaign of patent litigation allowed Corning to enter the new business, not
just collect royalties for it. Breaking into the new business involved personal
risks for CEO Amory Houghton at a time when money was scarce and other
Corning businesses were crying for cash for far less risky investments. According to family tradition it was the CEO's duty to support long-term innovation,
and Amory Houghton took the necessary risks, though he said later that if he
hadn't been a family member the board would have fired him.

One of the personal risks Houghton took was to persuade two distinguished
veterans of Corning's established businesses, Dave Duke, who had led the Celcor project, and Al Dawson, who had led Corning's television glass business and
then its electronics division, to build the fiber optics business from scratch, giving up their senior executive positions in established product businesses. Duke
managed the fiber startup and Dawson took on the cable-manufacturing
alliance, Siecor. Siecor was yet another way in which Corning reestablished its
prewar pattern of innovation. The huge German electrical concern, Siemens,
had entered into one of the five Joint Development Alliances, and it had early
indicated its intention to work with Corning to build the fiber optics business
around the world. Siecor, Inc., and Siecor, GMBH, were two related alliances
formed to combine the Siemens expertise in cable making with the Corning
expertise in optical fiber. These alliances continued for twenty years and formed
the backbone of Corning's ability to supply not just optical fiber, which could
be predicted to become a commodity, but cable, which could be tailored for
individual customer need.

In addition to the internal risks involved in developing the business, there
were major timing risks. Corning had no control over when high-volume
demand might emerge. AT&T, however, which started producing fiber in the
late 1970s, still had a captive market among telephone operating companies and
could design experiments to use production quantities of the material. To break
into the new business Corning had to anticipate demand and to be ready to produce large quantities at a reasonable cost. CEO Amory Houghton found himself
signing one requisition for a costly new pilot production plant at the very time in
the summer of 1975 when the list of Corning managers to be laid off was sitting
on his desk. Despite the personal pressures of low morale and fear of failure,
Houghton continued to support, and override opposition to, every new plant
request for fiber. Often the paperwork for these plants was making its way

through Corning's formal approval processes while the plant was already under construction.

The decisive break for Corning came when AT&T was broken up by the courts in 1982 — effective 1984 — and MCI prepared to compete for AT&T's long-distance service by moving aggressively into fiber optics transmission. In 1982, fifteen years after the start of the fiber research project, MCI came to Corning with a large order on an urgent timetable. MCI not only wanted massive quantities for fiber, they wanted it priced well below the prevailing price at that time. MCI also wanted Corning's latest laboratory version, its single-mode fiber, which was significantly higher performing than the multimode version on which AT&T had standardized, but which had never been produced in quantity. Once again, project leaders Duke and Dawson took the huge risk of committing to the contract on the spot.

By 1984 Corning was positioned to be the largest supplier of optical fiber and, through Siecor, optical cable in the world. The newly liberated phone companies — the Baby Bells — welcomed an alternative supplier to Western Electric, which until the breakup they had been required to use as their supplier. Corning secured its dominant position by continuing to "turn up the clock speed" on its competitors, as it had so recently learned to do with Celcor. Though there were several times when financial analysts predicted that optical fiber would go through the premature downturn that Corning's earlier growth product, television glass, had gone through, Corning's ability to keep advancing product performance and Corning and Siecor's efforts to improve the quality of production contributed to growing the market and maintaining Corning's high share of it.

Optical fiber mirrored some of Corning's earlier patterns of innovation, but it also improved upon previous patterns. Unlike the days of television glass, Corning did not rely on relationships alone to maintain its profitable position, but kept aggressively improving the product beyond customer requests, or at least well ahead of them. Moreover, it did not rely on just two major products to support its business. It continued to keep alive other smaller business opportunities, on the basis of active, in-use technologies. It also developed two complementary networks over which to share and develop technologies — an outside network of alliances, as in the days of the associations, and an inside network of laboratories. This second, inside network had its start with a new laboratory at Corning's international headquarters in France, the Avon Laboratory.[42] Over time the internal network expanded to include an optical laboratory in St. Petersburg, Russia, and a television display glass laboratory in Shizuoka, Japan, as well as, for a short time, a few laboratories added with acquisitions.

Both its outside and its inside networks worked best when Corning followed the principles it had long observed—attempting to manage for a high degree of creative autonomy at its various sites. Elsewhere I have argued that these outside sources of technology were key to Corning's continued ability to find and pursue innovative ideas in the 1980s and 1990s, none more important than the Avon Laboratory in France. However, it was the availability of live technologies waiting to be developed into products, and shared with the other laboratories, that gave the new laboratories their chance to benefit the whole. Corning was hardly alone in developing network relationships, but its unusually long history of making them work, and of using them in a way that allowed them to innovate rather than just modify, was unusual.

CONCLUSION: SCHUMPETER'S PREDICTION

Toward the end of his career, Joseph Schumpeter believed that the day of the innovating entrepreneur, the active intentional risk-taker for ideas, had ended. Schumpeter believed that large companies had developed the capacity to institutionalize all the elements they needed to generate innovation internally, and further that the capacity to innovate might well be systematized to the point that large corporations could control both customer preferences and their own environments in their own interests. At the same time, Schumpeter feared that much could be lost in the process. When large companies essentially took the place of the customer, and taste became endogenous, he believed that capitalism could decline beyond hope of revival.

Had Schumpeter lived twenty years longer he might have seen that his prediction of the end of the entrepreneur, or at least the end of the need for the entrepreneurial function, was premature. As Corning's experience in the mid-twentieth century shows, creative response and the creative responder might not have been evenly visible at all times, but they were at work; and they would come to be sorely needed by an economy dominated by large, risk-averse companies, especially when their first experience of intense international competition drove them to focus narrowly on productivity and quality.

The three indicators Schumpeter cited as central to creative response all point to factors that lie at the heart of what entrepreneurial risk is about. Creative response is hard to find in the context of conventional managerial hierarchies because it involves risks that bureaucracies are structured and managed to avoid or minimize. Though Corning's survival in the 1970s hardly offers a testable formula for creative response, it does suggest ways in which creative responders make their choices differently from others. Risk accompanied many of these choices: financial risk for the company or for individual investors to

110   be sure, but also personal risk—of failure or of personal disappointment—and organizational risk—of disruption of management systems, or plans painstakingly crafted. At Corning, accepting these risks, and striving for the rewards that accompanied them, was the essence of innovation. Without such patterns of creative response, found at many companies and repeated over many business cycles, Schumpeter's prediction of the decline of the capitalist system would almost certainly have been fulfilled. Disruptive though they are, especially to industrial organization, creative responders provide a vital balance to the capitalist system, and are one of the most important resources the system as a whole can use for self-renewal.

NOTES

1. See Thomas K. McCraw, *Prophet of Innovation: Joseph Schumpeter and Creative Destruction* (Cambridge, Mass.: Harvard University Press, 2007); and Thomas K. McCraw, "Schumpeter's Business Cycles as Business History," *Business History Review* 80 (Summer 2006): 231–61. See also Louis Galambos, "End of the Century Reflections: Weber and Schumpeter with Karl Marx Lurking in the Background," *Industrial and Corporate Change* 5 (Jan. 1996): 925–93; and Nathan Rosenberg, *Exploring the Black Box: Technology Economics and History* (New York: Cambridge University Press, 1994), 47–61.

2. It was to this paragon of conscious risk-taking that Schumpeter's contemporary Herbert Zassenhaus referred when he called Schumpeter's entrepreneur, "a social miracle . . . an event beyond the laws of nature and society." Quoted in McCraw, *Prophet of Innovation*, 12.

3. For Schumpeter's distinction between "objective rationality," which does not have to be conscious, and "subjective rationality," which is a highly conscious working out of an individual, particular, and contextual view of economic rationality, see Joseph A. Schumpeter, "The Meaning of Rationality in the Social Sciences," in *Joseph A. Schumpeter: The Economics and Sociology of Capitalism*, ed. Richard Swedberg (New York: Cambridge University Press, 1991), 327–30.

4. See Joseph A. Schumpeter, "Comments on a Plan for the Study of Entrepreneurship" in Swedberg, *Joseph A. Schumpeter*, 406–24. Note more recently Clayton M. Christensen's similar categories of disruptive and sustaining innovation in "The Rules of Innovation," *Technology Review* 105 (June 2002): 33–38. Christensen's categories assume that companies exist within reference industries. Schumpeterian theory, by comparison, leaves open the possibility that creative response transcends the industrial frame of reference.

5. This argument is drawn from Margaret B. W. Graham and Alec T. Shuldiner, *Corning and the Craft of Innovation* (New York: Oxford University Press, 2001).

6. Naomi R. Lamoreaux, Daniel M. G. Raff, and Peter Temin have argued that integrated managerial hierarchies have given way to organizations with more open and permeable boundaries ("Beyond Markets and Hierarchies: Toward a New Synthesis of American Business History," *American Historical Review* 108 (Apr. 2003): 404–33), but

I am here suggesting that open and permeable boundaries were an alternative employed by innovating entrepreneurial firms in an earlier era as well, as a more flexible alternative to combinations.

7. Margaret B. W. Graham, "Financing Fiber: Corning's Invasion of the Telecommunications Market," in *Financing Innovation in the United States: 1870 to the Present*, eds. Naomi R. Lamoreaux and Kenneth L. Sokoloff (Cambridge, Mass.: MIT Press, 2007); and Jeffrey Hecht, *City of Light* (New York: Oxford University Press, 1999).

8. David J. Teece and Gary Pisano, "The Dynamic Capabilities of Firms: An Introduction," in *Technology, Organization, and Competitiveness*, eds. Josef Chytry, Giovanni Dosi, and David J. Teece (New York: Oxford University Press, 1998), 193–212.

9. Direct ownership by the family fell below the 10 percent level requiring disclosure in the 1980s, but a Houghton family member continued to serve in the senior management of the company for most of the rest of the twentieth century.

10. Davis Dyer and Daniel P. Gross, *The Generations of Corning: The Life and Times of a Global Corporation* (New York: Oxford University Press, 2001), 109, 115–16.

11. Dyer and Gross, *The Generations of Corning*, 47.

12. Graham and Shuldiner, *Corning and the Craft of Innovation*, chapter 3.

13. See the advertisement "The Magic and Mystery of Glass," listing a large array of Corning's specialty glass products, that ran in the *Saturday Evening Post* in November 1917.

14. Alanson Houghton, Corning's third-generation president and later U.S. ambassador to the Weimar Republic, did graduate work in Germany after graduating from Harvard. There he made contacts with German Junker families, scientists, and industrialists.

15. Under Amory Houghton, from 1895 to 1905, Corning had entered into its own set of clandestine agreements to control the market for railway signalware, railway lanterns, and electric lightbulb production. The next generation needed an alternative to these agreements, which were clearly banned under the Sherman Antitrust Act of 1890. See Dyer and Gross, *The Generations of Corning*.

16. Graham and Shuldiner, *Corning and the Craft of Innovation*, 80.

17. Day's brief stint as Corning's head of manufacturing after World War I reinforced his belief in this as a competitive advantage. Notwithstanding the sophisticated process knowledge that he had encountered in European companies and industrial laboratories, Day concluded that the seasoned U.S. workforce was superior to a European workforce in ingenuity and ability to adapt to new mechanical processes. He believed that the legions of unskilled Eastern European laborers being imported to staff the high-volume glass establishments in Pittsburgh and Wheeling in the 1920s could be no match for Corning's experienced workforce, with decades of skilled practice and a wartime's worth of maximum effort under its belt.

18. General practice would have suggested that the Mechanical Department receive greater funding, but at Corning equal budgets had symbolic value as well as substance.

19. For the importance of what would now be called "absorptive capacity" enabling firms to use knowledge generated outside the firm, see Wesley M. Cohen and Daniel A. Levinthal, "Absorptive Capacity: A New Perspective on Learning and Innovation," *Administrative Science Quarterly* 35 (Mar. 1990): 128–52.

112

20. George McCaulay, quoted in Graham and Shuldiner, *Corning and the Craft of Innovation*, 112.

21. J. F. Hyde, "Method of Making a Transparent Article of Silica," U.S. Patent 2,272,342, filed on August 27, 1934, issued on February 10, 1942.

22. For modern interpretations of this strategy of association, see Stefano Brusoni, Keith Pavitt, and Andrea Prencipe, "Knowledge Specialization, Organization Coupling, and the Boundaries of the Firm: Why Do Firms Know More Than They Make?" *Administrative Science Quarterly* 46 (Dec. 2001): 597–621.

23. Although the non-compete agreements clearly violated the letter of the Sherman Antitrust legislation, Corning executives and other observers liked to point out that the result of this practice had been steady reduction in the cost of glass products along with constantly improving quality and continued investment in innovation. Graham and Shuldiner, *Corning and the Craft of Innovation*, 282.

24. Detailed discussion of the conviction of the Glass Trust, Corning, Hartford Empire, and the glass container companies can be found in Graham and Shuldiner, *Corning and the Craft of Innovation*, Chap. 8, 279ff. Although the Supreme Court weakened the penalties imposed by the lower court, all of the companies were required to license their technology at reasonable royalties, and Corning ended up signing consent decrees in three different matters.

25. Graham and Shuldiner, *Corning and the Craft of Innovation*, 144.

26. Ellis W. Hawley, *The New Deal and the Problem of Monopoly* (Princeton: Princeton University Press, 1966), chapter 22. See also Spencer Weber Waller, "The Antitrust Legacy of Thurman Arnold," *St. John's Law Review* 78 (Summer 2004): 569–613.

27. Graham and Shuldiner, *Corning and the Craft of Innovation*, Chap. 7.

28. For an account of Condon's career before World War II, see Thomas C. Lassman, "Industrial Research Transformed: Edward Condon at the Westinghouse Electric and Manufacturing Company, 1935–1942," *Technology and Culture* 44 (Apr. 2003): 306–39. Lassman points out that at Westinghouse Condon had been charged with putting together a fundamental research program in theoretical physics, which included putting in an atom smasher, and creating a Westinghouse Fellows program.

29. Graham and Shuldiner, *Corning and the Craft of Innovation*, Chap. 6.

30. See Margaret B. W. Graham, *RCA and the VideoDisc: The Business of Research* (New York: Cambridge University Press, 1986).

31. Graham and Shuldiner, *Corning and the Craft of Innovation*, 272–73, 305–6.

32. Here again a respect for the mix of craft and science was an important factor. Manufacturing managers whom Corning sent to help improve Signetics's approach to semiconductor production encountered resistance on this issue. Graham and Shuldiner, *Corning and the Craft of Innovation*, 304–7, 349.

33. Among numerous examples of high-tech companies that acquired unsuitable subsidiaries for stock in the mid-1970s, two that stand out for having similar negative outcomes are Northern Telecom and Xerox.

34. The widespread belief that top-down management of technology-based projects should almost automatically result in swifter and more efficient use of resources and more dramatic results was characteristic of the postwar "linear model" of innovation. David A. Hounshell, "The Evolution of Industrial Research in the United States," in

Engines of Innovation: U.S. Industrial Research at the End of an Era, eds. Richard S.
Rosenbloom and William Spencer (Boston: Harvard Business School Press, 1996),
13–85. Many large companies with significant R&D establishments and budgets to
match fell victim to this blockbuster mentality—RCA, Alcoa, Eastman Kodak, Xerox,
and Polaroid are among the many documented cases. See Martin Kenney and Richard
Florida, *The Breakthrough Illusion: Corporate America's Failure to Move from Innovation to Mass Production* (New York: Basic Books, 1990). See also Margaret B. W. Graham and Bettye H. Pruitt, *R&D for Industry: A Century of Technical Innovation at Alcoa*
(New York: Cambridge University Press, 1990).

35. See Graham and Shuldiner, *Corning and the Craft of Innovation,* 275, for how
difficult it was to get approval for advanced forms of dishware that would formerly have
been developed without question.

36. So many large professionally managed companies went overboard that two pro-
fessors at the Harvard Business School published a controversial article on the subject.
See William J. Abernathy and Robert H. Hayes, "Managing Our Way to Economic
Decline," *Harvard Business Review* 58 (July-Aug. 1980): 67–77.

37. See Graham, "Financing Fiber."

38. Graham and Shuldiner, in *Corning and the Craft of Innovation,* 371–72, quote
Joseph Littleton's memo to Amory Houghton in which he points out that Corning's success has been just as dependent on singles and doubles as it has on research-based long
balls. This memo came in 1979 when the general disaffection with research came to a
head in U.S. industry.

39. For the failure of malleable glass and the new product programs that followed
it, see Graham and Shuldiner, *Corning and the Craft of Innovation,* Chap. 9.

40. A Harvard Business School case study, "Corning Glass Works: Erwin Automotive
Plant, 1973" describes how hard this extrusion process was to balance, and how risky it
was to the company.

41. Graham and Shuldiner, *Corning and the Craft of Innovation,* Chaps. 9 and 10;
Dyer and Gross, *The Generations of Corning;* Jeffrey Hecht, *City of Light;* Les C. Gun-
derson and Donald B. Keck, "Optical Fibres: Where Light Outperforms Electrons,"
Technology Review 86 (May-June 1983), 32–44; Ira C. Magaziner and Mark Patinkin,
The Silent War: Inside the Global Business Battles Shaping America's Future (New York:
Random House, 1989); and Joseph G. Morone, *Winning in High-Tech Markets* (Boston:
Harvard Business School Press, 1993), 125–98. See also Graham, "Financing Fiber."

42. See Margaret B. W. Graham, "Less Transfer Than Transformation: The Formation of Corning's Avon Laboratory" (paper presented at the Johns Hopkins Conference on Organizing for Innovation, Baltimore, October 26, 2002); and Margaret B. W.
Graham, "From Satellite Laboratory to Partner: The Formation of Corning's
Fontainebleau Laboratory" (paper presented at the Business History Conference, Lowell, Massachusetts, June 26–28, 2003).

3

Probability Theory and the Challenge of Sustaining Innovation

Traffic Management at the Bell System, 1900–1929

Paul J. Miranti Jr.

This chapter explains how the Bell Telephone System sought to sustain the innovative application of probability theory in traffic management to enhance operational efficiency and to minimize the financial risk during the period 1900–1929. A near bankruptcy in 1906–1907 resulting from overinvestment in traffic facilities persuaded the new top management headed by Theodore N. Vail of the need for improved capital budgeting methods. Probability theory became a key element in overcoming uncertainty and quantifying risk in an integrative procedure known as the "fundamental plan" that equated market demand, system capabilities, and finance. The need to sustain this managerial innovation increased during the 1920s, when the firm began making the transition from manual to automatic message switching.

Mathematics incorporated attributes well suited for sustaining managerial innovation. It extended understanding by reducing complex socioeconomic and technological phenomena to simpler and more tractable quantitative expression such as formulas and graphic representations. This afforded the possibility of extending business understanding by applying the rules and axioms of mathematical science in the analysis of economic data. The insights derived from these cognitive endeavors often proved useful in enhancing the effectiveness of administrative processes.

The boundaries for maintaining creative innovation for probability remained largely circumscribed within the organizational confines of the Bell System. The restricted focus resulted primarily from the fact that few external organizations of any type sought to apply probability to the resolution of the unique problems of message switching. The locus of firm-specific learning changed in response to an evolving set of organizational priorities. Initially this activity centered on AT&T's staff engineering department during the 1890s and focused on the discovery of new quantitative methods for analyzing the most efficient design and location of local switching exchanges. This first venture proved unsuccessful primarily because of the insufficient mathematical learning of the staff, poor data, and misconceptions about the factors that drove telephone demand. The second thrust that occurred during the first decade of the twentieth century proved much more successful after the firm succeeded in developing new ways to apply probability to the challenges of local office organization and to analyze the factors contributing to local market growth. Top management further extended the importance of this new knowledge through its integration as a key component in a complex sequence of procedures that involved local office planning, capital budgeting, and production scheduling. By the mid-1920s, however, the boundaries of creative innovation had changed again in two ways. First, the exchange layout capabilities derivable in exchange design from probability found a major new outlet as an adjunct procedure in the broad adaptation of automatic switching facilities throughout the telecommunication network. The second involved the inclusion of a specialized unit for studying probabilistic applications in traffic operations in the Bell Telephone Laboratories. The new unit concentrated on extending innovation though new channels that included conducting research on improved methods, monitoring external professional developments, promoting greater understanding through educational programs, and representing the firm before regulatory authorities.

The institutional context in which the firm operated also provided incentives for maintaining a strong commitment to innovation in probabilistic management. The changing nature of the market and of the structure of regulation represented two principal factors that affected innovation. Although the market experienced two dissimilar phases, the demand for new ways to accumulate and analyze data remained high. In the first the firm sought new information about market dynamics and ways to control costs as it prepared to face greater competition after the expiration of the Bell patents in the 1890s. The need for this information actually increased during the first decade of the twentieth century as the firm became increasingly dominant in telecommunications. Strong market power seriously reduced the usefulness of market pricing signals as an aide to managerial allocation decisions. To overcome the loss of price data, the firm, like a command economy, became increasingly dependent in its planning on a growing body of institutional arrangements for gathering and analyzing accounting and statistical information about its business environment. With respect to regulation, the development of probabilistic management paralleled the rise of state regulation that began during the Progressive Era. In this latter case, probability theory represented a powerful tool rooted in science that both

116   enhanced the firm's image of authority in the public mind and provided powerful evidence to justify its policies in actions before oversight boards.

Although scholars have emphasized how statistical knowledge extended
understanding of the physical and social worlds, they have said little about its
role in maintaining innovation. Some have concentrated, instead, on showing
how statistical thinking was useful in fields as diverse as economics, navigation,
geodesy, genetics, and thermodynamics.[1] Others have evaluated the central
role of statistics in the evolution of econometrics.[2] Still others have stressed the
role of this knowledge in fostering new thinking about social matters.[3] Historians have also analyzed in great depth the role of probability theory in advancing military operations research during World War II.[4] Although AT&T's role
in promoting statistical analysis was chronicled in its A History of Engineering
and Science in the Bell System, that publication primarily focused on technical development.[5] Nor have management historians examined the business-
political nexus that fostered operations research.[6]

The following five sections analyze how the Bell System sustained the innovative application of probability theory in traffic management from 1900
through 1929. The first section evaluates how AT&T's staff engineering department unsuccessfully sought to control costs and enhance competitiveness by
trying to gain through quantitative analysis a better understanding of its operations and markets. The next section explains how a new approach to statistical
analysis, known as the commercial survey, strengthened planning in the years
following the firm's near-fatal financial crisis in 1906. The third section assesses
how, during a period of rising public oversight, probabilistic analysis enhanced
the company's ability to satisfy regulatory standards of economy and efficiency
by reducing uncertainty in capital budgeting and production planning. The
fourth section discusses how the formation of a department in the Bell Laboratories improved the firm's ability to remain innovative in the application of
probability theory to the problems of traffic management. The conclusion evaluates the significance of the Bell experience and what it tells us about the
nature of twentieth-century innovation.

QUANTITATIVE INNOVATION AND MARKET COMPETITION, 1892–1902

The initial efforts to develop innovative ways to apply statistics to achieve
greater efficiency in traffic operations began in the mid-1890s as new competitors entered the telecommunications market after the expiration of the Bell
patents. Although the firm had differentiated its services by beginning to
develop long-distance service, competition for local and toll (inter-exchange)
service became more intense. Smaller local rivals sought to secure market share

through sharp pricing competition. Such competition made Bell management more sensitive to the need to control costs. A major focus of this effort centered on the construction and operation of message switching offices and call trunking facilities that connected individual exchanges. However, the company experienced severe difficulties in building efficient local networks. Lacking effective guides for estimating market growth, it too often had incurred high costs from inefficiently allocating resources in local markets. Moreover, this translated into high costs and uneven quality of service that by the first decade of the twentieth century was creating problems in dealing with the rising number of state regulatory bodies who increasingly demanded high levels of economical and efficient service.[7]

Thomas B. Doolittle, a pioneer in exchange development from Connecticut and a close personal associate of Thomas J. Watson of Western Electric, first espoused the innovative use of statistical analysis to guide toll exchange development in 1892.[8] In Doolittle's view, three factors would affect the success of toll exchange development. First, it required that the associated companies upgrade their transmission facilities by investing in state-of-the-art equipment such as metallic circuits and solid-back transmitters to ensure high-fidelity communications. Second, he wanted the service to be marketed aggressively by motivated canvassers to ensure an adequate customer base. And third, he needed substantial amounts of data, much of it never before collected, about subsidiary companies' operations in order to prepare feasibility studies for exchange construction.[9]

Doolittle began to evaluate data collected from the firm's operating subsidiaries after the engineering department formed its toll data bureau in 1892. Each unit reported the number of toll calls from private subscribers and from public pay facilities for each city, town, or village in their service area identified in the 1890 census.[10] The associate companies also submitted maps that identified the locations of their transmission facilities.[11] Doolittle used this information to construct ratios that equated population—which he incorrectly believed was the driver of service demand—to revenues and number of calls on a daily per-capita basis for each census district.[12] Extrapolating local growth trends from census data, Doolittle drew graphs that projected future toll revenue and traffic levels and then used these to estimate profits and returns on invested capital.[13] The usefulness of this work, however, was marginal because of the staleness of the census population data.[14] Nor was the information effective in making reliable projections, because it failed to factor the crucial effect of consumer income on demand.[15]

Later, Doolittle developed a second system of analysis to estimate the number of trunk lines necessary to accommodate toll service at local telephone offices. The procedures were outlined in an Engineering Department document titled "Traffic Estimate," derived from an evaluation of toll call patterns for the Chicago exchange from 1894 to 1897.[16] Using peg count information for all originating calls from local offices in the exchange, AT&T engineers evaluated changing traffic patterns on a monthly and daily basis. From this data Doolittle developed an algorithm for predicting the number of trunk lines necessary for efficiently accommodating local office toll traffic:

$$X = .75 \underbrace{\frac{(ST-S^*T^*)}{ST}}_{(A)} \times \underbrace{\frac{S^*T^*}{110}}_{(B)} + \underbrace{M\,(N-1)\,(1)^{17}}_{(C)}$$

Where:

.75 = adjustment factor based on corporate experience for reducing local office toll calls as a percentage of total local office calls

X = total number of trunk lines for a local office

S and S^* = number of subscribers in total exchange and local office, respectively

T and T^* = average daily originating calls in exchange and local office, respectively

110 = maximum daily number of toll calls per trunk line

M = arbitrary coefficient of .8 indicating excess number of actual trunk lines above theoretical projection to accommodate business variation and short-term growth trends

N = number of local offices making up an exchange

However, significant differences emerged in Chicago between the values that the formula predicted and actual experience.[18] A major shortcoming of the algorithm was that its two constants, .75 and M, fluctuated significantly in practice. They changed in response to variations in size of subscriber base, geographic distribution of subscribers, and local economic characteristics. This clearly implied that the patterns of consumer telephone usage were sensitive to more subtle and complex forces than were discernible from the analysis of raw calling data.

QUANTITATIVE INNOVATION AND PATTERNS OF CONSUMER CHOICE: THE COMMERCIAL SURVEY

In 1902 W. F. Patten and Malcolm C. Rorty of AT&T's engineering department achieved a breakthrough with the development of a method of statistical analysis of the poorly understood problem of consumer preference in the selection of telephone service.[19] The new approach known as the "commercial survey"

largely supplanted Doolittle's procedures. The new model concentrated primarily on the study of both macro and micro elements of urban centers that
promised the greatest profit potential because of their high traffic density.[20] The
pioneering engineers first divided the cities served by the firm into two broad
categories: (1) those who were economically self-sufficient and (2) those whose
economic future was contingent on adjacent regional development. They then
further classified them into seven categories depending on rates of population
growth and nature of local economic activity. They projected demand for slow-
growing cities simply by extending linear extrapolations of the past decade's
trend. They projected demand for rapidly growing cities through the extrapo-
lation of growth curves that incorporated geometric slopes derived from the
analysis of the previous two decades' worth of experience.[21]

On the micro level, they segmented markets served by local switching offices
into several household and business income categories, which they believed
was a more important driver of demand than population. Although they thought
that all residences of the first class should have telephone service, the maximum
ratio for second-class households was 50 percent, and for those in the lowest tier
it was only 5 percent. However, the ratio for central business districts was
expected to be equivalent to the square root of the city's population. They estimated that the ratio of business to residential service fluctuated in most U.S.
cities between 40 and 50 percent. Maps plotted the different sectors within an
urban economy and identified the location of major transmission resources.
Besides facilitating capital investment planning, these maps proved useful in
determining the mix of local services and rate structures.[22]

These methods assumed greater importance after the installation of a new
senior management team under the leadership of Theodore N. Vail following
the firm's near-bankruptcy. Vail recognized that the debacle had resulted in
part because of unrealistic estimates of demand growth that encouraged an
overexpansion in new plant investment. Commercial surveys soon became
relegated under Vail's extensive reorganization of corporate administration to
the newly formed Commercial Department that had the responsibility for promoting demand for telephone service and for controlling traffic operations.
The new unit began to refine Patten and Rorty's technique by incorporating
innovative practices developed at some of the operating subsidiaries. New York
Telephone, for example, surmounted the contemporary dearth of local income
information by using rental data as a proxy. Under the leadership of its president, J. J. Carty, planners at the associated company developed a bottom-up
approach to growth analysis that focused on the relative income distributions
within local office territories.[23] They assumed that the "telephone habit" represented a luxury afforded primarily by upper- and middle-class families.[24] They
believed, probably on the basis of the findings of a 1903 study by the U.S.

Bureau of Labor Statistics,[25] that family income on average was 4.5 times greater than the family's rent expense. They built up their estimates of potential telephone service by subdividing urban markets into segments based on a six-level income to rent scale. In 1915, for example, they thought that families with incomes less than $1,100 per annum were poor candidates for telephone service, whereas those with incomes in excess of $2,400 were good candidates for the highest level of private service.[26]

The classification of business, however, involved only three categories. The highest category included large office or factory buildings that were candidates for private branch exchanges (PBXs). (By the 1920s PBXs connected to 19 percent of the total telephones served and accounted for 17 percent of local telephone volume in the Bell System.) At the opposite end of the spectrum of prosperity were small, marginal shops that were candidates for coin-box service.[27]

Commercial surveying eventually became formalized in a twelve-step process that, besides estimating telephone-line growth, also projected differing revenue outcomes depending on alternate rate plans for particular markets.[28] This information also helped traffic engineers to estimate new plant and equipment requirements for accommodating growth.[29] The treasurer's department also started to use the revenue projection in its cash planning.[30]

The formation of a statistics division at AT&T's corporate headquarters affected commercial survey activity.[31] This unit, formed in 1909 by President Vail and initially directed by Malcolm Rorty, had expanded by 1921 to six sections: economic and financial statistics, special statistical analysis, statistical methods, mathematical statistics, telephone statistics, and public utility statistics.[32] Two of these sections had important influence on commercial survey work. The first, economic and financial statistics, emphasized business-cycle analysis to evaluate the likely impact of changing economic trends on both national and state income levels.[33] The second, the statistical methods section, was formed in 1919 to provide guidance on how advances in statistical methodology might be applied to the telephone business.[34]

The statistical methods section also experimented with using more rigorous statistical analytical techniques in commercial survey work. Efforts to apply statistical measures of dispersion and skew to create a uniform national framework evaluating the connections between rental data, residential income, and telephone service demand proved unsatisfactory. These efforts were impeded by the high degree of variability in both inter- and intra-city data relating to rents, which was far greater than such other determinants of household disposable income as food or clothing.[35] However, W. C. Helmle discovered a means for making useful rental data comparisons through the analysis of logarithmic skew distributions. This revealed a high degree of constancy in indices

of rental dispersions within particular cities in the face of price-level changes.[36]
Thus the Helmle study highlighted the uniqueness of local telephone markets
and the importance of detailed local analysis in planning and forecasting. In
addition, the study suggested that the relative distribution of income within a
community—the factor thought most critical in determining the level of telephone demand—tended to remain stable over the intermediate time horizons
encompassed in commercial survey forecasts.[37] This finding also affirmed the
soundness of the rental variable as an income proxy.

SUSTAINING MANAGEMENT INNOVATION:
OVERCOMING UNCERTAINTY IN TRAFFIC PLANNING
AND OPERATIONS WITH PROBABILITY THEORY

The Bell System eventually discovered the utility of probability theory for sur-
mounting uncertainties inherent in the process of local exchange design and
operation. A key question involved the determination of the level of plant
investment necessary to ensure that the chances of customer service denial
remained at some acceptable level during the normal peak traffic times at
11 A.M. and 1 P.M. daily. Although exchange daily call averages could support
rough approximations, historically such approaches had proven too inaccurate
and had led to the misallocation of transmission resources. The 1906 crisis had
persuaded management of the need for better estimating to minimize financial risk. The successful development of new methods more firmly grounded
in mathematical science would also help the company affirm its compliance
with regulatory economy and efficiency standards.

The idea that probability theory could assist traffic planning first began to
crystallize when G. T. Blood, a member of Doolittle's toll records bureau,
observed that the pattern of exchange busy signals seemed to conform to that of
a binomial distribution.[38] However, it took five years before anyone recognized
the significance of Blood's observation and thus to develop a useful application for probability theory for traffic planning. In 1903, Malcolm Rorty, now
the traffic manager of Pittsburgh's Central District and Printing Telegraph
Company, prepared along with C. N. Fosdick of the Central Union Telegraph
Company and W. O. Pennell of the Missouri and Kansas Telephone Company
a set of graphs based on the binomial formula that projected the probability of
service denial at various levels of call volume and trunk line availability. The
relative slowness of the pace of development stemmed from the lack of formal
training for contemporary engineers in statistics. Rorty and his associates, for
example, relied on the expositions on probabilities and the binomial that
appeared in the *Encyclopedia Britannica* to guide their graphic analyses.[39]

This work eventually became the basis for what was known as the "binomial trunking loss formula" that remained a basic statistical tool of traffic analysis well into the 1960s.

$$P(c,n,p) = \sum_{x=c}^{x-n} \binom{n}{x} p^x (1-p)^{n-x}$$

Where:

n = number of subscribers

c = total number of trunks in a network

x = number of trunks in full use at a given time

p = probability that a subscriber is on the line

P = probability of service denial because of unavailability of a trunk line[40]

Through the use of this formula, Bell engineers could calculate precisely the station equipment requirements necessary to ensure that the probability of service denial would not exceed what eventually became the company's standard of 1 percent.

The joint study provided AT&T management with another useful insight. It showed that investment in additional trunking lines led to a greater proportional increase in efficiency in message transmission. This insight helped to persuade AT&T's management of the benefits that could derive from a more capital-intensive and automated telecommunications system.[41]

The usefulness of probability theory increased in 1908 with the development of an improved trunking formula by E. C. Molina, a gifted, self-educated mathematician who was employed then as an Engineering Department circuit designer. In a successful effort to develop a simpler means for approximating the binomial summation, Molina independently derived what later was recognized to be the Poisson Exponential Binomial Limit first discovered by French mathematician Simeon Poisson in the 1830s.[42]

$$P(c,a) = \sum_{x=c}^{\infty} \frac{a^x e^{-a}}{x!}$$

The Poisson formula offered many advantages over the binomial. First, it was an expression that was particularly accurate in dealing with large populations of events and low probabilities, parameters frequently encountered in traffic planning. In addition, the Poisson incorporated one less variable than the binomial, thus contributing to computational efficiency. The Poisson also yielded more conservative results than the binomial. Thus it automatically induced traffic planners to incorporate slightly higher amounts of equipment than in the case of the binomial.[43]

SUSTAINING INNOVATION THROUGH SYSTEM INTEGRATION AND AUTOMATION

By 1911 the Bell System had begun to sustain its innovations in probabilistic traffic operations through the system's inclusion in the firm's evolving managerial hierarchy. This had three aspects. First, it involved the development of routine procedures and other institutional arrangements for using the new knowledge in the design and operation of the firm's basic operating unit, the local telephone exchange. Second, the firm developed procedures for integrating the planning information for local exchanges submitted annually by the regional operating subsidiaries into the firmwide capital budget and the production planning schedules for its manufacturing subsidiary, Western Electric. Third, the company also established a unit in the Bell Telephone Laboratories to extend and disseminate this specialized knowledge.

That same year the Bell System promulgated a new set of procedures known as "fundamental plans" for use by subsidiaries in local exchange planning.[44] Relying on the probabilistic methods worked out by Molina and others to confront the problems of operational uncertainty, the fundamental plans equated the physical capabilities for new or expanded telephone exchange to expected call traffic. Although these plans projected conditions for up to twenty years, capital investment decisions were more usually geared to one-, five-, and ten-year time horizons.[45] The preparation of fundamental plans for the associated companies integrated the knowledge of AT&T's three basic classes of specialists, representing traffic, plant, and commercial operations. Traffic engineers defined exchange operational procedures and the requisite amount of capital investment. They also made decisions about the layout and arrangement of equipment to ensure the most efficient operating results. The plant function drew on the special skills of both equipment and outside plant engineers. The equipment engineers selected and installed the equipment and circuit arrangements that would satisfy the goals laid out in the traffic engineer's studies. The outside plant engineers established connectivity between the local office and the external telephone grid.[46] Commercial engineers completed the process through the creation of rate plans that provided the basis for projecting financial performance.[47]

In 1912 the Bell System introduced a new method originally developed by Edgar S. Bloom at its Pacific Telephone and Telegraph subsidiary for the regional operating companies to incorporate information from their fundamental plans in their capital budget or "provisional estimate."[48] Corporate headquarters required the subsidiaries to report this information to its centralized treasury department, which had overall responsibility for firmwide financial planning. It required the associated companies to project the costs of their capital outlays and estimated increases in revenue for the next fiscal year. To avoid the danger of overspending, it also imposed the discipline that capital outlays could not exceed estimated revenue increases. The provisional estimate also required the calculation of returns on investment to judge the efficiency of capital resource allocation.[49]

The effects of the probabilistic analysis at the local-office level filtered into production planning through the provisional estimate. The scope of this document gradually expanded over time to help ensure greater effectiveness in coordinating planning between the Bell System's many operating elements. In the case of manufacturing, the key supplement was Schedule G, which communicated information about the physical quantities of key system inputs including cable, poles, and central office switching equipment. Western Electric used this data to prepare its sales budget and its quarterly production schedules.[50]

In addition to these operating arrangements, the Bell System sustained its commitment to innovative applications of probability theory in traffic management through the formation of a specialized department in the Bell Laboratories. The firm concentrated its experts in this field from both AT&T and Western Electric in Department 312.[51] This afforded several advantages. First, it brought together the differing perspectives of engineers who confronted traffic management problems from the complementary perspectives of equipment manufacturing and systems operation. Second, a single departmental focus diminished the likelihood of the costly and confidence-shaking debates between experts about technical matters. Molina, for example, became embroiled in such a controversy with Thornton C. Fry of Western Electric over the effectiveness of the Poisson method in practice. Third, the new unit became dedicated to performing research relating to the ways that the use of probability theory might be better applied in traffic management. These findings often became communicated within the firm through research articles appearing in the *Bell System Technical Journal*.[52] Fourth, the department also monitored external developments in the field that might affect the firm's interests. Fifth, it became a center for the development of educational materials about the use of probability theory for its field engineering staff. Sixth, it provided consultation services to regional operating companies experiencing difficulties in resolving probability problems. Seventh, the firm also called on its experts to provide testimony before state regulatory authorities about how their probabilistic methodologies contributed to the achievement of economy and efficiency in telephone operations. For example, the firm in justifying the fairness of its management charge to its New York Telephone affiliate noted the intra-corporate consulting and educational services rendered by Molina's department.[53] In addition, the relative advantages of the Poisson formula and the Erlang formula developed by the Royal Danish Postal Administration in developing accurate estimates of demand patterns became part of the discourse in this inquiry. Eighth, the researchers also promoted innovative continuity through their penchant for looking at each extension of their knowledge as a contribution to theory. This conception of their work as part of a broader intellectual continuum helped to define the baseline of understanding in traffic management and to identify the remaining problems that required resolution in order to round out the understanding of their specialized body of knowledge.

The integration of probability theory in traffic management facilitated the expansion of automated message switching during the 1920s. This investment and the use of the new quantitative techniques had dramatic impacts on operations. During World War I, for example, there were about 100,000 operators, or about 15 per 1,000 telephones; by World War II the numbers of operators had increased to about 140,000, or about 7 per 1,000 telephones.[54] Between 1912 and 1929 the efficiency of local exchange operations increased 6.5-fold.[55] The total number of telephones served by the Bell System increased from about 4 million in 1910 to over 14 million in 1928.[56] The total number of telephones served by automatic dialing exchanges increased from several hundred thousand in 1920, or 2.7 percent of the total, to over 3.5 million, or 27 percent, in 1929. Moreover, prior to the onset of the Great Depression, the firm had contemplated an expansion to over 8 million, or 47 percent of the total, by 1933.[57] The average time necessary to complete a toll call fell from 7 minutes in 1920 to 1.5 minutes in 1928.[58] In these and other ways the sustaining of innovative methods of applying probability theory to resolve management problems contributed strongly to the economy and efficiency of the Bell System.

CONCLUSION

What then does this experience tell us about sustaining innovation in the Bell System prior to the Great Depression?

The discovery of the usefulness of probability theory in traffic operations in part represented a product of the Bell System's management culture. Engineers and scientists dominated the management cadre. Through their formal training they had become imbued with belief in the centrality of mathematical analysis to the resolution of technological problems. Probability theory's

126   effectiveness in addressing uncertainty enhanced its attractiveness to a management that favored solutions grounded in mathematical science.

Because of its successful integration in critical operational routines, the probability theory ensured continued future perfection and refinement in traffic management. There were no viable alternative means for confronting the uncertainties that permeated the running of complex messaging systems. The resultant drive to sustain innovation led to both the discovery of more efficient formulas and modes of analysis and the development of new ways to embed this knowledge in a changing institutional and organizational context.

The ability of the Bell System to sustain innovation in this field also derived in part from its status as a regulated utility that could charge developmental expenses against its rate base. Private companies operating in more competitive markets did not have access to this form of subsidization. Although some oversight bodies challenged the inclusion of such costs, the firm generally experienced full recovery because of its ability to argue convincingly about the reasonableness of the outlays in achieving economical and efficient service.

The internalization of the drive to sustain innovation reflected the underdevelopment of the U.S. research community during this era. Neither federal nor state governments supported foundations to pursue research in the industrial application of probability theory. Universities also lacked resources and experienced personnel necessary for providing useful counsel to giant industrial organizations. The most important centers of inquiry remained the telephone systems that had a pressing need to advance the horizons of understanding for these applications. Not surprisingly, the other great set of such formulas for message switching derived from the work of Erlang at the Danish telecommunications system. Moreover, one of the first textbooks to discuss these applications comprehensibly was Thornton Fry's *Probability and Its Engineering Uses*, which did not appear until 1928.[59]

The Bell System sustained innovation because of its ability to develop institutional frameworks to exploit what Alfred Chandler has termed "firm-specific learning."[60] This had four dimensions. First, the company succeeded in solving the underlying analytical problem of applying probability theory to many crucial problems encountered in traffic operations. This understanding in itself was insufficient to exercise long-term influence on the firm's business activities. The company also needed a second type of innovation that involved the effective deployment of the new knowledge within the enterprise's organizational context so as to improve operational efficiency. The third dimension involved the creation of monitoring capabilities to assess the overall performance in practice of both the mathematical and organizational structures. The fourth related to the establishment of capacities for studying the problems identified in the process of operational feedback and initiating programs for their resolution.

The sustaining of innovation strengthened the position of the Bell System in dealing with public oversight authorities. As Louis Galambos has persuasively argued in his insightful works on the organizational society and triocracy, government during this earlier era frequently deferred to the superior knowledge of private business groups that impinged on complex public-interest controversies.[61] Few state regulatory boards could muster the company's heavy expertise in adjudicating issues that pertained to the use of probability in traffic management. In addition, as Roland Marchand has noted, such competency also contributed to a growing public image of the firm as a leading agent for promoting progress through applied mathematics and science.[62] The creation of a favorable aura that highlighted the firm's civic-mindedness and its great technical capacities also attracted adherents to its positions in public policy debates.

How then might the history of this era have been different if AT&T did not develop probabilistic management?

First, this would have retarded efforts to strengthen the control of the central office within the Bell System over its subsidiary activities. Second, it would have compelled management to rely on more descriptive and less analytical statistical and accounting data in managing these affairs. This latter option would have weakened efforts to control traffic investment and operating activities. Third, lower operating effectiveness would have hampered the firm's ability to make the transition to more automatic switching and, doubtless, would have led to serious service bottlenecks, particularly in major metropolitan centers with high traffic density. Fourth, a consequent higher reliance on manual switching would have increased the firm's personnel requirements and the proportion of variable costs as part of its total cost structure. This latter circumstance would have reduced the firm's ability to achieve economies of scale from heightened service volumes in a high fixed-cost environment. Fifth, diminished managerial and planning effectiveness would have increased the firm's financial risk, as in the 1906 panic when insolvency threatened because of overbuilding. Finally, the combination of these problems would have made it much more difficult to reach accommodation with state regulatory bodies.

NOTES

1. Stephen M. Stigler, *The History of Statistics: The Measurement of Uncertainty Before 1900* (Cambridge, Mass.: Harvard University Press, 1986).

2. See R. J. Epstein, *A History of Econometrics* (Amsterdam: North Holland, 1987); and Mary S. Morgan, *The History of Econometric Ideas* (Cambridge, U.K.: Cambridge University Press, 1990). See also the excellent collection of essays on this topic in David F. Hendry and Mary S. Morgan, eds., *The Foundations of Econometric Analysis* (Cambridge, U.K.: Cambridge University Press, 1995).

3. Robert M. Porter, *The Rise of Statistical Thinking, 1820–1900* (Princeton: Princeton University Press, 1986).

4. For military applications of probability theory in operational research (OR), see E. A. Johnson and E. A. Katcher, *Mines Against Japan* (Silver Spring, Md.: Naval Ordnance Laboratory, 1973); M. W. Kirby and R. Capey, "The Air Defense of Great Britain, 1920–1940: An Operational Perspective," *Journal of the Operational Research Society* 48 (June 1997): 555–68; M. W. Kirby and R. Capey, "The Area Bombing of Germany in World War II," *Journal of Operational Research Society* 48 (July 1997): 661–77; Harold Lardner, "The Origin of Operational Research," *Operations Research* 32 (Mar.-Apr. 1984): 465–75; C. W. MacArthur, *Operations Analysis in the U.S. Army Eighth Air Force in World War II* (Providence: American Mathematical Society, 1990); J. F. McCloskey, "British Operational Research in World War II," *Operations Research* 35 (May-June 1987): 453–70; J. F. McCloskey, "US Operational Research in World War II," *Operations Research* 35 (Nov.-Dec. 1987): 910–25; H. J. Miser, ed., *Operations Analysis in the Eighth Air Force: Four Contemporary Accounts* (Linthicum, Md.: Institute for Operations Research and Management Science, 1997); Philip M. Morse and George E. Kimball, *Methods of Operations Research*, rev. ed. (New York: The Technology Press of the Massachusetts Institute of Technology and John Wiley & Sons, 1954); K. R. Tidman, *The Operations Evaluation Group: A History of Naval Operations Analysis* (Annapolis, Md.: Naval Institute Press, 1984); and C. H. Waddington, *OR in World War II: Operations Research Against the U Boat* (London: Elek Science, 1973).

5. See, for example, M. D. Fagen, ed., *A History of Engineering and Science in the Bell System: The Early Years (1875–1925)* (New York: AT&T Bell Laboratories, 1975), 538–44, 859–82 passim, 923–34; and M. D. Fagen, ed., *A History of Engineering and Science in the Bell System: Communication Sciences (1925–1980)* (New York: AT&T Bell Laboratories, 1984), chapter 1 passim.

6. Claude S. George Jr., *History of Management Thought* (Englewood Cliffs, N.J.: Prentice-Hall, 1968), chapter 11.

7. "Economy" essentially meant at a reasonable cost whereas "efficiency" related to the qualitative nature of the service.

8. See Robert W. Garnet, *The Telephone Enterprise: The Evolution of the Bell System's Horizontal Structure, 1876–1900* (Baltimore: Johns Hopkins University Press, 1985), 21–22 and 71. For discussion of early strategic evolution, see George David Smith, *The Anatomy of a Business Strategy: Bell, Western Electric and the American Telephone Industry* (Baltimore: Johns Hopkins University Press, 1985).

9. For a more detailed discussion of the strategy and the difficulties of implementing it in the South see Kenneth Lipartito, *The Bell System and Regional Business: The Telephone in the South, 1877–1920* (Baltimore: Johns Hopkins University Press, 1989), 117–48. See also the letter of T. B. Doolittle to Joseph P. Davis, January 17, 1898, *1897 Annual Report of the Engineering Department*, Box 250-06-18, AT&T Archives, Warren, New Jersey (hereafter "ATT").

10. Joseph P. Davis to John E. Hudson, Jan. 30, 1895, *1894 Annual Report of the Engineering Department*, 146–67, Box 250-06-18, ATT.

11. T. B. Doolittle to Joseph P. Davis, Jan. 17, 1898, *1897 Annual Report of the Engineering Department*, 3–8, Box 250-06-18, ATT.

12. T. B. Doolittle to Joseph P. Davis, Jan. 7, 1895, *1894 Annual Report of the Engineering Department*, 155–66, Box 250-06-18, ATT.

13. See, for example, an unsigned report titled "Toll Traffic, Boston," Apr. 1901, Box 14-05-03, ATT. See also reports, T. B. Doolittle, "Re. Cost of 1000 Subscriber Exchange," May 13, 1896, Box 01-04-02, ATT; and T. B. Doolittle, "Subject: Cost of Establishing an Exchange of 100 Subscribers," Apr. 17, 1895, Box 01-04-02, ATT.

14. T. B. Doolittle to Joseph P. Davis, Jan. 16, 1901, *1900 Annual Report of the Engineering Department*, 2–7, Box 250-06-18, ATT.

15. Hammond V. Hayes to F. P. Fish, Dec. 31, 1906, *1906 Annual Report of the Engineering Department*, 1–12, Box 250-06-18, ATT.

16. "Operating Cost-Traffic Estimate" (hereafter "Traffic Estimate"), no date (circa 1898), Box 23-01-01, ATT.

17. See "Traffic Estimate," 21–23. The expression at A is the percentage of local office toll calls adjusted for an "acquaintance factor," which takes into consideration that toll call potentials are rarely achieved because of the unique communication patterns of differing subscriber groups. The expression at B is the total number of trunk lines necessary to accommodate daily local office trunk volume assuming that each trunk can carry a maximum of 110 calls per day. Expression C is the margin of safety for local office trunk lines to accommodate peak variations in demand and growth.

18. "Traffic Estimate," 27–30.

19. Rorty joined New York Telephone in 1897 and transferred to AT&T's Engineering Department, where he was employed until 1903. Rorty had a distinguished career at AT&T, eventually rising to the rank of vice president before joining International Telephone and Telegraph Company in 1923 at that same rank. Rorty was a prolific inventor, enjoyed wide intellectual interests, and served as president (1922–23) of the National Bureau of Economic Research and as president (1930) of the American Statistical Association. See "Malcolm Churchill Rorty" in *National Cyclopedia of American Biography* (Ann Arbor, Mich.: University Microfilms, 1967), Vol. 27, 269–70.

20. Joseph P. Davis to Frederick P. Fish, dated Aug. 28, 1902, and attachment attributed to W. F. Patten and Malcolm Rorty titled "Note on Development Plan," Box 137-09-01-14, ATT.

21. "Note on Development Plan," 1–2.

22. "Note on Development Plan," 3–7.

23. For a general discussion of Carty's approach to commercial survey work while at New York Telephone, see R. A. Davis, "Method for Making Development Studies and What They Are Expected to Accomplish," New England Telephone and Telegraph, *Commercial Conferences* (Boston: 1909–1910), 109–20, Box 185-03-02, ATT.

24. Davis, "Method for Making Development Studies," 110.

25. See W. C. Helmle, "The Relation Between Rents and Incomes and the Distribution of Rental Values," in *Bell System Technical Journal* (Nov. 1922), 84–5 and 89 for discussion of U.S. Bureau of Labor Statistics 1903 report.

26. Davis, "Method for Making Development Studies," 111–13, 114–15.

27. Davis, "Method for Making Development Studies," 113, 115–16.

28. See E. L. Stone Jr., "Commercial Surveys," in *Traffic Engineering Conference* (New York: 1922), 123–32, Box 185-09-03, ATT.

130

29. Stone, "Commercial Surveys," 123.

30. Stone, "Commercial Surveys," 127.

31. See S. L. Andrews, "The Work of the Chief Statistician's Division with Special Reference to Certain Economic Studies Bearing Upon the Business Outlook," *General Accounting Conference* (New York: May 1921) in Box 185-03-01, ATT.

32. Andrews, "Work of the Chief Statistician's Division," 3–16.

33. Andrews, "Work of the Chief Statistician's Division," 13–29.

34. Andrews, "Work of the Chief Statistician's Division," 10–13. See also A. H. Richardson, "Discussion of Graphic Methods of Presenting Telephone Information," *General Accounting Conference* (New York: May 1921), Box 185-03-01, ATT.

35. See Helmle, "Relation Between Rents and Incomes," 82–90.

36. Helmle, "Relation Between Rents and Incomes," 90–8.

37. Helmle, "Relation Between Rents and Incomes," 79–104.

38. Fagen, *History of Engineering and Science in the Bell System: The Early Years*, 539.

39. M. C. Rorty to Joseph P. Davis and attachment, Oct. 22, 1903, Box 1360, ATT. Memorandum by R. I. Wilkinson, "Great Debate on Probability Formula for Traffic Engineering in 1920," Mar. 19, 1965, Box 85-09-03, ATT.

40. The binomial trunking formula and its use is discussed in Bell Telephone Laboratories, *Probability and Statistics Fundamentals—Application to Traffic and Design Problems* (New York: Bell Telephone Laboratories, n.d.), chapter 9, 7–9.

41. "History of Development of Panel Machine Switching System," 3, Box 85-06-02, ATT.

42. Wilkinson, "Great Debate on Probability Formulas for Traffic Engineering," 1, Mar. 19, 1965, Box 85-09-03, ATT. See also transcript of interview with E. C. Molina, Dec. 5, 1962, 25–8, Box 106-10-03, ATT.

43. Bell Telephone Laboratories, *Probability and Statistics Fundamentals*, chapter 9, 10–17; Thornton C. Fry, *Probability and Its Engineering Uses* (New York: D. Van Nostrand, 1928), 232–35, 237–40.

44. For general discussion of fundamental plans, see D. M. Rice, "Application of Fundamental Plans," *Plant and Engineering Conference* 2, Shawnee, Pennsylvania, May 1923, 543–53, Box 185-07-03, ATT.

45. For discussion of planning horizons, see E. N. Renshaw, "Comparative Cost Studies and Estimates," in *Building and Equipment Conference* (New York: 1923), chapter 7, Box 185-03-02-02, ATT.

46. The roles and responsibilities of both traffic and plant engineers are described by W. E. Farnham, "The Relationship Between the Work of the Traffic Engineer and the Equipment Engineer," in *Building and Equipment Conference* (New York: 1923), chapter 6, Box 185-03-02-02, ATT.

47. For discussion of rate plans, see F. N. Bratney, "Rate Plan Studies and Their Relation to Methods of Planning for the Future," *Plant and Engineering Conference* 2, Shawnee, Pennsylvania, May 1923, 581–93, Box 185-07-03, ATT.

48. For the origins of the provisional plan, see H. B. Thayer to N. C. Kingsbury, Apr. 6, 1912, and attached report attributed to Edgar S. Bloom, "The Provisional Estimate," Box 125-07-03-24, ATT.

49. Bloom, "Provisional Estimate," 1–6, Box 125-07-03-24, ATT. For an example of how provisional estimate information was used in evaluation see H. B. Thayer to P. L. Spalding of New England Telephone and Telegraph Company, Aug. 12, 1913, Box 126-05-03-12, ATT.

50. J. W. Cogan, "Analysis of Monthly Report No. 22 and Correlation of Construction Program with Western Electric Sales" (paper presented at the AT&T Accounting Conference, Briarcliff Manor, New York, 1923), 1–11. For more detailed discussion of the coordination problems in manufacturing, see Nandini Chandar and Paul J. Miranti, "Networks and Uncertainty: Forecasting, Budgeting and Production Planning at the Bell System" (paper presented at the Fifth Accounting History International Conference, Banff, Canada, August 10, 2007).

51. Memorandum, R. I. Wilkinson to W. S. Hayward, "Technical Developments in the Traffic Studies Area of Bell Telephone Laboratories, 1924–1965," and appendices, May 3, 1965, Box 85-06-02, ATT.

52. See, for example, E. C. Molina and R. P. Crowell, "Deviation of Random Samples from Average Conditions and Significance to Traffic Men," *Bell System Technical Journal* 4 (Jan. 1924): 88–99.

53. Transmittal sheet, E. C. Molina to Paul H. Burns, June 17, 1929; and memorandum, "Probability Studies, New York Rate Case B-1.14"; and transmittal sheet, E. C. Molina to E. L. Blackman, May 14, 1928; and memorandum, "Instruction Course in Probabilities, New York Rate Case A-1.14," Box 481-02-01, ATT.

54. Fagen, *History of Engineering and Science in the Bell System: The Early Years*, 550n73.

55. Bancroft Gherardi and F. B. Jewett, "Telephone Communication System of the United States," *Bell System Technical Journal* (Jan. 1930): 9.

56. Gherardi and Jewett, "Telephone Communication System," from table on 88.

57. Gherardi and Jewett, "Telephone Communication System," from table on 22.

58. Gherardi and Jewett, "Telephone Communication System," from table on 34.

59. Fry, *Probability and Its Engineering Uses*, especially chapters VIII and X.

60. For a discussion of Chandler's notion of firm-specific learning and its importance in maintaining competitiveness, see Alfred Chandler, *Inventing the Electronic Century: The Epic Story of the Consumer Electronics and Computer Industries* (New York: Free Press, 2001); and *Shaping the Industrial Century: The Remarkable Story of the Evolution of the Modern Chemical and Pharmaceutical Industries* (Cambridge, Mass.: Harvard University Press, 2005).

61. Louis Galambos, *America at Middle Age: A New History of the United States in the Twentieth Century* (New York: New Press, 1982); and Louis Galambos, "Technology, Political Economy, and Professionalization: Central Themes of the Organizational Society," *Business History Review* 57 (Winter 1983): 471–93. See also Louis Galambos and Joseph Pratt, *The Rise of the Corporate Commonwealth: U.S. Business and Public Policy in the Twentieth Century* (New York: Basic Books, 1987).

62. Roland Marchand, *Creating the Corporate Soul: The Rise of Public Relations and Corporate Imagery in American Big Business* (Berkeley, Calif.: University of California Press, 1998), chapter 2.

4

RETHINKING THE INVENTION FACTORY

Bell Laboratories in Perspective

Kenneth Lipartito

Bell Telephone Laboratories is the most famous and most successful corporate research organization in history. At its height in the 1980s it employed over twenty-five thousand people and spent, in real (2005) terms, over \$4 billion per year.[1] Attracting top Ph.D.s from the nation's finest graduate programs, it housed departments in physics and chemistry that held weekly seminars and colloquia, and its scientists presented thousands of papers each year.[2] Among its most famous inventions were the transistor, sound motion pictures, television, and the laser, as well as key contributions to fiber optics, communications satellites, data networking, light-emitting diodes, and, almost as an afterthought, the cellular network design for modern mobile phones. Its scientists have won seven Noble Prizes, including one for the discovery of the cosmic background radiation that established the Big Bang theory of the universe's origins. Earning some twenty-five thousand patents in seventy years, researchers kept up the remarkable pace of one patentable idea per day.

Yet for all of its success, Bell Labs has gradually lost favor as a model for industrial research. Corporations today face tough competitors; rapid, unpredictable shifts in technology; and constant pressure from financial markets for short-term profits. The polite environment of long-term, scientific research that once characterized Bell Labs is out of place in this world. Several of the Labs' big projects failed the test in the market place, most notably its vaunted Picturephone system of video telephony developed in the early 1960s. The transistor, clearly a research triumph, was developed by other firms into integrated circuits and microprocessors.[3] One critic has argued that Bell Labs only flourished so long as it was part of a regulated monopoly, and that it was unable to compete in a deregulated market.[4]

As I shall argue, however, the story of Bell Labs offers important lessons for firm research organizations. Contrary to the beliefs of celebrants and skeptics alike, the Labs owed its success not to pure research or an academic culture but to the careful integration of research, development, manufacturing, and operations. Notions of a past predicated on pure research are largely a myth. But like all myths, this one had the power to shape the action and behavior of those who believed it. One of the first victims of the myth of corporate research was, ironically, Bell Labs itself.

RESEARCH AND THE FIRM

As economist Nathan Rosenberg has noted, nearly everything about research suggests that firms should not do it. Research yields results too general to have direct commercial application and too hard to expropriate and keep out of competitors' hands. For these reasons, Rosenberg observes, much basic research takes place outside of profit-making firms, in universities and government laboratories. But not all. Some firms do their own, and many that do have long histories of research.[5]

AT&T is one of these firms, but the closer one looks, the greater becomes the mystery of why. As a regulated monopolist, for example, AT&T should have been complacent, uninterested in innovative activity. Of course, monopoly status also gave AT&T certain advantages when it came to research. Most notably, it could charge the costs of its research to its customers, in contrast to competitive firms, which must absorb the costs of unfruitful research. But even these advantages were less solid than first they appear. AT&T was closely watched by the FCC and other government bodies, who accused it of overcharging customers for research and frittering away dollars on ill-conceived or overdesigned hardware.[6] As a result, Bell Labs had to justify, on a cost-benefit basis, the value of its work. At the same time, peer firms such as RCA, IBM, and Ford had stable sources of income that supported research and development even absent a monopoly.

AT&T also appeared to lack the motivations for research common in other firms. At General Electric, Du Pont, IBM, and Corning, for example, research followed a strategy of diversification and new product development. AT&T was a centralized firm largely limited to a single service. Efforts to diversify into radio, sound motion pictures, and television were either abandoned, blocked by regulators, or reduced to minor activities. Nor did international markets provide much reason for research. AT&T withdrew from international competition in 1925. In a 1954 consent decree, the corporation also agreed to stay out of nontelephone markets and to license its patents to all who wanted them. The company

134   continued to give its laboratories millions for research in a wide range of areas, but in many cases was unable to commercialize the outcome of this work.

One reason for research does apply to AT&T, however. Only by actively participating in the generation of new knowledge can firms learn about and understand important breakthroughs. A firm does not have to make new discoveries to benefit from research that helps keep it abreast of the latest developments. Even a protected monopoly would understand this, that new products or technologies could threaten its secure position.

Research as learning is one way to mitigate the risk and reduce the uncertainty that surrounds all innovation. In fact, research can be used to transform uncertainty into calculable risks. When scientists have a good understanding of the basic physical characteristics of their firm's products, they may be able to predict future areas in which research is likely to be rewarded, or which will challenge those existing products. Once uncertainty becomes risk, it can be more easily worked into the firm's calculus of decisions. Knowing when threatening breakthroughs (or exploitable new opportunities) are likely to occur affords firms a measure of protection against market change. Seen this way, firm-level research moves from questionable luxury to strategic resource.[7]

Risk and uncertainty, I shall argue, are the proper framework for understanding the motivation of research at Bell Labs. It is the starting point for unpacking the reasons why AT&T invested in basic as well as applied research, even when it held a monopoly. But in large, complex firms, motivation is rarely unitary. Whereas scientists may appreciate the value of research, production engineers, salespeople, and accountants may not. By definition uncertainty does not yield a clean expected value or payoff. It is therefore easy to ignore. Even risk, more readily quantified, can be under- or overestimated or distorted by subjective factors, such as risk tolerance. Although statistics show that firms that engage in research generally have a better survival rate than those which do not, most of the big basic breakthroughs still come from outside of corporations.[8] So risk and uncertainty may be necessary, but they are not sufficient explanations for industrial research at AT&T. A second important factor is the relationship between research and corporate culture.[9]

Most institutions will tend to avoid speculative, costly, cutting-edge work. For this reason, research must be embedded culturally into the firm if it is to thrive. It must also have a place in the corporate structure. Here leadership is crucial. Leadership in this case means the skillful articulation of reason, narrative, and corporate identity to create a vision of research's payoff to the firm. It is a fine balancing act that maintains the ties between the laboratory and other parts of the organization, without sacrificing the gains of learning and broad exploration (crucial to reducing risk and uncertainty) to short-term needs and profitability. Although it seems clear in retrospect that research serves the firm precisely by exploring the uncertain and unknown, the costs of "not knowing what you don't know" are too easily ignored in the face of immediate pressures. Connecting research to core areas without strangling creative work is the task of a good research manager.

Good leaders come and go, making it difficult, perhaps impossible, to maintain the ideal balance over time between scientifically oriented research and commercial mandates. That is why a final factor is crucial—history. Over time, the active, creative balancing done by key leaders will be reduced to routine, with tendency for the original inspiring vision to languish. Here, reflexivity becomes important. How firms read their own history, how they interpret the sources of success and the meaning of their own traditions is an important tool in maintaining a successful research culture.

RESEARCH AND LEADERSHIP AT AT&T

The tradition of research at AT&T began to evolve even before the formation of Bell Labs. Particularly important were a number of experiences stemming from a period of intense competition in the American telephone industry between 1894 and 1907. The modern "Bell System," consisting of AT&T the corporate parent and operator of the long-distance lines, regional Bell Operating Companies throughout the nation, and manufacturer Western Electric, was born out of this competitive era. The industry structure proved durable and stable, reinforced by later regulation. Competition, which cut AT&T's market share from nearly 100 percent to barely 50 percent, almost drove the firm to bankruptcy. Yet crucially, this intense competitive experience convinced AT&T president Theodore Vail of the long-term importance of research.

As Louis Galambos has persuasively argued, Vail instilled in AT&T a deep commitment to research. This commitment was keyed to Vail's overall corporate strategy stressing "universal service," or the provision of telephone service nationwide, through one organization, gradually reaching all households. Research, Vail argued, was a tool to lower costs and improve quality, needed to make what was still an expensive and limited service universal.[10]

Vail's research strategy was something of a departure in the company. Before, AT&T had used research largely defensively, protecting its monopoly through patents. This defensive strategy would not completely disappear, as Leonard Reich has demonstrated, but under Vail the shift toward a more progressive use of research occurred.[11] In Vail, AT&T found its first leader committed to a culture of innovation, with research not simply a tactical weapon for immediate gain but a resource that could serve for almost any future scenario.

One of the issues Vail broached was the value of a long-term perspective on technological innovation that included a fair amount of basic research. Earlier, AT&T had stifled such creative work as too speculative.[12] Ph.D. scientists working at the firm understood their brief as limited to clearly practical work. At the same time, they learned to stay out of the manufacturing plants, where bench engineering reigned supreme.[13] Scientists, even those working in industry, still tended to believe that the "inventive temperament" was distinct from the more routine work connected with engineering and manufacturing, and hence probably not suitable for an industrial organization anyway.[14]

The crucial learning experience for Vail was the effort to build a national long-distance telephone network. One important component of that system was the loading coil, which greatly increased the range and quality of telephone calls. Developing the loading coil required an appreciation of theoretical physics. AT&T's George Campbell did pioneering work on this device, though he was finally beaten to the patent office by Columbia University's Michael Pupin. AT&T acquired Pupin's patent, but then had trouble coordinating research and manufacturing to make the most out of the invention.[15]

These issues motivated Vail to reorganize the company's research staff after 1907. Although AT&T was still locked in a fierce competitive struggle, Vail gave greater scope and authority to his research personnel, freeing them from narrow practical concerns. He consolidated research in New York, near headquarters, established a new staff department called Development and Research, and placed the entire research structure under a new chief engineer, J. J. Carty. Carty held a staff position and so was not linked to any operating company or to the firm's manufacturing arm, Western Electric. Scientists engaged in research (as opposed to development) were encouraged to perform high-level work. At the same time, though, Vail connected research more firmly to the company mission by having all requests for equipment go through Carty rather than directly to the factory.[16]

The new structure was put to the test when AT&T committed itself to building a transcontinental telephone line. This venture required technology that did not yet exist. Carty quickly turned to his research scientists.[17] He purchased a new electronic device, the audion, from independent inventor Lee de Forest. Refining de Forest's invention, company scientists designed electronic equipment to "repeat" or amplify telephone signals. The transcontinental line opened ahead of schedule in 1912.

The experience with the transcontinental line offered several important lessons that would shape research at AT&T and influence the creation of Bell Labs in 1925. First, it showed that even a well-equipped and well-funded corporate laboratory could not eliminate uncertainty. Efforts by AT&T engineers to build a mechanical repeater failed, while the crucial electronic technology was invented out of house. But research could provide the knowledge and expertise to learn about outside inventions and reduce them to practice. De Forest's audion needed substantial refinement, which AT&T scientists quickly understood.[18] Thus even without making the initial breakthrough, corporate science could help identify, acquire, and adapt unforeseen innovations to telephone needs. This approach also required significant coordination with Western Electric manufacturing.[19] Learning taking place within the company could prepare the way to exploit unexpected developments in the external environment. In this regard, the lessons taught by competition and the transcontinental line episode paralleled those being developed by AT&T for managing demand and investment, as explored by Paul Miranti in Chapter 3 in this volume.

The episode also demonstrated that corporate research could be a very powerful symbolic device. The cross-country line added little to AT&T's bottom line, but it earned the firm great prestige.[20] The astute Carty wasted no opportunity to tout achievement (while slighting the contribution of de Forest's audion). The line's opening was masterfully staged, with an aging Alexander Graham Bell in New York calling his old assistant Thomas Watson on the West Coast. Within AT&T, the transcontinental line served as proof that Vail's initiative was correct. Corporate research was worth the expense.[21]

Over the next decade, top personnel at AT&T refined their understanding of what they had learned. Vail and Carty had initially believed that inventive genius was unique and individual, and so could not be replicated in an organized setting such as a corporate laboratory. Star performers were born, not made. Most of them would not be happy within the confines of business anyway. But now it seemed that one did not have to rely on cranky individualists to have an inventive organization. The distinction between breakthrough and development of technology was not so sharp as it had once seemed. Applied research was simply basic research focused on particular problems, according to Carty.[22] Significant creative work needed to be done on rough inventions such as the audion to make them serve complex systems, work that required personnel trained at the frontiers of knowledge, though not necessarily geniuses. Though brilliant individuals were important, Vail acknowledged, "all laboratory and experimental work . . . must be co-ordinated and carried on in connection with the practical operation of an overall system."[23]

Soon there was evidence to back up the claim that organized research could yield fundamental innovations. Work on the transcontinental line, for example, led to "carrier" transmission, or the imposition of multiple channels on a single circuit.[24] Other surprising breakthroughs followed in the same unforeseen manner. In dealing with problems of nonlinear distortion in an amplified line, AT&T researcher H. S. Black began feeding the circuit's output back through the amplifier, resulting in an entirely new area of research on feedback

138 amplification. As time went on, it appeared that a single technical and commercial problem—long-distance transmission across the nation—had spawned
all sorts of new, unexpected, and in some cases quite basic results.

The emphasis of research now shifted at AT&T. Patents and other assets
that promoted corporate survival remained important by-products of corporate
research. But "defensive" research was less important than creative investigations that solved crucial system problems, lowered costs, or led to new products and services.[25] Rather than trying to control or block the unexpected and
unforeseen, AT&T would prepare for change by undertaking research for the
purpose of learning. Dealing with uncertainty, pursuing development after an
invention, and, most important, having a research structure that could connect basic research with engineering and manufacturing, were the new lessons
of the company's emerging research culture.

By 1919, the year Vail retired as president, AT&T had a modest-sized
research force of some one hundred scientists spending about one-quarter of a
million dollars per year. This made it roughly the same size as the research
operation at General Electric, with about half the budget.[26] On January 1, 1925,
a new entity, Bell Telephone Laboratories, came into existence. With nearly
four thousand employees, it was the largest corporate research organization in
the world.[27] Repeating Vail's and Carty's ideas about the value of coordination,
AT&T President Walter Gifford saw this central laboratory as the place where
"highly trained individual minds are fused into one composite mind."[28] Drawing staff from both Western Electric's factories and AT&T's own Department
of Development and Research, Bell Labs was designed to foster cooperation
between headquarters and plant. The person who would make all this happen
was Frank Baldwin Jewett.

FRANK JEWETT AND THE INSTITUTIONALIZATION
OF SCIENCE AT BELL LABS

From his office in a remote part of Bell Laboratories, Lloyd Espenschied surveyed his new boss. He did not like what he saw. The leader of the company's
new research organization, he wrote in private, was a second-rate scientist with
"quite limited imagination," a glad hander, "conventional and practical." Even
his strengths provoked Espenschied's ire. Jewett "appreciated so highly the
teamwork side of research . . . that he came to regard inventing as incidental
to development rather than the spearhead of it." That was unfortunate, Espenschied concluded, for "in this age of huge collective enterprise, we need, more
than ever, superman, and alas, we are all born very ordinary indeed."[29]

It would be easy to dismiss Espenschied as another jealous corporate func-
tionary, or an old-timer who preferred the individual and personal touch to the

new imperatives of modern organization. Though he had done important research on radio and coaxial cable, he had lost much of his purpose and place as AT&T's research organization grew. But his criticism was not that of nostalgia, for Espenschied appreciated and even applauded purpose and order. He favored a Romantic version of modernism, however, with a charismatic leader building an organic community of followers around his genius.[30] Jewett, in contrast, offered a different place for the individual in research, one that emphasized the new science of organization. He would both create the organizational structure of the modern research lab and, just as important, find it a place in the corporate order.

The roots of Jewett's ideas went back to his mentor, J. J. Carty. As we have seen, Carty brought to research the philosophy that organized labor, not individual genius, made progress. "Great ideas must come from the mind of an individual genius" he understood, "but to apply the idea to such a complex entity as a telephone system . . . no individual unaided can bring it to a successful practical application."[31] A sort of evolutionary functionalism guided Carty and those at the firm who followed him. As Jewett wrote, "[i]n the importance Carty attached to cooperation, he was a true disciple of Herbert Spencer," the nineteenth-century social thinker who applied Darwinian notions to society.[32] In this tradition, organization was seen as good because it "assigns to the individual the tasks and duties to which he is best fitted . . . [to] get the most out of the efforts of all."[33]

Drawing on Carty's ideas of functionalism, Jewett would coordinate the specialized labor in the laboratory and keep it central to the corporate mission. Even more than Vail, Jewett created the structure, strategy, and culture that ensured research and innovation would remain key features of company behavior. Unlike his predecessors, though, he would place more emphasis on the individual. Rather than replacing individuals with organization, as Espenschied had feared, organization, for Jewett, served as a tool for multiplying the powers of individual geniuses.

Jewett's background was perfect for these tasks. Receiving his Ph.D. from the University of Chicago under Albert Michelson, he taught physics and electrical engineering at MIT before entering the business world. At AT&T he headed up the corporate office of engineering research, but in 1912 went over to manufacturer Western Electric, where he eventually became chief engineer.[34] With his academic training in science and experience in manufacturing, Jewett came to understand the need for mechanisms to connect these two without reducing the one to the other.

Under Jewett's leadership, Bell Labs organized into two main groups. The "commercial group" provided all the legal, financial, accounting, and personnel services that a giant research body required. This arrangement reflected a

140   basic concept endorsed by Jewett—the Labs was "in the last analysis . . . a commercial organization."[35] It was to be a profit center, not a cost center. Its primary activities, however, were to be found in the other half, the "technical group." Here engineers and scientists engaged in both research and applied development activities. Among the technical group, the Research Department, staffed by academically trained scientists, did the most basic work. But it stood in close proximity to engineers in the Apparatus Development Department, the Systems Development Department, the Inspection Engineering Department, and the Patent Department. In time patents were moved from the technical to the commercial wing, and a new Transmission Development Department was added. The latter did research on the area of telephony most closely connected to academic science.[36]

    Size and centralization offered distinct advantages. Problems could be attacked from multiple angles, bringing together disciplines ranging from chemistry and physics to metallurgy, mathematics, and statistics. Research directors broke down problems and assigned them to specialists, coordinating the lines of attack. Centralization also freed scientists and engineers from more immediate commercial concerns to pursue speculative research opportunities.[37] In these respects, Bell Labs followed the pattern of other new industrial research organizations, centralized so as to remove scientific activity from day-to-day operating concerns.[38] But Bell Labs quickly developed organizational lines that connected research to engineering and manufacturing. Engineering groups at the Labs, such as Systems Development and Apparatus Development, quickly obtained the fruits of research for their design projects. Both research and development personnel also connected to the regional Bell Operating Companies ("the field") through another unit, AT&T's Department of Development and Research (D&R).

    Not folded into Bell Labs, D&R was a corporate unit that advised headquarters on technical issues. Whereas the basic units of the Labs organized around disciplines and apparatus, D&R organized around functions. It housed transmission, switching, and outside plant departments.[39] In a complex interplay of organizations, it mediated between the Labs and the field. D&R received information from another unit, the Department of Operations and Engineering, which had close contact with the Bell Operating Companies strewn across the nation. With this information, D&R set standards for equipment and apparatus, and passed them on to the Labs. Ideally the entire structure connected field operations to manufacturing through the Labs.[40]

    As Espenschied recognized, Jewett was chosen to head Bell Labs because he was willing to work through this complex bureaucratic structure, not around it. Here he contrasted sharply with others heading industrial research organizations. Willis Whitney at General Electric, and even J. J. Carty, had relied more on personality than impersonal structures to connect research to corporate needs. Carty was a general, commanding his troops as in a military campaign. Whitney was a patriarch who smoothed over relations between scientists and corporate types with his familial, antibureaucratic style. Jewett had little charisma, but personal style and authority were not as important as they once were. Leaders such as Whitney, romantic generalists who disliked bureaucracy, grew less effective at promoting the place of research in the corporation. Those who understood how to use bureaucracy, such as Jewett, became more successful.[41]

As a Ph.D. scientist, Jewett appreciated the value of fundamental scientific work. But he had also seen how difficult it was to move the fruits of research over to manufacturing.[42] He was astute enough an organizational politician to know that over time, expensive research might lose favor in a corporate setting. With its monopoly position secure, AT&T could grow complacent. Without a direct competitive challenge, even committed researchers and engineers might become lazy or lose touch with customers.

Within his first decade as Bell Labs president Jewett had produced organizational solutions to both of these issues. One of his first moves was to bring the all-important right to set equipment standards inside the Labs. It was crucial, Jewett maintained, that development and research be located together, rather than leaving the development and engineering work in the hands of Western Electric.[43] Western Electric could perform its own research to improve manufacturing processes, but otherwise it had to rely on the Labs when it came to equipment, and the Labs would have final say on quality and performance.[44] In this way standards would be set with "due regard to the betterment of service throughout the entire system," something that the Labs' scientists were best positioned to determine.[45] If, in the course of manufacturing, small changes in equipment were needed, the Labs would recheck the results.[46] Similarly, the Labs sought to take a more direct role with field operations, investigating on its own complaints about equipment. In this way researchers gained more intimate knowledge of field conditions that helped them in design work back at the Labs.[47]

These ties between research and operations went both ways. Carty had demonstrated how even in a narrowly focused project such as the transcontinental line, deep research was of value. Jewett generalized from this experience, giving his research scientists freedom to pursue opportunities arising "out of the growth of science itself. . . ."[48] This, in Jewett's view, would prevent AT&T from falling prey to complacency or stagnation, or being blindsided by an unexpected technological change. Jewett took away from the audion and transcontinental line episodes the role of uncertainty in technological change. The

142 danger that industrial research entities faced was that the practical demands of everyday work, which the Labs was obligated to perform, would push aside other things that did not seem to have immediate direct benefits. "It becomes increasingly easy to pass from seeking knowledge because it is useful," Jewett warned, "to seeking what we may consider from *a priori* reasons to be useful knowledge."[49] However, pursuit of basic research did not make the Labs a staff or service operation. Rather, it reflected the "line" nature of science in industry. Researchers manipulated science, in Jewett's mind the basic raw material of the telephone business.[50] The Labs was thus a full part of the vertical structure of the Bell System, and just as much a producer of profits as was Western Electric. Jewett, by implication, was more than an advisor to executive personnel on technical matters; he was a peer executive.

The research organization needed to be connected to the company mission, but not so tightly connected that the more open-ended work disappeared. Jewett's most important contribution here was the articulation of a powerful rhetoric about science, technology, and markets. Drawing on precedent, he forged a new synthesis, a new vision of science in the corporate setting. It would define Bell Labs and other pioneering industrial research organizations for half a century.

The Jewett Synthesis began by redefining the place of creative individuals in innovation. Earlier research directors at AT&T had thought inventive genius beyond control or organization. Following Carty, however, Jewett argued that organization improved on individual genius by capturing multiple points of view.[51] Individual inventors would still turn up the unexpected, of course. But if no one could control or predict when breakthroughs occurred, with the Labs AT&T stood well equipped to learn about and make use of these unpredictable advances.

Next Jewett dealt with the related issue of basic science in a corporate setting. Just as individual genius had been thought incompatible with an industrial laboratory, so too had science been assumed to be at odds with business. Distinguishing between "pure" and what he called "fundamental" research, however, Jewett shifted the debate from the type of knowledge to the site of production. Pure research went on in universities and other nonprofit settings. It uncovered the laws of the physical universe and produced new knowledge about physical reality. But so too did the "fundamental" research at Bell Labs. The only difference was in purpose. The purpose of fundamental research was to "achieve a result," rather than to "extend the boundaries of knowledge."[52] Pure research sprang from a love of knowledge. Fundamental corporate research served business needs, though otherwise it was the same as pure research. Jewett did not seek to build a "university-in-exile," but rather to overcome traditional objections that science and business were incompatible.[53]

The rhetorical trick was Jewett's stripping science of any social or other inherent values beyond pursuit of truth. This opened the door for researchers at Bell Labs to serve the corporation while remaining members of the scientific community. Truth, after all, was truth, regardless of where it was produced. Precisely because science had no fixed or *a priori* values it could be bent to the needs of the firm (or any other institution). It could serve the practical needs of the telephone. Older beliefs that basic research could not serve "applied" ends, or that value-free science would be debased by connection to a commercial product, vanished through Jewett's rhetorical move. What also disappeared from view, however, were the ways in which Jewett's very definition of science stripped it of any independence from business interests. "So far as the physical machinery of electrical communication is concerned," Jewett argued, choices made by future engineers "will be determined . . . primarily by considerations of economics or policy and not by those of a limiting science."[54] Science itself, clear and objective, could provide no independent check on or critique of corporate activity. In a private organization questions of value would be settled strictly by business criteria.[55]

Jewett's argument for basic research provided a way to bring scientists into a profit-making setting without their losing their professional identity. Simply turning scientists into engineers or technicians would not secure the benefits of fundamental knowledge. To keep high-level Ph.D.s engaged, Jewett argued that AT&T should fund cutting-edge work even when the payoff was unclear or remote. He also encouraged scientists to publish and attend professional conferences. Between 1925 and 1928, the Labs' output in scholarly journals was the ninth largest in the nation, behind only the top research universities.[56] Carty had once imagined the research laboratory as almost a Taylorite factory of intellect. Jewett tempered this more extreme formulation with greater respect for the professional reputation and the personal perspective taken by the individual scientist. But creativity could not be limited to the "prejudice" of one person's viewpoint, which was why scientists in pursuit of truth should respect the value of organization. Bell Labs thus preserved a role for individual difference within the well-organized lab, sidestepping outworn arguments about the corruption of pure science by the corporation.[57]

Institutional collaboration through the central research laboratory also made possible fruitful interaction between basic research and practical issues of engineering and manufacturing. New knowledge, if it were truly new, could not be fully predicted. But scientists in close contact with design and manufacturing engineers would come to see avenues and opportunities available to serve the telephone network. They possessed the skills, training, and professional contacts to pursue such opportunities when they arose. For example, scientific knowledge could "predict" or lead researchers to pursue certain likely paths

144   that would result in a solution to a technical problem. Conversely, work on an engineering problem would raise questions that required basic research, opening up whole new lines of inquiry. According to Oliver Buckley, head of research at Bell Labs, the problem was not in finding something for researchers to do but to "select from all the promising leads that are uncovered."[58] The built-in goals and the economics of the Bell System would provide focus to make these choices.[59]

Research and manufacturing had to reach the customer as well. This may seem surprising in a monopoly. AT&T management saw customers as a fairly homogenous group, largely as men who valued quality, standardization, and the decreasing cost of service. If managers held a rather limited view of their public, they still understood their mission as providing a service to users and not simply perfecting system operations in purely technical or abstract terms. Bell Labs had to remain in touch with the field if it were to participate in this project.[60] In 1932 the corporate D&R Department was absorbed into Bell Labs. This was partly a response to the Depression, which put a premium on reducing costs and streamlining operations. But Jewett appreciated how the move strengthened his position. With D&R gone no other corporate body "check[ed] the Laboratories' performance."[61] More important, by taking over the fieldwork done by D&R, Bell Labs moved one step closer to Jewett's organizational vision of a Labs positioned right in the vertical chain of production.

Bell Labs now had both disciplinary and functional departments, and both fed results to development departments that paralleled the engineering groups at Western Electric. Indeed, it was hard to tell where research ended and development began. Oliver Buckley argued that a quarter of "research" was really "development work in areas where fundamental research and development are closely connected." Here was the Labs' strength, "continuous operation" from research to application. Though personnel were free to pursue fundamental work, the Labs was not set up "separate and apart" from daily operations of "commercial design and economic consideration," as were other industrial research entities.[62] Research and development departments were in "close proximity," and information flowed between them casually and informally. The research worker served as a consultant to the development engineer, and researchers had a good understanding of the field operations of the apparatus they were working on.[63]

As a monopoly, AT&T had the option of simply charging the public the full cost of all this work. Indeed, unlike other firms, it had a huge source of revenue to fund its laboratories—a 4.5 percent fee collected from the local operating companies, and through them, the consumer. Nonetheless, this source of funds did not relieve Bell Labs of the obligation to be self-financing. Jewett

had to charge as much of his work as possible to another part of the company. 145
Western Electric, for example, paid substantially for the fruits of laboratory research, between $14 and $20 million per year in the 1920s.[64] This sort of internal market was used at competitive firms GE and Du Pont to pay for research as well. It provided a way of connecting research activities back to the basic economic functions of the firm.[65]

At AT&T, such financial and organizational arrangements were, in essence, embodiments of Jewett's belief that fundamental and applied science could not be logically separated. Research always linked to development; science always had a practical payoff.[66] Under Jewett, basic research had been redefined, away from individual genius or the sort of work that had gone on in the nineteenth-century university, replaced by organized science.[67] The financing of the Labs, the way it organized research to attack problems, and the role it played in decisions about standards and service quality provided a powerful internal argument for supporting scientific research at the corporation. The structure and operation of Bell Labs was, at the same time, an argument against notions that progress only came from science in a university setting removed from commercial and financial concerns. Fundamental research could thrive and pay for itself in a profit-making context.

Jewett's faith in smooth functional progress through organized science soon faced a major challenge: the Great Depression. In the late 1920s AT&T was spending some $20 million on research annually, one tenth of all private-sector R&D dollars in the United States, and it employed 5,300 people.[68] With the Depression, demand for telephone service slackened, and for a time the number of homes with telephones dropped. Western Electric laid off thousands of factory workers. By 1934, the Labs's intracompany revenues had fallen by 30 percent.[69] In response, Jewett argued for keeping up research work. As he wrote to company president Walter Gifford, "[t]he future is promising for more science and technology. It would be wise to maintain our strength in this field and play for the long term."[70] When this appeal failed, Jewett altered his justification for research from the promise of the future to the immediate benefits of cost reduction.[71] Accordingly, work at the Labs shifted firmly toward incremental improvements in existing technology to support AT&T's primary mission and markets.[72] In the end, neither argument worked, and the Labs let go 1,500 people as the organization's payroll shrank 35 percent. There was no new hiring until 1936.[73]

Still, for the most part the Depression years confirmed rather than undermined the Jewett Synthesis. By the mid-1930s, Jewett was again arguing that innovation through science would lead the nation out of the Depression and the Labs's budget was on the rise.[74] Jewett's organizational innovations and tight links

146   between science and engineering weathered the Depression quite well. In contrast, the organizationally skeptical Willis Whitney of GE suffered massive budget cuts. His personal style of leadership and the "family"-like atmosphere did not survive the crisis, and he retired from the directorship of GE's labs.[75]

CONTINUITY AND CHALLENGE IN THE POSTWAR ERA

Although Bell Labs came out of the Depression in strong shape, its mission and culture began to change, particularly following Jewett's retirement. The change was gradual, and at first hard to perceive. More emphasis in the post–World War II era went toward basic research. Nobel Prizes and breakthrough inventions added luster to the Labs, and also provided powerful symbols to defend the Bell monopoly against detractors. Oddly enough, the economic crisis of the 1930s may have actually reinforced this shift toward basic research, because engineering and manufacturing work had to be put on hold during the economic downturn. Jewett had protected key staff by turning to longer-term research projects.[76] After the economy recovered, the Labs rapidly added new, university-trained personnel. But the Depression years had undercut the faith of science graduates that corporations were a viable alternative to academic life. Increased government spending on science also raised prospects for employment outside of the private sector. To strengthen the scientific departments at Bell Labs, new hires were permitted to foster a more academic atmosphere, with colloquia and increased opportunities for publication.[77]

One visible symbol of change appeared in 1941. Bell Labs opened a major new research center in Murray Hill, New Jersey, a sprawling new campus-like complex of the sort that would typify postwar industrial research facilities. The facilities at Murray Hill were soon matched by others dotted among the rolling hills of central New Jersey.

Frank Jewett's successor, Oliver Buckley, was even more committed to basic research and to the role of science in promoting innovation and creativity in large organizations. Buckley also understood better than Jewett the new role of government in sponsoring basic research, sitting on the committee that recommended creation of the National Science Foundation and eventually serving as science advisor to President Truman.[78]

The new trend continued under Buckley's successor, Mervin Kelly. A strong researcher, Kelly believed that the Labs needed personnel capable of handling cutting-edge theory. While serving under Jewett in the 1930s, he had advocated hiring theoretical physicists and was able to create a strong physics department of Ph.D.s versed in quantum mechanics. This group included William Shockley and the other inventors of the transistor.[79] The hiring of physicists with advanced degrees was not itself a radical departure. Following Jewett's principle that science and economics could work in tandem to guide research agendas, Kelly reasoned that solid state physics would be the likely area to look for solutions to the problems of speed, power, and reliability that bedeviled vacuum tube electronics. Indeed, it was only when it seemed that this area of physics had a clear payoff to telecommunications that Kelly was able to hire the people he wanted.[80] Once economics and business criteria had defined the problem to be solved, research, even the deeper branches of theory, could swing into action. Theory guided experimenters by determining where the likely payoffs of empirical work would be. In this case, the results—the transistor—stimulated work that led back down to basic research, which was needed to gain a fuller understanding of the technology.[81] Thus work on the transistor followed the tradition of basic and applied science feeding back and forth to each other that had been established with the audion and transcontinental line. But this project turned out to be the last example of such a fine balance between basic research and practical commercial demands.

The success of the transistor provided a compelling argument for a new emphasis on theoretical knowledge and strong disciplinary departments at Bell Labs. Under Kelly's leadership, a multidisciplinary team had brought to fruition a practical solution to a key telephone problem by starting from theory. The old interplay between science, technology, and business was still there, but in contrast to the Jewett years, science and theory were taking leading roles. Bell Labs remained an industrial laboratory, but a new term, *mission oriented research*, suggested a softening of the line between pure research of an academic type and fundamental industrial R&D.

One example of this shift was in personnel. The Labs sought out the best graduates in a variety of scientific disciplines, and disciplinary lines became more important to the organization of research. Top researchers, particularly those with a theoretical bent, wanted strong links back to their academic fields. That meant having the autonomy to define research topics, take sabbaticals, travel to professional meetings, attend seminars, and publish peer-reviewed articles.[82] Jewett had already permitted such professional ties at the Labs. But the balance was shifting more toward what Jewett had defined as "pure" science in an academic mode as the Labs competed with universities for scientists.

A pillar of the Jewett Synthesis had been that breakthroughs and new knowledge had unpredictable sources. This pillar too began to bend a bit, as success with the transistor pushed Bell Labs personnel into a belief that science could not just solve, but define problems. Thorough understanding of the sciences underlying telecommunications, it was now believed, would tell Labs scientists where to focus their research programs.[83] This was different from Jewett's

148   more modest belief that scientific research could help the firm cope with uncertainty and prepare for the unexpected, but that "business criteria" would determine what to do.

Still, for all the new emphasis on science, many aspects of the original Bell Labs were carried over into the postwar era. New directors came to appreciate the need to balance research, development, and manufacturing in an industrial setting. Bell Labs's Murray Hill facility, for example, actually encouraged science-engineering interaction, despite its campus-like quality. This contrasted with the new technical and research centers going up at IBM and GM, which were more exclusive and academic in physical design and conception.[84] Labs director Mervin Kelly also understood, like Jewett, that cutting-edge research had to connect with engineering and manufacturing. Kelly recognized, for example, that despite the Labs's pathbreaking work in electronics, knowledge of this field was quickly escaping to other firms. RADAR development during the war and greater public funding of science at institutions such as MIT's Rad Lab allowed other firms and institutions to master electronics. In 1956, settlement of an antitrust case against Western Electric required the company to license its patents to whomever wanted them. "We have been a conservative and non-competitive organization," wrote Kelly " . . . but our basic technology is becoming increasingly similar to that of a high volume, annual model, highly competitive, young, vigorous and growing industry."[85] If AT&T did not show progress in manufacturing, nontelecommunications firms would make it abundantly clear that the monopoly was not as efficient as it should be.

"Scientifically controlled" improvements in processes and apparatus required new links between research and manufacturing. This included hiring well-trained design engineers and installing them at the factory. Western Electric was still too oriented toward mechanical technology. New engineers would help translate the breakthrough work of the Labs, especially in electronics, into products and processes at the manufacturing end. Scientific knowledge alone would not be sufficient; there had to be strong expertise and tacit knowledge to turn scientific breakthroughs into new products and low-cost, high-quality apparatus.[86] This necessitated, Kelly realized, putting engineers and researchers together on specific projects and moving from a functional to a line-of-business structure.[87]

To ensure that basic research did not stray too far from business concerns Kelly used organizational and geographical decentralization.[88] Western Electric provided space for development engineers from the Labs to set up shop, bringing the fruits of research directly to the site of production. By the late 1960s, there were sixteen Western Electric manufacturing works and twelve decentralized laboratories in or near them. One-quarter of Bell Labs's personnel worked in these field laboratories.[89] The branch labs retained a high degree of autonomy from the central labs in development and design work for their manufacturing plants. To strengthen connections between manufacturing and research, Western Electric engineers also took stints in the branch labs. This was a direct descendent of policies developed in the Jewett era, when face-to-face dialogue, site visits, correspondence, and other "informal" methods had helped to bridge the research-manufacturing gap.[90]

Supporting these moves was another major structural development, the formation of a new department of systems engineering in 1955. Systems engineering is a way of organizing the multiple functions of a large engineering project or technical system.[91] It uses formal models and measures to integrate system components in optimal fashion, and to relate the system to its environment by taking into account economic and other factors that influence technical choices. In Jewett's time, systems-engineering functions had been scattered throughout Bell Labs, notably in the work done by the D&R Department. Jewett himself had recognized that choices about system components had to be made on the basis of economic factors such as labor cost and demand. Indeed, the entire "Spenserian" apparatus of thought that Carty had first brought to the Labs was oriented to thinking in terms of specialization, functional division, and optimal integration. This way of thinking became formalized in the postwar period in systems engineering.[92]

By combining technical and economic criteria and relating the system to the market, systems engineering seemed a perfect method for rationalizing innovation. Beginning with the basic truths of science, one could construct an optimal system of technology even in a monopoly through systems engineering that linked research, development, manufacturing, and field operations. It was another example of how, beyond the rhetoric about scientific research, Bell Labs continued to be directed by and to acknowledge the importance of business and economic considerations in its work. But systems engineering was also a highly formalized method, one that substituted internally generated information for market signals. With a dominant, near-monopoly position, AT&T had few market signals to rely on. The formalized methods of systems engineering provided only a partial substitute for market signals.

THE END OF THE JEWETT SYNTHESIS

In the postwar period, Bell Labs continued to conduct basic research—extremely good basic research, judging by the prizes and publications. It continued to integrate this research into the chain of production. Although the transistor showed how science and theory could take a leading role, sometimes practical

150   work paid off in new scientific knowledge, as when Arno Penzias and Robert
Wilson used the Labs's sensitive horn antenna to research noise interference
with microwave transmission and ended up confirming the Big Bang theory.
Either way, science had been made a basic part of the telephone system.
Indeed, the more stable the system and the less competitive the environment,
the more science could be expected to provide that spark of creativity that the
monopoly needed. Contrary to critics who feared corporate science would
merely serve narrow interests, the Labs was performing almost a public service
with its broad research agenda. It was the culmination of Jewett's insight that
one could never control science, and hence never truly stabilize technology.
There would always be something unexpected beyond the horizon. The wis-
est strategy for survival was to take an active role in grasping the new through
basic research.

It would be tempting to end the story here, circa 1970, when all was right
with Bell Labs's world. But subsequent events have not been kind to the crown
jewel of American industrial research. By the early 2000s, it was a beleaguered
research center of a few hundred, not twenty-five thousand, employees. Development, once 70 percent of its work and a key link between basic research and
business and technology, had been stripped out of the Labs and completely
decentralized. Not surprisingly, the result has been to leave the research scientists somewhat adrift, and vulnerable to management skepticism about the
worth of their contributions.[93]

Various explanations have been offered for this dramatic decline. For some,
it simply reflects fundamental changes in science and technology that undermined Bell Labs's unique competence and left it behind in the innovation race.
This answer, however, raises the question of why, after negotiating other big
shifts in science and technology, the Labs failed to keep pace? Jewett's whole
concept, after all, had been a laboratory that could deal with unexpected external shifts in science and technology. A second explanation is that the entire
concept of a centralized research laboratory has grown obsolete. Such institutions are too disconnected from commercial needs, a view that is undercut by
the well-established connections Bell Labs had long made between research,
manufacturing, and the field.[94] Others argue that the dismantling of the old
Bell Labs was a mistake borne of radical deregulation and short-term corporate
thinking. Certainly the direct cause lies with the breakup of the Bell System
beginning in 1984. But again, questions remain. If the Labs really was a jewel,
how could anyone have made such a mistake and let such a valuable asset
depreciate?

I argue instead that what mattered to Bell Labs's fate was history. This
accounts for the Labs's ability to negotiate external shifts in one period but not

in another. By history I mean in particular the way in which those managing and working at Bell Labs read and reflected on their own past, and the lessons they chose to take from it.[95]

The problem began with the ideology of science and technology that emerged at the Labs after World War II. As we have seen, Frank Jewett constructed an ideology that served the Labs well in its early struggle to establish the place of research at AT&T. Since science itself was value neutral, Jewett argued, the pursuit of truth could go on inside as well as outside of the firm. But over time, and with the flush of success that came with Nobel Prizes, this belief in the efficacy of science for industry grew so inflated that it actually knocked away one of the pillars of the Jewett Synthesis. Far from being corrupted or constrained by industry, science at Bell Labs became too free of its partners—technology and economics.

For all the efforts to maintain continuity with the past, some things changed decisively at Bell Labs after World War II. The memory of competition, which had partly spurred the pursuit of innovation at AT&T, was all but lost. Even Mervin Kelly, a perceptive leader who well understood the value of connecting research and manufacturing, succumbed to the postwar temptation toward detached scientific research. While professing the strategic value of decentralization, Kelly also argued that researchers assigned to factories were scientists who assisted factory engineers in working with new science-infused technologies. As Steven Usselman notes, Kelly's recommendation to centralize research at IBM was rejected.[96] In like manner, Kelly advocated that leadership of the Labs be in the hands of strong scientists, and that the Labs president stay on site at Murray Hill, rather than moving to corporate headquarters. Science became a separate realm within the corporation.

This was a subtle but crucial reworking of Jewett's belief that scientific autonomy made it possible to connect science and corporate needs. If science were a pursuit of truth that could never quite be standardized, the smart firm respected science and stood ready to absorb its results. Conversely, high-quality researchers continued to pursue their projects, but had no objection to doing so in ways that served corporate ends—truth being truth in the final analysis.

Through the early 1960s, notions of progressive, autonomous, and value-free science fit perfectly with the Cold War emphasis on the efficacy of America's free, liberal, pluralistic institutions. Professional scientists, operating on the basis of objective and nonideological inquiry, would apply their expertise freely in a market economy. This sort of Parsonian structural-functionalism, which respected professional norms but also firmly integrated professionals into the social system (and in this case, the Bell System), was an intellectual outgrowth of the Spenserian functionalism that had convinced Carty and Jewett that they could create an effective scientific establishment in a large corporation. A few antitrust attacks aside, AT&T was largely left alone to perfect the workings of the telephone network, with science as the most basic tool of the task. It was a neat circle. Autonomous, value-free science ensured that research at Bell Labs would produce only objective results that would be applied to the needs of the Bell System, and the success and stability of the Bell System justified the status quo, including the role of science and the Labs.

At precisely the moment when this happy, functional system seemed to be at the height of its powers, however, confidence in the autonomy and beneficence of science came under attack. The general questioning of science began in the era of the Vietnam War, when critics began to condemn Cold War–related research on college campuses and in corporations as part of the military-industrial complex. At a particularly fraught moment in the 1970s, in fact, the head of Bell Labs, William O. Baker, felt he had to defend his institution and science in general from these critics. Speaking on "leadership and authority," Baker rejected what he believed were naive calls for "more socially responsible science." Science needed no external values or responsibilities beyond the pursuit of truth. He defended Jewett's original notion that science and business could be partners and that science met its social responsibilities by working for private corporations.[97] But the space Jewett had carved out for corporate science was collapsing rapidly. The pursuit of truth was no longer a sufficient defense for what Bell Labs did.

In response to such skepticism, corporate scientists began to emphasize their independence. Independence from corporate objectives, not pursuit of truth regardless of context, defined the professionalism of the scientist working for industry now. The glory of Bell Labs became its Nobel Prizes and tradition of almost academically pure research. This restating of the Labs's history missed or overlooked the way in which Jewett had defined fundamental research as connected to engineering and economics. By stressing independence, scientists were cutting themselves off from the corporation that supported them, and removing research from external market signals. The defense against critics of corporate science was to deny links between scientific research at Bell Labs and the corporate mission.

None of this is to deny that basic material changes contributed to the erosion of research at AT&T. New competition, falling profits, and structural reforms all helped to sweep aside in the 1980s and 1990s the long tradition of mission-oriented research at Bell Labs. But the challenge to the key ideas that Jewett and his followers had articulated to justify industrial research had begun even earlier. Bell Labs, even in the better-funded and more stable era before the deregulation of the 1980s, had always required the validation that basic research paid dividends by connecting the firm to the unpredictable and neu-

tral learnings of science. Without such an argument, there was no way to jus- $\qquad$
tify the expenditure of corporate resources to the long-term prospects of
research with uncertain payoffs. As the notion of value-free science collapsed,
so did the justification. Or to put it another way, if corporate science was indeed
tainted by its mission, as critics charged, then corporate managers could feel
justified in tossing aside any respect for science as pursuit of truth in favor of
research narrowly focused on corporate needs. Efforts to protect basic research
by reasserting scientific objectivity only cut off the Labs from engineering and
manufacturing, destroying the links here that Jewett had so carefully forged.

CONCLUSION

Schumpeter's claim that innovation could be routinized rested in part on the
work being done within large corporate laboratories.[98] But as we have seen,
even in the largest, most famous example, Bell Labs, the reality belied any simple notion that innovation could be strictly routinized. In fact, Bell Labs was
at its best when it found ways to connect internally generated research and
knowledge to its external environment, through links that brought laboratory
scientists, factory engineers, and field engineers closer together. "Fundamental" research in a corporate setting could be both highly scientific and highly
relevant to corporate needs.

The large corporation's main advantage in the process of innovation lies not
in routinization but in coordinating a production chain running from basic
research to manufacturing and services. Even the best corporate laboratory cannot eliminate the unexpected, though it can scan the external environment
and pick up on or anticipate breakthroughs. Exploiting breakthroughs requires
the sort of institutional mechanisms present in large firms.[99] In its heyday Bell
Labs was extremely adept at shepherding long-term projects that began with
fairly esoteric science through a series of intermediate steps that ended with a
commercially useful outcome.

This work required institutional and cultural mechanisms for cultivating
innovation in the firm. Frank Jewett's arguments for the objectivity and long-
term payoff of scientific investigation, coupled with the relationships he built
between the Labs and engineering and manufacturing organizations, provided
an excellent environment for doing basic research. The irony of history is that
as this model of science came under attack for being naive and overly ideal-
ized, Bell Labs detached itself from the rest of the firm's value chain. Finance-
oriented managers skeptical of science in an industrial setting thus lost sight of
the alternative tradition founded by Jewett, which had successfully connected
basic and applied research in ways that did not subordinate one to the other.
The current skepticism about basic research in industry imagines a world of

154   detached corporate laboratories that existed only for a moment in a much longer history of a successful balancing of the commitment to the long term with the need for a commercial payoff. This historical misperception perpetuates the error that basic research is a luxury firms cannot afford. Reducing the scope of innovation in this way is a recipe for reducing innovation, period.

NOTES

1. *New York Times*, June 20, 1995, sec. C. For figures on expenditure and employment at the Labs, see *New York Times*, July 2, 1991, sec. D.

2. A claim made strongly in Jeremy Bernstein, *Three Degrees Above Zero* (New York: Charles Scribner's Sons, 1984). See also *New York Times*, May 27, 1982, sec. B. Far better than most universities, Bernstein argues, the Labs provided a cooperative and yeasty interdisciplinary environment for its employees. Portraits from the 1970s and 1980s drew it as a cloistered, campus-like world and noted the counterculture quirkiness of its high-level researchers.

3. Kenneth Lipartito, "Picturephone and the Information Age: The Social Meaning of Failure," *Technology and Culture* 44 (Jan. 2004): 50–81.

4. T. A. Heppenheimer, "What Made Bell Labs Great," *Invention and Technology* 12 (Summer 1996): 46–57. For a restatement of these charges, see also *New York Times*, Mar. 9, 1987, sec. D.

5. Nathan Rosenberg, "Why Do Firms Do Basic Research (With Their Own Money)?" *Research Policy* 19 (Apr. 1990): 165–74.

6. For example, see O. B. Blackwell to Mr. Clark, Feb. 18, 1944, Box 77-04-01, AT&T Archives, Warren, New Jersey (hereafter ATT).

7. Bernstein, *Three Degrees Above Zero*, 215; Leonard S. Reich, *The Making of American Industrial Research: Science and Business at GE and Bell, 1876–1926* (New York: Cambridge University Press, 1985), 172; Nathan Rosenberg, *Inside the Black Box: Technology and Economics* (New York: Cambridge University Press, 1982), 154; Rosenberg, "Why Do Firms Do Basic Research," 170.

8. David Mowery, "Industrial Research and Firm Size, Survival, and Growth in American Manufacturing, 1921–1946: An Assessment," *Journal of Economic History* 43 (Dec. 1983): 953–80.

9. Naomi R. Lamoreaux, "Reframing the Past: Thoughts About Business Leadership and Decision Making Under Uncertainty," *Enterprise and Society* 2 (Dec. 2001): 632–59.

10. Louis Galambos, "Theodore N. Vail and the Role of Innovation in the Modern Bell System," *Business History Review* 66 (Spring 1992): 95–126.

11. Leonard Reich places more emphasis on the narrower aspects of corporate strategy, in particular competitors and the threat posed by radio. Reich, *The Making of American Industrial Research*.

12. Vail left the company in 1887, in part in disgust at the refusal of the Boston owners to take a longer-term view, before his later return. Reich, *The Making of American Industrial Research*, 144–45.

13. Stephen B. Adams and Orville R. Butler, *Manufacturing the Future: A History*
of Western Electric (New York: Cambridge University Press, 1999).

14. Reich, *The Making of American Industrial Research*, 145–49.

15. Reich, *The Making of American Industrial Research*, 148.

16. J. J. Carty to E. J. Hall, July 17, 1907, Box 48-08-01-10, ATT.

17. Reich, *The Making of American Industrial Research*, 158–59.

18. M. D. Fagen, ed., *A History of Engineering and Science in the Bell System: The Early Years (1875–1925)* (New York: AT&T Bell Laboratories, 1975), 256–61.

19. H. D. Arnold to E. H. Colpitts, "Operation of Some Electric Valves and the Audion," Jan. 17, 1914, Box 43-05-03, ATT; Reich, *The Making of American Industrial Research*, 159; Leonard S. Reich, "Research, Patents, and the Struggle to Control Radio: A Study of Big Business and the Uses of Industrial Research," *Business History Review* 51 (Summer 1977): 208–35.

20. Events like these could also be directed at government antitrust policy, a use of science that many big firms found effective at this time. See David Hounshell, "Continuity and Change in the Management of Industrial Research: The Du Pont Company, 1902–1980," in *Technology and Enterprise in Historical Perspective*, eds. Giovanni Dosi, Renato Giannetti, and Pier Angelo Toninelli (New York: Oxford University Press, 1992), 239.

21. Reich, *The Making of American Industrial Research*, 161–64, 169.

22. John J. Carty, "The Relation of Pure Science to Industrial Research," *Science* 44 (Oct. 13, 1916): 511–18.

23. Theodore Vail to John Moon, Dec. 30, 1915, Box 1080, ATT.

24. On the development of carrier, see *Statement of O. E. Buckley*, 1944, Box 45-01-01-01, ATT.

25. Reich makes much of the patent motivation for research after 1907 ("Research, Patents, and the Struggle to Control Radio," 231), but patents had always been important. On the decline of patents as a motivation at General Electric, see George Wise, *Willis R. Whitney, General Electric, and the Origins of U.S. Industrial Research* (New York: Columbia University Press, 1985), 275.

26. On numbers and name, see Reich, *The Making of American Industrial Research*, 96, 182, 185–86.

27. Talk by Mr. Craft, Bell System Education Conference, 1926, Box 185-06-02, ATT.

28. Walter Gifford, *Addresses, Papers and Interviews of Walter Gifford* (New York: AT&T, 1928).

29. Frank Jewett Biographical File, Box 60-05-03, ATT.

30. Espenschied, though never fired, was reduced to *persona non-grata* for his pro-fascist sympathies and his role in the Nazi-linked German American Bund.

31. *J. J. Carty—Speeches and Demonstrations, 1889–1926*, "World Communications," an address at the University Club of New York, Feb. 10, 1923, Box 1073, ATT. On Carty's commitment to collectivity, in a functional, Spencerian sense, see *J. J. Carty—Speeches and Demonstrations, 1889–1926*, World Communications Congress, 1926, Box 1073, ATT.

32. *J. J. Carty, Biography*, Box 1073, ATT.

33. Vail address before the National Geographic Society, *Vail Speeches, 1913–1919*, Box 1080, ATT. The coinventor of the transistor, William Shockley, would spend his

156   later years trying to prove that race and IQ were linked and that blacks as a group had lower IQs. Shockley termed his theory "retrogressive evolution," a twist on Carty's Spenserian ideas. See *New York Times*, Aug. 14, 1989, D9. The Spencerian view held sway at Bell Labs at least until the 1970s. See specific reference to it in J. A. Morton, *Organizing for Innovation* (New York: McGraw-Hill, 1971), 94.

    34. *ATT Headquarters Bulletin*, 13 (Sep. 1939): 10–11.

    35. Talk by Mr. Craft, Bell System Education Conference, 1926, Box 185-06-02, ATT.

    36. Talk by Mr. Craft. On the change, see BTL—Organizational Chart, 1934, Box 73-06-02-04, ATT. On the design of industrial research laboratories, see Scott G. Knowles and Stuart W. Leslie, "'Industrial Versailles': Eero Saarinen's Corporate Campuses for GM, IBM and AT&T," *Isis* 92 (Mar. 2001): 1–33.

    37. Frank Jewett to David Houston, Mar. 9, 1926, Box 75-05-01-03, ATT. On the importance of gaining a measure of independence from manufacturing at GE, see Wise, *Willis R. Whitney*, 96, 179. Achieving independence without separating too formally from the basic corporate mission was a common problem. See, for example, Georg Meyer-Thurow, "The Industrialization of Invention: A Case Study from the German Chemical Industry," *Isis* 73 (Sep. 1982): 372–75. On a classic example of what happened when this connection failed, see Stuart W. Leslie, *Boss Kettering: Wizard of General Motors* (New York: Columbia University Press, 1983).

    38. This was the case at GE and at Du Pont, where consolidation and centralization took place in the labs during the first decade of the twentieth century. Hounshell, "Continuity and Change," 231–60.

    39. Organization, AT&T Development and Research Department, Frank Jewett to David F. Houston, Mar. 9, 1926, Box 73-05-01-03, ATT.

    40. Frank Jewett to E. B. Craft, Nov. 16, 1925, Box 48-08-01-10, ATT.

    41. On the declining effectiveness of Whitney in the 1920s, see Wise, *Willis R. Whitney*, 250–81.

    42. On faith in the long-term payoff of science, see Ulrich Marsch, "Strategies for Success: Research Organization in German Chemical Companies and IG Farben Until 1936," *History and Technology* 12 (Issue 1, 1994): 23–77, which discusses the leadership of Carl Duisberg in supporting R&D at that company. On Pierre Du Pont in a similar capacity, see David A. Hounshell, "Du Pont and the Management of Large-Scale Research and Development," in *Big Science: The Growth of Large-Scale Research*, eds. Peter Galison and Bruce Hevly (Stanford: Stanford University Press, 1992), 237.

    43. Frank Jewett to E. B. Craft, Nov. 16, 1925, Box 48-08-01-10, ATT.

    44. Organization, AT&T Development and Research Department, Responsibilities and Duties, Box 73-05-01-03, ATT.

    45. Frank Jewett to David F. Houston, Mar. 9, 1926, Box 73-05-01-03, ATT.

    46. *Relations Between the Western Electric Co. Inc. and the Bell Telephone Laboratories, Inc., with Respect to the Design and Quality of Products Furnished by the Western Electric Company*, July 12, 1926, Box 48-08-01-10, ATT.

    47. Mr. Hosford to Mr. Kilpatrick, Feb. 24, 1926, Box 48-08-01-10, ATT.

    48. Bell System Education Conference, 1926, Dr. Jewett's Talk, Box 185-06-02, ATT. It was Jewett who convinced Carty that the route to success with the transcontinental line lay through an investment in the discipline of physics.

49. Frank Jewett, "The Motives of Pure and Applied Scientific Research," *Bell Telephone Quarterly* 5 (Jan. 1926): 12–22.

50. Frank Jewett, "Electrical Communications," *Bell Telephone Quarterly* 14 (July 1935): 167–95.

51. Robert Williams, "Functions of a Laboratory Organization," 1932, Box 55-06-01-04, ATT.

52. Bell System Education Conference, 1926, Dr. Jewett's Talk, Box 185-06-02, ATT.

53. Frank Jewett, "The Motives of Pure and Applied Scientific Research," 12–22; Jewett, "The Laboratory—A Potent Source of Progress in Industry," *Bell Telephone Quarterly* 18 (Jan. 1939): 13–22; and Lillian Hartmann Hoddeson, "The Entry of Quantum Theory of Solids into the Bell Telephone Laboratories, 1925–40: A Case-Study of the Industrial Application of Fundamental Science," *Minerva* 18 (Sep. 1980): 430.

54. Frank Jewett, "Electrical Communication: Past, Present, Future," *Bell Telephone Quarterly* 14 (July 1935): 178.

55. On consumer variability and how it can shape research and development, see Sally H. Clarke, "Negotiating Between the Firm and the Consumer: Bell Labs and the Development of the Modern Telephone," in *The Modern Worlds of Business and Industry*, ed. Karen Merrill (Turnhout, Belg.: David Brown, 1998), 161–82; and Kenneth Lipartito, "Constructing Telephone Networks in Britain and America," paper presented at the Conference on Constructing Markets, Shaping Production: The Historical Construction of Product Markets in Europe and America, Idöborg, Sweden, July 5–7, 2002.

56. Organization, Bell Telephone Labs, Mr. Mills to E. H. Colpitts, Nov. 7, 1934, and E. H. Colpitts to Frank Jewett, Jan. 7, 1935, Box 72-06-03-03, ATT; and Arturo Russo, "Fundamental Research at Bell Laboratories: The Discovery of Electron Diffraction," *Historical Studies in the Physical Sciences* 12, Part 1 (1981): 117–60.

57. A point repeated by AT&T president Walter Gifford, "Address at Cooper Union Commencement, June 6, 1928," in *Addresses Papers and Interviews of Walter Gifford* (New York: AT&T, 1928). See also Robert Williams, "Functions of a Laboratory Organization," 1932, Box 55-06-01-04, ATT.

58. Statement of O. E. Buckley before the FCC, Jan. 21, 1936, Box 45-01-01-01, ATT.

59. The Nobel Prize–winning work of Davidson and Germer at Bell Labs on the wave-particle duality of electrons was a "sideline" from work on the technical opportunities in the new field of electronics. Hoddeson, "The Entry of Quantum Theory," 433; Arturo Russo, "Fundamental Research at Bell Laboratories."

60. Jewett talk, Bell System Educational Conference, 1926, Box 185-06-02, ATT.

61. E. H. Colpitts to Frank Jewett, Oct. 18, 1928, Box 73-05-01-03, ATT. Also E. B. Craft to Frank Jewett, May 1, 1928, Box 73-05-01-03, ATT. These letters discuss the problems of overlap and efforts to reduce the D&R to a small staff function in the years before its final incorporation into Bell Labs.

62. Craft talk, Bell System Educational Conference, 1926, Box 185-06-02, ATT.

63. Statement of O. E. Buckley before the FCC, Jan. 21, 1936, Box 45-01-01-01, ATT. Buckley had reason to slant his testimony toward practical work and development, given the FCC's skepticism about consumers paying for research. But his statements comport with the organizational philosophy articulated by Jewett a decade or so earlier.

158   64. *Review of Bell Telephone Laboratories Activities*, Dec. 26, 1934, Box 45-01-01-01, ATT.

65. Organization—Bell Telephone Laboratories, Box 72-06-03-01, ATT. On GE, see Wise, *Willis R. Whitney*, 217–18, 248.

66. The ways that Jewett connected the most basic research and the most practical results can be seen in Frank Jewett, "Research Methods," *Bell Laboratories Record* 6 (Jan. 1926): 349–52.

67. On the changes in universities that parallel the organization of research in the private sector, see Michael Dennis, "Accounting for Research: New Histories of Corporate Laboratories and the Social History of American Science," *Social Studies in Science* 17 (Aug. 1987): 479–518.

68. Bell Labs vastly overshadowed public sector research as well. Herbert Hoover's ambitious Commerce Department spent a little over $2 million on research. David M. Hart, *Forged Consensus: Science, Technology, and Economic Policy in the United States, 1921–1953* (Princeton: Princeton University Press, 1998), 35; and Organization—BTL, Box-73-06-02-04, ATT.

69. *Review of Bell Telephone Laboratories Activities*, Dec. 12, 1934, Box 45-01-01-01, ATT.

70. Frank Jewett, Memo for Walter Gifford, June 9, 1930, Box 53, ATT.

71. *Economies Effected by Development Work of BTL*, 1935, Box 74-07-02-03, ATT.

72. *Review of Bell Telephone Laboratories Activities*, Dec. 26, 1934, Box 45-01-01-01, ATT.

73. Organization—BTL, Box 73-06-020-04, ATT; and Hoddeson, "Entry of Quantum Theory," 440–41.

74. Despite the hard times, many firms opened new research labs, and overall research expenditures fell only 10 percent during the Depression years. Mowery, "Industrial Research and Firm Size," 978–79; and Bell Telephone Laboratories Organization, Box 76-06-02-06, ATT.

75. On Whitney at GE, see Wise, *Willis R. Whitney*, 278–83, 301.

76. Bell Labs, Review of Expenses, Frank Jewett to J. J. Carty, June 2, 1930, Box 53, ATT.

77. Bell Telephone Laboratories Organization, Frank Jewett to Arthur Page, July 10, 1942, Box 76-06-02-06, ATT.

78. Hart, *Forged Consensus*, 75, 87, 161; and Robert Kargon and Elizabeth Hodes, "Karl Compton, Isaiah Bowman, and the Politics of Science in the Great Depression," *Isis* 76 (Sep. 1985): 300–18.

79. Bernstein, *Three Degrees Above Zero*, 130–36.

80. Hoddeson, "Entry of Quantum Theory," 434–44.

81. Lillian Hartmann Hoddeson, "The Discovery of the Point-Contact Transistor," *Historical Studies in the Physical Sciences* 12, Part 1 (1981): 41.

82. Hoddeson, "Discovery of the Point-Contact Transistor," 41, 53.

83. Morton, *Organizing for Innovation*, 48.

84. Knowles and Leslie, "'Industrial Versailles,'" 22. Saarinen's masterpiece for Bell Labs at Holmdel proved far less effective and more inward looking than the Murray Hill facility.

85. M. J. Kelly, "A First Record of Thoughts Concerning an Important Postwar
Problem of the Bell Telephone Laboratories and Western Electric Company," May 1,
1943, Box 85-05-01-01, ATT.

86. Kelly, "A First Record of Thoughts."

87. Kelly, "A First Record of Thoughts."

88. On Kelly and decentralization of research and development work, see Stuart
W. Leslie, "Blue Collar Science: Bringing the Transistor to Life in the Lehigh Valley,"
Historical Studies in the Physical Sciences and Biological Sciences 32, Part 2 (2001):
71–113.

89. H. W. Bode, *Synergy: Technical Integration and Technological Innovation in the
Bell System* (Murray Hill, N.J.: Bell Laboratories, 1971), 52–53, 105–106.

90. M. H. Cook, "Decentralization—A Feature of Today's Bell Laboratories," Dec.
1961, Box 86-05-01-03, ATT.

91. Stephen B. Johnson, "Three Approaches to Big Technology: Operations
Research, Systems Engineering, and Project Management," *Technology and Culture*
38 (Oct. 1997): 891–919.

92. By 1970, 15 percent of Bell Labs personnel were in systems engineering. Morton, *Organizing for Innovation*, 67–71.

93. Narain Gehani, *Bell Labs: Life in the Crown Jewel* (Summit, N.J.: Silicon Press,
2003); and A. Michael Noll, "Telecommunication Basic Research: An Uncertain Future
for the Bell Legacy," *Prometheus* 21 (June 2003): 177–93.

94. This belief that research had enjoyed great freedom but now had to be made to
pay appeared right after the Bell breakup. *New York Times*, Mar. 9, 1987, sec. D1.

95. On reflexivity and learning from history in organizations, see Jeffrey Fear,
"Thinking Historically About Organizational Learning," in *Handbook of Organizational
Learning and Knowledge*, eds. Meinholf Dierkes, Ariane Berthoin Antal, John Child,
and Ikujiro Nonaka (New York: Oxford University Press, 2001), 162–91.

96. See Steven W. Usselman, Chapter 8 in this volume, and also Steven W. Usselman, "Learning the Hard Way: IBM and the Sources of Innovation in Early Computing," in *Financing Innovation in the United States: 1870 to the Present*, eds. Naomi R.
Lamoreaux and Kenneth L. Sokoloff (Cambridge, Mass.: MIT Press, 2007): 317–63.

97. William Baker, "Leadership and Authority," Skytop Conference, Sep. 27–Oct. 1,
1976, Box 10-07-02-02, ATT.

98. On the application of this Schumperian perspective to corporate R&D, see
Zoltan J. Acs and David Audretsch, "Innovation in Large and Small Firms: An Empirical Analysis," *American Economic Review* 78 (Sep. 1988): 678–90; Susan Morris, "Organizing for Industrial Research: The Resource Networks of Small Enterprises," paper
presented at the Society for the History of Technology Annual Meeting, Toronto,
Canada, October 2002; and F. M. Scherer, *Industrial Market Structure and Economic
Performance* (Chicago: Rand McNally, 1970).

99. David Edgerton, "Industrial Research in the British Photographic Industry,
1879–1939," in *The Challenge of New Technology: Innovation in British Business Since
1850*, ed. Jonathan Liebenau (Brookfield, Vt.: Gower, 1988), 106–34; and Edwin Mansfield and Samuel Wagner, "Organizational and Strategic Factors Associated with Probabilities of Success in Industrial R&D," *Journal of Business* 48 (Apr. 1975): 179–98.

PART II
AMONG FIRMS

Introduction to Part II

Sally H. Clarke, Naomi R. Lamoreaux,
and Steven W. Usselman

AT&T's Bell Labs was once the epitome of corporate innovation. By the late twentieth century, however, this mantle had passed to Silicon Valley. Rather than shifting to another large enterprise with extensive investments in in-house R&D, leadership went to an agglomeration of firms clustered around a university that had consciously reinvented itself as a hub of innovation. AT&T had sited Bell Labs in a bucolic campus environment, at a distance from other firms and even from most of the company's other facilities, not to mention the vibrant urban environment of New York. By contrast, Stanford University made a deliberate effort to lure industry to the doorstep of its previously isolated campus and to extend its reach throughout the surrounding region, bringing industry personnel to campus and sending faculty and graduate students out to nearby enterprises. Once an academic backwater, Stanford saw its stature climb as the fertile regional knowledge environment it created attracted top scientific and technical talent from around the country. Bell Labs, conversely, suffered one budget cut after another until AT&T finally spun it off as part of the ill-fated Lucent Technologies venture.[1]

The innovator credited with leading Stanford to its new regional role was Frederick Terman, a professor of electrical engineering who rose to be dean of the school of engineering and later provost of the entire university. As Stephen Adams explains in "Stanford University and Frederick Terman's Blueprint for Innovation in the Knowledge Economy," from his earliest days as a professor Terman had urged his students to exploit their discoveries commercially by starting local firms. He himself had provided some of the initial capital that William Hewlett and David Packard needed to commercialize their audio-oscillator in the late 1930s. As Terman moved upward in the university, he

164   expanded these encouragements into a four-pronged program to build both
the region's economy and the university's role in it: Stanford Industrial Park
attracted investments from leading firms such as General Electric, Eastman-
Kodak, and Xerox, as well as from homegrown enterprises such as Varian Associates and Hewlett-Packard; the Stanford Research Institute provided industry
with consulting services; the Honors Cooperative Program trained employees
of local enterprises; and the Affiliates Program created job opportunities for
Stanford graduates in Valley firms. The "community of technical scholars" that
these programs created, Terman believed, would enable researchers in industry and academia to cross-fertilize each other by facilitating the flow of knowledge and ideas in the region.

Stanford's phenomenal rise and Silicon Valley's economic dynamism
attracted attention from around the globe. Regional boosters seeking to revi-
talize their economies and large firms struggling to remain innovative eagerly
sought Terman's advice about how to replicate the Silicon Valley model in
other locations. Few of these efforts panned out, however, in large part because
success required much more than a university. It required re-creating the whole
package of relationships—the cultural ambience—that made Silicon Valley
firms so able to benefit from Stanford's programs.[2] Adams's essay explores one
of these failed projects, Terman's proposal to build a new institution, Summit
University, at the instigation of Bell Labs and other large R&D-intensive firms
in New Jersey.

What is most intriguing about the Summit University proposal is the way it
caricatured the real Stanford. In Palo Alto, Terman's efforts to redirect resources
within the university away from the humanities and from undergraduate programs more generally, as well as to run academic departments like businesses
or bypass them entirely by forming interdisciplinary centers targeted at particular technologies, had often met with determined faculty opposition. So his
plan for Summit eliminated undergraduate teaching and the humanities altogether and instead focused resources on a small number of "steeples of excellence" built around areas of research he thought would be important to
businesses in the region. Whether this narrowing of focus would have enriched
the flow of information between industry and academia, or whether it would
have upset the balance that enabled Stanford to play so vital a role in its region,
is impossible to know because the university was never built. When push came
to shove, the New Jersey firms that had commissioned the plans were not willing to provide the necessary level of financial support—perhaps because they
were still so focused on building up their own internal R&D operations.

Although Silicon Valley was difficult to replicate, it was not the only example of a regional high-tech agglomeration. Another cluster of innovative activity

and enterprise that emerged during the same period was the Gulf Coast petro-
leum industry. Efforts to tap oil supplies beneath the Gulf of Mexico sprang to
life following World War II, as small independent firms that had been denied
access to onshore fields looked to share in the postwar bonanza. Over the
course of the next twenty-five years, these pioneers pressed steadily further offshore into deeper waters. By 1969, over a thousand platforms stood in the Gulf,
at depths up to three hundred feet. Collectively, they pumped forth about a
million barrels of oil annually.

As Joseph Pratt explains in "The Bold and the Foolhardy: Hurricanes and
the Early Offshore Oil Industry," this impressive engineering accomplishment
involved a highly collaborative effort. In pushing offshore, the pioneers confronted unprecedented conditions—most important, the threat of destructive
hurricanes, with their powerful winds and potentially damaging waves. Lacking reliable data on such phenomena, firms had much to gain by learning collectively. They regularly exchanged lessons derived from practical experience
and also drew upon an expanding network of technical expertise, including
specialized engineering design firms, private weather forecasters, academic
consultants, and clusters of oil-related manufacturers and service companies.
This "offshore fraternity," as Pratt aptly characterizes it, collectively transformed
the uncertainties surrounding open-water drilling into manageable risks. A recognizable "Gulf of Mexico System" took shape and even exported its know-
how to other offshore areas.

Like many of the other organizations discussed in this volume, the offshore
fraternity eventually fell victim to its own success. The insular regional cluster
possessed a certain inbred capacity for self-delusion. Lulled by years of unusual
calm, participants tended to underestimate the destructive potential of the most
powerful hurricanes until severe storms revealed serious cracks in the system.
In the wake of these disasters the fraternity redoubled its collaborative efforts
at learning, ultimately producing new generations of oil rigs and platforms.
This response paralleled similar episodes in the electronics industry of Silicon
Valley, which weathered disruptions such as the collapse of the market for computer memory. Whether, as AnnaLee Saxenian has argued, such regional clusters provided an especially resilient form of business organization, superior to
that of the tightly controlled corporation, remains an open question.[3] But
research on Silicon Valley, the Gulf Coast oil industry, and other similar technology clusters certainly casts into doubt the idea that large, national bureaucratic enterprises have been a more important source of innovation over the
long run than local, informal networks of firms.[4]

Over the years executives in large firms have come to recognize the vibrancy
of this alternative way of organizing innovation and have tried to learn from it,

166   turning to consulting firms and faculty in business schools for help in understanding how to import some of its lessons. The example that Christopher
McKenna discusses in "Mementos: Looking Backwards at the Honda Motorcycle
Case, 2003–1973" is a particularly good one because Japanese companies emerged
from an environment that seemed to combine elements of both models—large
firms with in-house R&D capabilities imbedded in broader networks of innovative enterprises.[5] Honda's ability to penetrate the U.S. market—previously the preserve of large American and British producers—made the episode particularly
galling and enthralling. The problem, as McKenna shows, is that observers drew
very different lessons from the case.

McKenna takes a decidedly unconventional approach in his essay. Using
Christopher Nolan's movie *Memento* both as an organizing device and a source
of object lessons for scholars, he moves backwards through time, examining
the stories that consultants and academics have constructed to explain Honda's
success. McKenna shows how each version built upon earlier accounts
(mementos) at the same time as it critiqued them until "in the end, the fragments bec[a]me the story itself, misdirecting subsequent attempts to analyze
the original event." The various accounts—and ultimately the original episode
itself—can only be understood historically, McKenna argues. One must peel
back the layers of interpretation, always keeping in mind the prevailing political and epistemological conflicts that informed each in turn. By demonstrat-
ing that "as mementos accumulate, so too does the certainty that only one
explanation is right," McKenna offers an important cautionary tale for anyone
seeking to gain perspective on current business practice from past or parallel
experiences. His creative study adds a cognitive layer to the problems business
executives must overcome as they struggle to keep their companies innovative.

NOTES

    1. On Bell Labs, see Chapter 4 by Kenneth Lipartito in this volume; and Peter
Temin with Louis Galambos, *The Fall of the Bell System: A Study in Prices and Politics*
(New York: Cambridge University Press, 1987). On Silicon Valley, see AnnaLee Sax-
enian, *Regional Advantage: Culture and Competition in Silicon Valley and Route 128*
(Cambridge: Harvard University Press, 1994); the essays in Martin Kenney, ed., *Under-
standing Silicon Valley: The Anatomy of an Entrepreneurial Region* (Stanford, Calif.:
Stanford University Press, 2000); and Christophe Lécuyer, *Making Silicon Valley: Inno-
vation and the Growth of High-Tech, 1930–1970* (Cambridge, Mass.: MIT Press, 2006).
    2. Stuart W. Leslie and Robert H. Kargon, "Selling Silicon Valley: Frederick Terman's
Model for Regional Advantage," *Business History Review* 70 (Winter 1996), 435–72.
    3. Saxenian, *Regional Advantage*.
    4. See, for examples, Michael J. Piore and Charles F. Sabel, *The Second Industrial
Divide: Possibilities for Prosperity* (New York: Basic Books, 1984); and the essays in Charles

F. Sabel and Jonathan Zeitlin, eds., *World of Possibilities: Flexibility and Mass Produc-*
tion in Western Industrialization (Cambridge, U.K.: Cambridge University Press, 1997).

    5. Toshihiro Nishiguchi, *Strategic Industrial Sourcing: The Japanese Advantage* (New York: Oxford University Press, 1994); William Mass and Andrew Robertson, "From Textiles to Automobiles: Mechanical and Organizational Innovation in the Toyoda Enterprises, 1895–1933," *Business and Economic History* 25 (Winter 1996): 1–37; Takahiro Fujimoto, *The Evolution of a Manufacturing System at Toyota* (New York: Oxford University Press, 1999); and Susan Helper, "Strategy and Irreversibility in Supplier Relations: The Case of the U.S. Auto Industry," *Business History Review* 65 (Winter 1991): 781–824.

5

STANFORD UNIVERSITY AND FREDERICK TERMAN'S BLUEPRINT FOR INNOVATION IN THE KNOWLEDGE ECONOMY

Stephen B. Adams

What role should a research university play in building and sustaining the knowledge economy? As a magnet for and developer of high-tech talent? As a node for technology transfer? As an incubator for startup companies? Some universities have done all of these things and have, by doing so, become anchors of high-tech regions—regions that spawn clusters of high-tech companies that interact with one another and with the university anchor to sustain high levels of innovation over time. The most notable example has been Stanford University, which became the anchor of Silicon Valley. The indispensable individual at Stanford during this process was Frederick Terman, who served as dean of Stanford's School of Engineering and then as provost during the two decades after World War II. During his career at Stanford, Terman helped build the School of Engineering into a world-class institution that either competed with or surpassed institutions such as the Massachusetts Institute of Technology (MIT); the University of California, Berkeley; and the California Institute of Technology and transformed the university into a catalyst for innovation and regional economic development. His planning and leadership at Stanford during the development of the high-tech region that grew nearby justly earned him the moniker "Father of Silicon Valley."[1] By examining the organizational transformation Terman accomplished at Stanford, we can learn how research universities go about nurturing and sustaining innovation in the high-tech industrial clusters that grow up around them.

Terman's years as a Stanford administrator (1945–1965) coincided with two decades of unprecedented activity of American universities in national life, including the rise of what Clark Kerr, the president of the University of California, dubbed the "multiversity." Throughout the country, fundamental questions about the future of the university were debated. Should research universities' activities extend beyond campus boundaries, as did those of the land-grant universities? What was the research university's proper role in spurring innovation, and how should that innovation be transferred to or harnessed by the outside world? Should the university more closely follow the Johns Hopkins model, with its emphasis on graduate education and research? Should those able to perform fundable research (that is, scientists) prosper, while those who were not (that is, humanities scholars) die on the vine? Where should power reside: with the faculty or with the administration? As an administrator at Stanford, Terman would confront all of these questions, and would do so using the principles not just of academia but also of business administration.

Terman's model had two organizing principles, which he termed *communities of technical scholars* and *steeples of excellence*. As with many enterprises, Terman's institutional approach began outside the walls of the organization with the recognition of a need or demand. Such an opportunity could come from industry or from government in the form of a research problem to address (and the funding to address it). The interplay between the university and the public and private sectors, and the transfer of knowledge (and movement of individuals) among these spheres, represented communities of technical scholars. The public and private sectors would not be interested in just any old university, however; they would want to fund the cutting edge. That is where Terman's steeples of excellence came in. This was the academic equivalent of a niche strategy — preferring superiority in a few fields to mere "coverage" of many. Terman's model was quite corporate, mirroring industry's approach to innovation, including top-down power relationships, devotion to efficiency, thorough and quantifiable performance measures, and allowing the marketplace to shape the services offered.

Terman's model of the university included some features common to the firms Alfred Chandler wrote about. Terman presumed an oligopolistic world of research universities in which a select few reap the lion's share of rewards. His model also posited the importance of managerial hierarchies in academia, as opposed to grassroots power of the faculty at the department level. He was not entirely sanguine, however, about the academic equivalent of the Chandlerian firm. Terman saw lack of strategic direction, inefficiencies, and bad habits ("barnacles") in existing universities. Terman's model also parted ways from Chandler's in that the external environment mattered so much. Much of Terman's innovation had to do with organizational boundaries: he did not hesitate to redraw external boundaries to foster closer relations with industry, and he matched the university's strategy and structure to the needs of the particular surrounding environment with which it interacted. Therefore, Terman recast an established California university (Stanford) in order to jump-start a high-tech region on a model somewhat different from that of the startup New Jersey university (Summit) he later proposed, which was aimed at sustaining innovation in a well-established high-tech region.

Terman's success was as an agent of change; he was forever changing the strategy, the structure, and the operation of universities. He therefore expressed a preference for the fresh start, whether in overhauling an established institution or in providing a blueprint for a new one. His skills in leveraging limited resources into impressive outcomes would be recognizable to entrepreneurs as a successful bootstrapping strategy. Little wonder, then, that Terman became associated with the rise of the "entrepreneurial" university and its approach to innovation.[2]

In this chapter, I will explore Terman's philosophy for building a university to power innovation and how he carried it out at Stanford. I will also describe Terman's approach to starting a university from scratch after his retirement from Stanford, and conclude with lessons regarding the nature of the research university in the knowledge economy.

OTHER MODELS

Before becoming a shaper of Stanford University, Frederick Terman was a product of it: class of 1920. When he began his studies there, he was already familiar with the university and the surrounding area. The son of Lewis Terman, a prominent Stanford psychology professor, Terman grew up in Palo Alto and was one of the radio buffs in the orbit of the Federal Telegraph Company (FTC), the region's first high-tech enterprise. FTC's relationship with Stanford University foreshadowed the sort of relationships Terman would later foster between Stanford and local entrepreneurs. Cyril Elwell, a recent Stanford graduate, had founded the firm in 1909 at the urging of the electrical engineering department chairman, Harris J. Ryan, and with the personal financial support of Stanford's president, David Starr Jordan.

Early in his career at Stanford, Terman would be influenced by two other academic models: MIT and Clark University (also in Massachusetts). Clark was where Lewis Terman had earned his Ph.D., and MIT was where Frederick Terman earned his. In the 1890s, Clark was the only nonsectarian all-graduate university in the United States. Clark followed the approach of Johns Hopkins University, emphasizing research rather than teaching. The Clark model involved striving for excellence in a few disciplines rather than "coverage" of many. As Lewis Terman noted, "There was no effort to make the courses of different professors dovetail."[3] Clark University's approach to internal operations would foreshadow Frederick Terman's concept of steeples of excellence.

After graduating from Stanford, Frederick Terman earned a Ph.D. in electrical engineering from MIT, which provided him with a model of university-industry cooperation and permeable boundaries between the two. Terman's mentor at MIT, Vannevar Bush, emphasized the practical and encouraged interaction between his students and local industry. Bush also helped start the company that became Raytheon, a leading maker of radio tubes. In 1922, Terman witnessed the graduation of the first class enrolled in MIT's cooperative program, a joint venture with General Electric established by the chairman of MIT's electrical engineering department, Dugald Jackson, in 1917. "Co-op" students earned simultaneous bachelor's and master's degrees at the end of five years, the final three of which involved periods when the students alternately worked at a local GE plant and attended classes at MIT.[4] MIT's permeable boundaries with industry would foreshadow Terman's idea of communities of scholars.

Imbued with principles of university-industry cooperation, in many ways Terman found an "already prepared environment" at Stanford University. Stanford was no ivory tower; it was founded in 1891 as an institution that would emphasize the practical and provide a service to America's western states. By the mid-1920s, the newly established engineering school was not a solitary real-world vessel in a sea of the abstract; Stanford also had a medical school, a law school, and a business school.[5]

When he began teaching at Stanford in 1925, Terman began to mimic the MIT model. He directed his work and that of his students toward commercial applications. Terman's text *Radio Engineering* was a best-seller in the field because of its emphasis on the practical. He not only brought real-world problems into the classroom but also arranged for his students to visit local electronics firms.[6] His relationships with local firms also included consulting, which was so widespread at MIT that when its president, Karl Compton, later suggested that the faculty could devote 20 percent of their time to it, that meant a reduction in such activity.[7]

One thing Terman could not mimic in the 1930s, however, was MIT's cooperative program. FTC, the largest high-tech employer in the area, had struggled in the 1920s before being acquired by International Telephone and Telegraph (ITT), which had moved FTC's operations to New Jersey. The departure of FTC meant that there were few local jobs for graduates of Terman's electrical engineering program at Stanford. Although FTC's employees had spun off their own local firms, from Magnavox to Fisher Research Laboratories to Litton Engineering Laboratories, Terman later estimated that in 1937 local electronics firms employed no more than one hundred engineers.[8] There was insufficient critical mass in the local electronics industry for his students to be able to count on getting jobs in the area, much less rising in the firm and recruiting more Stanford grads—as was the model at MIT and GE. Unable to land jobs on the peninsula south of San Francisco in the 1930s and 1940s, several of his students ended up at bigger firms in the East, such as AT&T, RCA, GE, and Westinghouse.

In the late 1930s, with so many of his students heading to eastern firms, Terman began to complain about a "brain drain" from the Bay Area. Why should Stanford, which sought to promote the economic development of the West, provide training for engineers who would take their skills elsewhere? Terman's top priority became attracting—and keeping—top technical minds at Stanford and in the surrounding area. Such was the situation when in 1939 Terman helped his former students William Hewlett and David Packard start a company, Hewlett-Packard, in the garage of Packard's home in Palo Alto. On the eve of World War II, the brain drain of his students and FTC's departure both reinforced the same lessons for Terman: the importance of attracting high-tech industry to the area—or developing it, if need be—and the role that Stanford must play in making it happen.

Terman also sought ways to build his institution. When he became chairman of Stanford's department of electrical engineering in 1937 and investigated best practices in technological education, Terman's systematic comparison of Stanford's most prominent rivals in electrical engineering singled out the California Institute of Technology (Cal Tech) for its emphasis on sponsored research and graduate education. During World War II, Vannevar Bush appointed Terman as the director of the Radio Research Laboratory (RRL) at Harvard, where he observed how MIT and Harvard operated in the world of government contracts. This was important because during the war MIT was the largest university defense contractor ($117 million), and Harvard was third ($31 million), while Stanford was what Stuart Leslie called a "benchwarmer."[9]

Terman returned to Stanford from RRL not only with a recognition of how important government funding could be and a plan for how to bring it to Stanford. He had conceptualized how Stanford should be transformed into an institution that could compete for that funding and work well with industry—in other words, an institution that could act as an effective anchor for a high-tech region. As dean of the engineering school—a position he assumed in 1945—he would also be in a sufficiently high-ranking administrative position to bring many of his ideas to fruition.

TERMAN'S INNOVATIVE MODEL

Terman's concept of a "community of technical scholars" hearkened back to the medieval communities that formed the basis of prominent European universities. This concept of community was not merely a glance backward, however; it

174   also grew out of prescient observations by Terman. At the heart of the modern community of scholars was a research university, and surrounding the university were research-based companies. "Industry is finding that for those activities that involve a high level of scientific and technological creativity," he wrote, "a location in a center of brains is more important than a location near markets, raw materials, transportation, or factory labor."[10] Terman's conception of a community of technical scholars was nothing less than a blueprint for what would later be called the "knowledge economy."

The university's central role in Terman's vision of a community of scholars had ramifications for its mission and contrasted sharply with the concept of the university as devoted to the pursuit of knowledge free from the socioeconomic influences of the day. Terman embraced a departure from such "little worlds largely isolated from the hurley-burley of everyday life," and the "splendid isolation" in which universities had operated prior to World War II. "Their ivory towers were sullied by the world around only if the institution happened to be located in the center of a large city, which most institutions were not."[11] Instead, Terman argued that universities represented "a natural resource just as are raw materials, transportation, climate, etc." As such, universities had become "major economic influences in the nation's industrial life, affecting the location of industry, population growth, and the character of communities."[12]

Clark Kerr, the president of the University of California system, would introduce the concept of the multiversity, the "community of the undergraduate and the community of the graduate; the community of the humanist, the community of the social scientist, and the community of the scientist; the communities of the professional schools; the community of all the nonacademic personnel; the community of the administrators."[13] Terman observed the same phenomenon, noting that rather than ivory towers, universities were "rapidly developing into more than mere places for learning."[14]

Many of the ideas Terman sought to implement were not new; he simply tried to apply ideas from the business world to an academic setting. He was not, of course, the first to do so. So many universities were either endowed by businessmen or had businessmen on their boards of trustees that there was long-standing tension between the academic and industry approaches. One observer had complained in the early twentieth century that "the men who control Harvard to-day are very little else than business men, running a large department store which dispenses education to the million[s]."[15]

In many ways, Terman's approach would reflect that of the companies with which Stanford sought relationships, and in order to better connect with them (and attract their resources) he would help redraw the boundaries between academia and industry. This was part of what Kerr would refer to as a "merging physically and psychologically" of academia and industry, and what Christophe Lécuyer calls the "permeable" university.[16]

In Terman's view, having high-tech companies in close proximity to the university was not just good for the wider community, it presented an opportunity to those inside the institution as well. Such proximity increased the chances of students landing local jobs. "The idea is to get lots of good Stanford people well placed in industry," he wrote in 1943. "And then as time goes on and they begin to work up to responsibility, see that they hire good Stanford men to work for them, and so on ad infinitum."[17]

Another major issue at research universities was the nature of the problems to be solved. By 1945, Terman favored faculty exploration of "profitable" areas of research, fields "which are going to have a big rather than a small future." For instance, "oil is an important western industry, and Stanford should be strong on all things relating to oil, including geology, heat transfer, chemical engineering as applied to oil, etc."[18] Because garnering financial support for such research was a primary goal, in Terman's model industry demand played a key role in determining which fields would be "big." So Terman was willing to have his people take their cues from the "marketplace" of industry and government.

Terman's concept of a community of scholars also fit Stanford University's financial predicament. During the 1930s and 1940s, the university was in nearly constant budget crisis. Therefore, anyone who could either attract resources or stretch the existing resources through an entrepreneur-like bootstrapping strategy would benefit the institution. Terman's various efforts would allow the engineering school to hire more faculty than otherwise would have been possible and, consequently, dramatically expand the school's research activities and improve its reputation. Having that well-placed Stanford grad would not just benefit recent grads; it also represented an opportunity for the institution to raise money.

In the ten years following World War II, Stanford would establish four innovative programs of outreach to industry. Stanford Research Institute (SRI) provided faculty know-how, in the form of consulting, to industry (and government). Stanford Industrial Park (later renamed Stanford Research Park) provided industry with proximity to Stanford's faculty, students, and ideas. The Honors Cooperative Program offered opportunities for employees of local organizations to obtain graduate degrees. The Affiliates Program provided member firms with a first shot at Stanford technology and graduates. The latter two were initiatives of Terman and his faculty. The Industrial Park was conceived as the business manager's solution to Stanford's predicament of being land rich and cash poor, and Terman recognized it as a vehicle to attract high-tech firms to the region and to Stanford's orbit.[19]

By 1945 Terman recognized that a community of technical scholars would enhance the institution that acted as a regional high-tech anchor, providing a magnet for contracts and grants. Yet Terman identified not just the rewards but also the challenges of doing this successfully: "In developing the possibilities of an existing embryo center of technical scholars, or in attempting to bring into being a new center," Terman noted, "one is involved in a highly competitive situation. Many others are trying to do the same thing."[20]

A university's creation of an internal-external community of technical scholars was nothing but a pipe dream, then, without an internal mechanism to ensure competitiveness. Stanford needed "a small faculty group of experts in a narrow area of knowledge." Terman imagined his solution in terms that equated the university with a church devoted to learning: "Academic excellence depends upon high but narrow steeples of academic excellence," Terman noted, "rather than upon coverage of more modest height extending solidly over a broad discipline."[21] In 1943, Terman had discussed the example of a particular star engineering professor at Stanford, who received "only about twice the salary of an associate professor, but his value to Stanford cannot be compared with any number of ordinary associate professors." Attempting to compare the star to the associates was, Terman suggested, "like asking how many men who can jump three feet must a track team have to equal a man who can jump six feet."[22]

Essentially, this was a niche strategy, an approach that has been widely adopted by business: it is better to dominate a few markets than to operate in a larger number of markets at a mediocre level.[23] Terman's strategy for Stanford paralleled what he observed in local industry in the 1930s and 1940s. The activities of electronics firms in the San Francisco Bay Area were circumscribed by RCA's patents and the company's efforts to defend them. The presence of RCA and the threat of litigation caused Bay Area firms to employ a technological niche strategy that kept them on the cutting edge and avoided direct competition with the giant.[24]

The beauty of academic steeples, in Terman's view, was in the eye of the beholder: "What counts is that the steeples be high for all to see and that they relate to something important."[25] To Terman, this meant emphasizing graduate education. He favored not only saddling less-grantworthy professors with a greater share of undergraduate teaching but also reducing the university's overall emphasis on undergraduate teaching. During his first four years as provost, Stanford's undergraduate enrollment would hold steady while the number of graduate students increased by 50 percent.[26] Even amidst America's crisis of technological confidence following the Soviet Union's launch of Sputnik in 1957, Terman remained calm, partly because he did not think undergraduate education was a vital competitive issue. So although he granted that "the Russians outproduce us by more than 2 to 1 in engineers at the bachelor degree level," he referred to this as a "so-called" shortage of engineers. The reason was that America equaled or exceeded Soviet output of Ph.D.s, and "it is those with graduate training on whom we are coming to depend disproportionately for leadership and for creative advances."[27] Only if the Soviets began to outproduce America in technical Ph.D.s would Terman worry.

One of the ultimate measures of professors' importance was their effectiveness in obtaining funding. Therefore it was crucial that certain outsiders viewed them as worthy: those in government or industry with the money to support research could influence the extent to which a particular steeple should be sustained. Terman would excel at the practices of "salary splitting" (convincing a firm to share the burden of paying a faculty member) and hiring faculty members who brought along funding. A community of scholars was Terman's conception of how the university should relate to the wider world; steeples of excellence captured his conception of how the university should operate internally so that the external world would want to fund that interaction.

ADMINISTERING THE MODEL

If a corporate administrator were asked to assess Stanford's situation, she or he would consider issues of strategy (what is Stanford's competitive advantage?), structure (what should the organization look like, and should decision making proceed top-down or bottom-up?), and staffing (what organizational capabilities should be emphasized, and what criteria should be used to make new hires?). Terman's ideas about the administration of the academic center of a community of technical scholars were little different from a for-profit administrator's approach.

The idea from which the rest of Terman's model flowed was his conception of where power resided and the direction in which it flowed. He believed that in a university, power should flow down from the top rather than up from the bottom. Terman was convinced that most of the power in a major university belonged in the hands of the administration, not the faculty. In this respect, Terman echoed arguments of the Stanford trustee Herbert Hoover from pre–World War I days.[28] Like Hoover, Terman was imbued with the Progressive-era sensibility that valued efficiency, which resulted in attempts to bring the principles of scientific management to the university in the early twentieth century and again in the 1930s, amidst efforts to solve budget problems.[29] That Terman would later do so in the School of Engineering was fitting because American management became "scientific" primarily in engineering-based industries.

178   The most basic principle of scientific management was the separation of planning from execution. In 1943, Terman argued that "the thing that counts in education is well thought out and well executed plans." Terman thought that these plans should be made by a university's administrators, because they were in a better position to see the big picture than were faculty members. He also believed that at most universities, administrators failed to use their positions to the best effect. He noted disdainfully that "universities are fundamentally inefficient" because they fail to take a long-term view.[30]

One example was in the area of faculty hiring. He found that the faculty was too willing to make a hire to meet an immediate need regardless of whether that hire was aligned with long-term strategy. In 1944, Terman objected to the approach a physics department search committee was using—which was to "be content to stop without competition with someone that has been recommended and who looks satisfactory." Instead, he suggested at the outset of the search "it should be decided in what directions it is desired for the Physics Department to go, and particularly on what it should concentrate." Then, and only then, should the committee "canvass the entire country to get the best man available for twenty years ahead."[31] Terman preferred using a twenty-year horizon "because in this length of time one has virtually complete turn-over of academic staff."[32]

He also observed that faculty members often made decisions without regard to the financial implications, creating the academic equivalent of unfunded mandates. Therefore, Terman applied the scientific management notion of hierarchy to universities, suggesting that the administration, not rank-and-file faculty, should do the planning and make the key decisions. This meant, among other things, limits on faculty autonomy and on the nexus of faculty power, the department.

The key issues in Terman's corporate model were productivity and the ability to measure it. Just as his father had attempted to quantify intelligence through the Stanford-Binet IQ test, Terman quantified everything he could.[33] Terman began counting academic accomplishments during his first year on the Stanford faculty. In April 1926, he presented to Theodore Hoover, the dean of the engineering school, information on "the amount of productive research that the American colleges are carrying on in the field of electrical engineering."[34] He would later count research contracts per faculty member and Ph.D.s produced per faculty member. His own productivity had been impressive: during his first six years of teaching, he had supervised 33 advanced degrees (out of 172 in the entire engineering school).[35]

He also came up with his own way to count faculty teaching loads: instead of counting the number of classes a professor taught, Terman counted the number of students the professor taught. Large lecture classes represented a more efficient use of faculty resources; Terman discouraged small classes. Here he echoed his father, who had made a similar suggestion: running the university like a factory.[36] For faculty members who did not attract funding, who had few industrial or governmental patrons, and who produced few Ph.D.s, there was one way to be productive. Terman favored having departments such as classics operate as "service departments" by teaching large numbers of undergraduates. This represented an academic equivalent of "job breakdown," in which one group concentrated on teaching and another on research.[37]

Terman believed departments should be run like profit centers. Too much power at the departmental level often meant too much support for less-productive faculty, he thought. To Terman, productivity meant research—fundable research. The faculty members Terman sought were less interested in teaching than in attracting patronage for research, just as Terman was. He warned new faculty not to spend too much time on teaching, once they had reached a certain level of competence. To demonstrate this, he would draw a learning curve for teaching, and arbitrarily cut it off at a certain point, indicating that any further time spent would be counterproductive.[38]

Terman believed that departments (and individual faculty members) should be self-sustaining, having each tub stand on its own bottom. They should be able to attract their own sources of funding rather than depend on the university to provide money. In effect, Terman was promoting a new form of faculty member: an academic entrepreneur. In the academic world Terman envisioned, the faculty member's career and the university's financial health would depend on the entrepreneurial search for research funds. For Terman, this was not just a theory: he used the individual's capacity to be "self-sustaining financially" as a factor in recommending appointments of engineering faculty. In 1945, in suggesting the hiring of a faculty member, Terman noted that "if the national research program develops as anticipated, [the prospective hire] can probably after the first six months be self-sustaining financially."[39]

Terman also attempted to redraw boundaries within the university. Weakening faculty departments, Terman concluded, would promote interdisciplinary research. Terman believed that academic experts should gather on the basis of the problems to be solved rather than on the disciplinary basis of their training.[40] In the late 1930s, he had seen the possibilities for increasing the connection between electrical engineering and physics. In American universities, such interdisciplinary work was rare; departments acted as independent silos. Bucking that trend, Terman sent two of his electrical engineering graduate students to work with William Hansen of the physics department, and then he worked with Hansen and other members of the physics department to institutionalize those connections in a microwave laboratory.[41] One result of this collaborative effort was the founding of Varian Associates.[42]

Several departments at Stanford had critics of the organizing principles Terman adopted for running the university and for dealing with the wider world. One such department was physics. During the 1940s, Stanford's president, Donald Tresidder, who periodically sought Terman's counsel—particularly on how the university could better serve industry—fired David L. Webster from the position of department chair, and asked for (and received) the resignation of Webster's replacement, Paul Kirkpatrick, from the same post. Both Webster and Kirkpatrick lost power struggles over questions of industrial sponsorship of faculty research, the relative value of teaching and research, and the extent of departmental autonomy.[43]

During the 1940s, in both physics and geology, faculty objected to appointments given to individuals from industry without teaching experience. Indeed, Eliot Blackwelder of geology saw such moves as part of a troubling larger pattern, bringing scientific management–style division of labor to the university: "Is U. [university] to be run like a business—Bd. [Board of Trustees] decides, employees execute."[44]

EXPANDING THE MODEL BEYOND ENGINEERING

Although its professional schools were a testament to the practical bent of Stanford's early leadership, and the university was far from the Ivy League ivory tower Terman criticized, Stanford had developed a large contingent that embraced the view that the quality of a university's humanities program and the share of resources directed to it would determine the university's reputation. This group included faculty who objected to the approach Terman used as dean of the engineering school.

When he became provost in 1955, many faculty members in other schools (especially in the School of Humanities and Sciences) feared that he would impose the same model universitywide. They had good reason. Tresidder's successor, President J. E. Wallace Sterling, had been so impressed with what Terman had accomplished in the engineering school that, upon naming him provost, Sterling asked Terman to do the same for the other schools.[45] This may seem rather odd because Sterling had a history Ph.D. (Stanford, 1938), but in the intervening years before assuming the presidency of Stanford, Sterling had taught at, of all places, Cal Tech.[46]

Almost immediately after Terman's appointment, Charles Park, the dean of the School of Mineral Science, objected to Terman's measures toward increasing efficiency and production. Among these measures were attempts to eliminate small classes, insistence that certain departments award more doctorates per year, and efforts to increase levels of government patronage and to steer the school into certain fields of research. Dean Park resigned in early 1956.[47]

A 1959 report commissioned by Philip Rhinelander, the dean of the School of Humanities and Social Sciences, warned that "Stanford may be losing its liberal arts character in exchange for a more technical emphasis." The report bemoaned the fact that during the 1950s, only 4 percent of a sample of Stanford graduates went into "intellectual" careers (university teaching, writing, the arts), as opposed to 70 percent who "entered business, engineering, or one of the other technical trades."[48] To Terman, this was not necessarily a bad thing; those going into writing and the arts were not likely to be part of the community of technical scholars he sought. The previous year, Terman had noted with approval that more than 40 percent of entering freshmen in the previous three years had been engineers—well above the national average of 17 percent, and "almost as high as Cal Tech's average."[49] Rhinelander's report suggested cutting "formal" teaching loads in half and establishing funds for humanities professors unable to garner research grants. When Terman highlighted a section of his copy of the report that read, "The faculty hopes that the central place of the humanities within the University will once again be affirmed," it is not likely that the reason was because he agreed with it. After a couple of years of conflict with Terman, Dean Rhinelander resigned in 1961.

Such battles reflected the difficult situation in which Terman found himself as Stanford's provost. On the one hand, as Stanford's humanities and social sciences grew in prestige, Stanford was more and more spoken of in the same breath with eastern powerhouses such as Harvard, Yale, and Princeton. On the other hand, as an engineering educator and onetime dean of Stanford's engineering school, Terman worried that the means to the end of greater university prestige might harm the engineering and physical sciences programs. In 1943, Terman had suggested that at "old established endowed schools . . . arts and pure sciences seem to prosper much more than the applied sciences. Harvard, Yale, Columbia, and Princeton are good examples." Given the power and resources vested in those areas, "this has prevented the applied science groups from having a free hand to develop in their own way."[50] One thing Terman did in 1959 to ensure that this would not happen at Stanford was to appoint a vice provost who would handle the social sciences and humanities, while the physical sciences continued to report directly to Terman.[51]

So how did things change at Stanford during the Terman years? Terman's two organizing principles resulted in resounding success. As a builder of "steeples of excellence" in the School of Engineering, Terman succeeded rapidly and dramatically—both by his measures and by those of others. In electrical engineering, Stanford had produced one Ph.D. before Terman arrived in 1925. By the 1960s, Stanford was producing on average more than forty per

182   year, about 30 percent more than MIT, the number two producer, and Stanford had top-ranked programs in electrical engineering, aeronautics, materials
science, and physics.[52] Meanwhile, the reputation of the rest of the university
had risen as well. Rankings of the quality of the faculty in 1964 showed that
Stanford's humanities, social sciences, physical sciences, biological sciences,
and engineering were all among the top ten in the nation. Overall, Stanford
ranked third, behind only Harvard and UC Berkeley.[53] So Terman may have
favored fields that were self-sustaining and in which faculty established relationships with industry, yet his tenure as provost lifted all boats.

As the site of a community of technical scholars, Santa Clara Valley had
gained a critical mass of scientific and engineering talent by the mid-1960s,
much of which moved freely across the permeable boundary between the university and industry. Large high-tech firms, from Lockheed to GE to IBM, had
located facilities in the valley in the 1950s. Startups had also sprouted in the
valley, including Hewlett-Packard, which would become part of the Dow Jones
averages, and Fairchild, which would spawn virtually the entire semiconductor industry. Stanford Research Park, featuring both startups and satellite operations of established firms, was rapidly growing and had become a model that
would inspire replicators around the world.

Terman had clear ideas about the nature of the university, and it would
appear that he had succeeded in implementing enough of them to turn Stanford University into an approximation of his ideal. But what would Terman's
ideal university look like if it responded to a dramatically different external
environment?

SUSTAINING A HIGH-TECH REGION

After his retirement from Stanford in 1965, Terman accepted a consulting assignment from a consortium of New Jersey businessmen to provide a blueprint for
a new university to spur regional high-tech development and technological
innovation. Terman was a logical person to ask: as dean of the engineering
school and then as provost of the university, he had helped make Stanford the
educational anchor of the noted center of high-tech activity that would later be
called Silicon Valley. The external environment in New Jersey was quite different from that in the Santa Clara Valley, however, and, reflecting the different environment, Terman's proposal for the new university differed from what
he had done at Stanford.

The Silicon Valley of Fred Terman began with a supply of university-trained
technical talent that exceeded local industry demand and a complaint from
academia about regional brain drain. The problem that Terman—along with

Summit University, the regional academic anchor he proposed—was to solve
in New Jersey was the reverse: industry demand for technical talent that
exceeded university production and a complaint from industry about regional
brain drought. Bell Labs and other New Jersey research institutions were concerned about their pipeline of university talent. In 1960, Terman used Bell Labs
as an example of "growth industries . . . located in an area that does not have
educational opportunities."[54] There were not enough highly qualified graduates of local universities to fill the high-tech jobs available in New Jersey. Of
the nearly five thousand Ph.D.s employed in New Jersey's research laboratories, two-thirds were out-of-state "imports."[55] New Jersey's universities could
not fill the pipeline, so local companies were forced to incur the expense of
recruitment and relocation of employees from outside the region.

The starting point Terman chose for Summit University was Cal Tech, not
Stanford. Terman had good reason to use Cal Tech as a model. During the
1920s, Terman's formative years as an engineering scholar, what had once been
tiny Throop Polytechnic Institute in Pasadena had become the California Institute of Technology, a leading light in the physical sciences.[56] Cal Tech's transformation was the closest to a fresh start among American technical universities
during the first half of the twentieth century, and Terman brought an entrepreneur's enthusiasm to the idea of the startup and its advantages over an established organization.

Terman found in Summit University the "ideal situation" for another fresh
start. One reason was that by design the university would have a limited scope,
and each of the pieces of the whole would be "interrelated with each other."
Therefore, Summit University could "concentrate *all* of its resources on a small
number of very important steeples" (he called them "steeples of strength" in
his report). Summit University's focused approach promised a competitive
advantage, which would "contrast with the established universities, which all
are trying to do too many things, and as a result typically do only a limited number of them exceptionally well."[57]

Almost every section of Terman's report reflects such a corporate attitude
toward university administration. Although it is common in the consulting profession to tailor reports to the perceived preferences of the client, in this case
there was a genuine confluence of views between consultant and client. For
most of his career, Terman had been trying to bring a corporate model to academia—and this crusade no doubt made him an especially congenial consultant when representatives of AT&T and the other New Jersey firms wanted
someone to write a blueprint for Summit University.

In the area of productivity, Terman believed Summit University could have
a competitive advantage because "present standards of faculty productivity are

184   unnecessarily low."[58] Similarly, at Summit University, research specialties would
       "provide interdisciplinary reinforcement whenever possible." Researchers in
       chemistry, for instance, "would also emphasize quantum effect in molecules
       in view of related competence in physics and in nearby industrial research
       laboratories."[59]

       Terman's proposal called for researchers to "emphasize those areas of basic
       science and fundamental technology that are paralleled by the research inter-
       ests in the industrial research laboratories located in northern New Jersey."[60]
       For instance, in chemistry, researchers would pay "particular attention to the
       polymer field in order to reflect the industrial research activity in the area."[61]
       Here is where Terman's steeples of strength would be reinforced by a commu-
       nity of technical scholars.

       Another area where Summit University departed from the Stanford model
       was in its treatment of the humanities. In this respect, the proposal for Summit
       University foreshadowed a corporate trend decades away: the outsourcing of
       functions not core to the institution. Although Terman acknowledged the need
       to "make limited offerings in the humanities and behavioral sciences," they
       would be minimal. Instead, Summit University would rely on other institutions
       to provide education in the humanities and social sciences. Rather than dis-
       tracting the faculty from its focus on research, such general education demands
       would "be met by an arrangement with neighboring institutions for use of their
       faculty, rather than by faculty members of the Summit University staff."[62]

       With respect to graduate education, Summit University would be tailor-
       made for Terman's preferences: it would be exclusively a graduate institution.
       Therefore, some of the teaching and resource allocation disputes he had faced
       at Stanford would go away. As Terman noted in the report, this new institution
       would "not have the distraction of undergraduate students."[63] Whereas the fac-
       ulty at Stanford who did not consider undergraduates a "distraction" battled
       the provost for resources, at Summit the administration would not have to fight
       for the proper emphasis on graduate education.

       More than twenty years before, Terman had noted that "even California Tech,
       which had been beautifully set up, has cracks in its armor (an example is electri-
       cal engineering) and these will become increasingly numerous and large as Cal-
       ifornia Tech becomes older and increasingly smug."[64] One crack in Cal Tech's
       armor was its undergraduates, and the associated institutional requirements. By
       contrast, Summit intended to provide "fewer distractions, and fewer peripheral
       activities."[65] Summit would be a startup, and that alone was a selling point.
       "Because of its age," noted Terman, Cal Tech had "developed some barnacles
       that add to cost without giving proportionate educational return."[66] Terman was
       convinced that he could establish a university that would "set an example" for Cal
       Tech and other "older institutions" in terms of faculty productivity.

Summit University provided Terman with a chance to create an ideal academic institution from scratch. Problems faced by established institutions could disappear. Too much emphasis (resources spent) on humanities? Summit University would deemphasize the humanities and count on other local universities to pick up the slack. The distraction of undergraduates? Get rid of them. Problems posed by too many tenured faculty? Establish a university that relied on adjuncts to a great degree. Too much power in the hands of individual departments, which acted like silos rather than seeking interdisciplinary projects and solutions? Create a university with a powerful administration and an interdisciplinary bent.

Terman had long fought against the ivory tower as an impractical model of education, one that was not held sufficiently accountable in the marketplace of industry and government patronage. Terman had battled at Stanford on behalf of individuals, departments, and schools that could bring in external funding. Ironically, his proposal for Summit University, Terman's "dream . . . of a new and unique educational institution,"[67] died on the vine because of an inability to sell the project to key industry patrons.

Terman's clients from Bell Labs seemed to think that Summit University could start small: they estimated $650,000 in annual operating expenses. One of the goals of Terman's report was to disabuse them of the idea of starting small and scaling up. He estimated annual operating costs of $4.8 million and startup costs of $15 million.[68] Even as a regulated monopoly with enormous resources, Bell Labs could not pull this off alone. The price tag did not kill the project; instead, it was the inability of Bell Labs to find enough partners willing to contribute. In New Jersey, there were two primary high-tech industries: telecommunications and chemicals-pharmaceuticals. The deal breaker for Summit University was the pharmaceutical industry, which was not interested in providing major financial support for Summit University. In the end, Bell Labs would have had to foot 90 percent of the bill, which would have left the new university open to criticism as a captive of AT&T.

CONCLUSION

The Terman model presented more than one paradox. It was both forward- and backward-looking. In recognizing and explaining the nature of the coming knowledge economy—and the possibilities the new economy would present to universities—Terman was well ahead of Alvin Toffler and other futurists in anticipating life in the twenty-first century. At the same time, the model Terman promoted had a retro quality: it bore a striking similarity to what Clark University had been when Terman's father studied and taught there in the late nineteenth and early twentieth centuries.

Fred Terman and Clark Kerr both anticipated the increasing significance of the research university during the second half of the twentieth century. Terman echoed what Kerr referred to as the "knowledge industry," forecasting that it would be "the focal point of national growth." Kerr agreed with Terman on the university's position—both symbolically and physically: "The university is at the center of the knowledge process."[69] They also described the oligopolistic result of competition among research universities. Although they agreed on the university's place in the greater scheme of things, however, their models for the university were quite different.

Terman's model was both complex and simple. His conception of the community of scholars and the relationships universities would have with the outside world represented part of the increasing complexity Clark Kerr alluded to when comparing the multiversity to what preceded it. However, Terman's steeples of excellence represented a stripped-down version of the multiversity, which might leave undergraduates, the humanities, and non-self-sustaining faculty members behind. This again echoed his father's approach to psychology, using IQ tests as a simple way to gauge intelligence and to choose whom society should leave behind.

What are the implications of the Terman model and Stanford's version of it? In many ways, Terman's ideas for an academic anchor in a high-tech region mirror the oligopolistic aspect of American industry. The coin of Terman's realm, as it had been for his father, was "brains." It was through attracting the best of them to the university, developing them there, and then establishing the necessary local industry to hire them that a high-tech region (or "community of technical scholars") would be formed. As Terman set out to establish such a community around Stanford, he quickly grasped the grim realities of the external environment: that government contracts and grants—the lifeblood of post–World War II high-tech regions—would land in relatively few sets of hands. Terman sought to create an institution that reflected this reality, focusing on the few who could become magnets for the patronage of government and industry. In this world, not all inputs were created equal: graduates eclipsed undergraduates, engineering and the natural sciences eclipsed the humanities and the social sciences, and self-sustaining departments and faculty eclipsed all others.

In many respects, the oligopolistic community of technical scholars came to pass. Certainly Silicon Valley's engine of innovation resulted in a tremendous creation of wealth. The members of the Silicon Valley community, those who received the spoils, were far from representative of the population as a whole in education level, in ethnicity, and in gender. Ironically, however, the academic anchor for this region did not follow Terman's path toward a Summit University–like extreme. As in many universities, the division of labor widened between the winners (who did the research) and the losers (often adjuncts, who did the teaching). Yet at Stanford, undergraduates—and their ancillary activities, which Terman dismissed—thrived rather than disappeared. The humanities thrived at Stanford as well. In short, Clark Kerr's multiversity in all its complexity appears to have found a home in the organization of the academic anchor of the world's foremost high-tech region. It is possible that the creative tension and vibrant complexity of Kerr's multiversity (including, to use the expression Terman applied to the business world, the "hurley-burley" of the marketplace of ideas) may turn out to be a means complementary to Terman's focused, efficient, corporate approach to creating steeples of excellence and communities of scholars, together reflecting and sustaining the innovative vibrancy of the surrounding region.

NOTES

The author thanks, for their assistance and comments, Margaret Kimball at the Stanford University Archives, the volume editors, two anonymous reviewers, and participants in an interactive session at the 2005 Academy of Management Conference; and, for research funding, the Salisbury University Foundation.

1. Accounts of the rise of Silicon Valley vary in tone, some giving a triumphal narrative of the evolution of Stanford's organization as the region's academic anchor, others a story of declension with Terman at center. AnnaLee Saxenian, perhaps the most prolific academic writer on Silicon Valley, gives Terman much credit for shaping both Stanford and the region. See, for instance, AnnaLee Saxenian, *Regional Advantage: Culture and Competition in Silicon Valley and Route 128* (Cambridge, Mass.: Harvard University Press, 1994). The most sympathetic account is C. Stewart Gillmor, *Fred Terman at Stanford: Building a Discipline, a University, and Silicon Valley* (Stanford, Calif.: Stanford University Press, 2004). Gillmor provides excellent detail regarding Terman's style and the programs he initiated. In *The Cold War and American Science: The Military-Industrial-Academic Complex at MIT and Stanford* (New York: Columbia University Press, 1993), Stuart W. Leslie focuses his Stanford section on the influence of government-sponsored research. A cautionary account of Stanford's evolution—and Terman's role in it—is Rebecca Lowen's *Creating the Cold War University: The Transformation of Stanford* (Berkeley and Los Angeles: University of California Press, 1997). Her account of the impact of the military-industrial-academic triocracy on the mission of the university lends an air of inevitability to Stanford's transformation. In a study focusing on makers of power grid tubes, microwave tubes, and semiconductors, Christophe Lécuyer suggests that Stanford's role was not as great as the conventional wisdom suggests, and that Stanford was as much a beneficiary of technological advance as it was a contributor. See Christophe Lécuyer, *Making Silicon Valley: Innovation and the Growth of High-Tech, 1930–1970* (Cambridge, Mass.: MIT Press, 2006).

2. On the entrepreneurial university, see Henry Etzkowitz, "The Making of an Entrepreneurial University: The Traffic Among MIT, Industry, and the Military,

188 1860–1960," in *Science, Technology, and the Military*, ed. Everett Mendelsohn, Merritt Roe Smith, and Peter Weingart (London: Kluwer, 1988), 515–40.

3. Laurence R. Veysey, *The Emergence of the American University* (Chicago: University of Chicago Press, 1965), 166–71.

4. Frederick E. Terman, "A Brief History of Engineering Education," *Proceedings of the IEEE* 64 (Sep. 1976): 1403.

5. Stephen B. Adams, "Regionalism in Stanford's Contribution to the Rise of Silicon Valley," *Enterprise and Society* 4 (Sep. 2003): 521–43, especially 523–27; Stuart W. Leslie, "The Biggest 'Angel' of Them All," in *Understanding Silicon Valley: The Anatomy of an Entrepreneurial Region*, ed. Martin Kenney (Stanford, Calif.: Stanford University Press, 2000), 51–52; and Edgar Eugene Robinson and Paul Carroll Edwards, eds., *The Memoirs of Ray Lyman Wilbur, 1875–1949* (Stanford, Calif.: Stanford University Press, 1960), 294–97.

6. Leslie, "The Biggest 'Angel' of Them All," 51–52.

7. David L. Kirp, *Shakespeare, Einstein, and the Bottom Line: The Marketing of Higher Education* (Cambridge, Mass.: Harvard University Press, 2003), 177.

8. Timothy J. Sturgeon, "How Silicon Valley Came to Be," in Kenney, *Understanding Silicon Valley*, 30; and Frederick Terman, "Bay Area Electronics—Then and Now," speech delivered at the WEMA 30th anniversary dinner, Nov. 20, 1973, Series VIII, Box 4, Folder 5, Stanford Archives and Special Collections 160 (hereafter cited as SC).

9. Leslie, *The Cold War and American Science*, 6, 12, 15.

10. Frederick Terman speech, "The Newly Emerging Community of Technical Scholars," p. 2, Nov. 5, 1963, Series VIII, Box 3, Folder 3, SC 160.

11. Terman speech, "The Newly Emerging Community of Technical Scholars," p. 6.

12. Terman speech, "The Newly Emerging Community of Technical Scholars," p. 4.

13. Clark Kerr, *The Uses of the University* (Cambridge, Mass.: Harvard University Press, 1963), 18–19.

14. Terman speech, "The Newly Emerging Community of Technical Scholars," p. 4.

15. Veysey, *The Emergence of the American University*, 346.

16. Kerr, *The Uses of the University*, 90–91; and Christophe Lécuyer, "Academic Science and Technology in the Service of Industry: MIT Creates a 'Permeable' Engineering School," *American Economic Review* 88, no. 23 (May 1998): 28–33.

17. Frederick Terman to Paul Davis, Dec. 29, 1943, Series I, Box 1, Folder 2, SC 160.

18. Terman to Paul Davis, Dec. 29, 1943.

19. Stephen B. Adams, "Stanford and Silicon Valley: Lessons on Becoming a High-Tech Region," *California Management Review* 48 (Fall 2005): 29–51.

20. Terman speech, "The Newly Emerging Community of Technical Scholars," pp. 10–13.

21. Terman speech, "Bay Area Electronics—Then and Now," p. 5.

22. Terman to Paul Davis, Dec. 29, 1943.

23. Stuart W. Leslie and Robert H. Kargon, "Selling Silicon Valley: Frederick Terman's Model for Regional Advantage," *Business History Review* 70 (Winter 1996): 439–40.

24. Adams, "Regionalism in Stanford's Contribution to the Rise of Silicon Valley," 527–29.

25. This was quoted from a comment Terman made in 1968. Everett M. Rogers and Judith K. Larsen, *Silicon Valley Fever: Growth of a High-Technology Culture* (New York: Basic Books, 1984), 36.

26. Frederick Terman speech to Stanford University Board of Trustees, Oct. 10, 1959, Series VIII, Box 2, Folder 3, SC 160.

27. Frederick Terman speech, "Some Observations on Engineering Education," July 8, 1958, Series VIII, Box 2, Folder 2, SC 160.

28. Lowen, *Creating the Cold War University*, 70; and George H. Nash, *Herbert Hoover and Stanford University* (Stanford, Calif.: Hoover Institution Press, 1988), 39.

29. Robert G. Sproul, "Opportunity Presented by Budgetary Limitations," *Chronicle of Higher Education* 5 (Jan. 1934): 7–13; and J. B. Speer, "The Functional Organization of the University," *Journal of Higher Education* 5 (Oct. 1934): 414–21.

30. Terman to Paul Davis, Dec. 29, 1943.

31. Frederick Terman to Paul Davis, Feb. 9, 1944, Series I, Box 1, Folder 2, SC 160.

32. Terman to Paul Davis, Dec. 29, 1943.

33. Mitchell Leslie, "The Vexing Legacy of Lewis Terman," *Stanford Magazine*, July–Aug. 2000, www.stanfordalumni.org/news/magazine/2000/julaug/articles/terman.html (accessed September 25, 2008).

34. Frederick Terman to Theodore Hoover, Apr. 26, 1926, Series II, Box 4, Folder 4, SC 160.

35. Leslie, *The Cold War and American Science*, 49.

36. Lowen, *Creating the Cold War University*, 71.

37. Lowen, *Creating the Cold War University*, 159.

38. Joseph Franzini, interview by Stephen Adams, June 11, 2002.

39. Frederick Terman to W. W. Hansen, Sept. 7, 1945, Series I, Box 1, Folder 6, SC 160; and Lowen, *Creating the Cold War University*, 89, 111, 175.

40. Lowen, *Creating the Cold War University*, 73.

41. Lowen, *Creating the Cold War University*, 84.

42. On the founding of Varian, see Timothy Lenoir with Christophe Lécuyer, "Instrument Makers and Discipline Builders: The Case of Nuclear Magnetic Resonance," in *Instituting Science: The Cultural Production of Scientific Disciplines*, ed. Timothy Lenoir (Stanford, Calif.: Stanford University Press, 1997), 239–92; Lécuyer, *Making Silicon Valley*, 53–128; and Henry Lowood, *From Steeples of Excellence to Silicon Valley: The Story of Varian Associates and Stanford Industrial Park* (Palo Alto, Calif.: Varian Associates, 1987).

43. Lowen, *Creating the Cold War University*, 82–88.

44. Lowen, *Creating the Cold War University*, 91.

45. Lowen, *Creating the Cold War University*, 156.

46. Lowen, *Creating the Cold War University*, 119.

47. Lowen, *Creating the Cold War University*, 159–61.

48. William M. McCord, "School of Humanities and Sciences Final Report—Faculty Survey" (revised Apr. 1, 1959), p. 23, Series III, Box 32, Folder 5, SC 160.

190   49. Frederick Terman speech, "Cooperation Between Industry and Education," May 12, 1958, Series VIII, Box 2, Folder 2, SC 160.

50. Terman to Paul Davis, Dec. 29, 1943.

51. Frederick Terman to department heads, Sept. 14, 1959, Series III, Box 2, Folder 1, SC 160.

52. Leslie, *The Cold War and American Science*, 12; and Terman, "A Brief History of Electrical Engineering Education," 1403.

53. Gillmor, *Fred Terman at Stanford*, 514–15.

54. Frederick Terman speech, "The Growth of Science in Industry," p. 7, Oct. 4, 1960, Series VIII, Box 2, Folder 3, SC 160.

55. Leslie and Kargon, "Selling Silicon Valley," 443.

56. Robert H. Kargon, "Temple to Science: Cooperative Research and the Birth of the California Institute of Technology," *Historical Studies in the Physical Sciences* 8 (1977): 3–31.

57. Frederick Terman, "Preliminary Report, Summit University: The Dimensions of the University," pp. 22–23, ca. 1966, Series IV, Box 17, Folder 12, SC 160. Emphasis in the original.

58. Terman, "Preliminary Report, Summit University," p. 32.

59. Terman, "Preliminary Report, Summit University," p. 3.

60. Terman, "Preliminary Report, Summit University," p. 1.

61. Terman, "Preliminary Report, Summit University," pp. 2–3.

62. Terman, "Preliminary Report, Summit University," pp. 13–14.

63. Terman, "Preliminary Report, Summit University," p. 6. Terman's emphasis on graduate education had been one of the hallmarks of his tenure as provost. In a 1960 speech, he had suggested that "by doubling or tripling our output of prospective college teachers for our western schools, Stanford can probably make a greater contribution to higher education in the west than by making a large increase in undergraduate enrollment." Frederick Terman speech, "Stanford Yesterday and Today," Oct. 7, 1960, Series VIII, Box 2, Folder 4, SC 160.

64. Terman to Paul Davis, Dec. 29, 1943.

65. Terman, "Preliminary Report, Summit University," pp. 29, 30.

66. Terman to Paul Davis, Dec. 29, 1943.

67. Terman, "Preliminary Report, Summit University," p. 1.

68. Leslie and Kargon, "Selling Silicon Valley," 446.

69. Kerr, *The Uses of the University*, 87–88.

6

The Bold and the Foolhardy

Hurricanes and the Early Offshore Oil Industry

Joseph A. Pratt

When the oil industry moved offshore into the Gulf of Mexico in the 1930s and 1940s, it plunged into an ocean of ignorance. Designers and builders of the earliest offshore drilling rigs, platforms, and pipelines had almost no data about basic wind, wave, and soil conditions in the Gulf. They knew still less about the engineers' nightmare out there beyond the horizon, the stunning force of hurricanes. A key challenge facing oil industry engineers in the first quarter century of offshore operations was to protect life and property while gaining the experience and information needed to reduce the vast uncertainties presented by high winds and waves driven by hurricanes. Only after such uncertainties had been reduced to manageable risks could the industry achieve orderly expansion in a demanding offshore environment.[1]

Those seeking to develop a technological system capable of finding and retrieving oil and natural gas from underneath the ocean faced formidable challenges in defining basic design criteria. Traditional engineering calculations could be adopted to estimate the environmental forces that would come to bear on offshore equipment and structures, but such calculations could be made only after the collection of fundamental data about these forces of nature. How strong would the winds blow? How high could hurricane-driven waves be expected to crest? How solid was the foundation provided by the soft, sandy bottom of the Gulf of Mexico, and how would this soil be affected by hurricanes? Underlying these questions was another, more practical one: How much were oil companies willing to spend in order to develop safe, durable offshore structures?

By the end of World War II, numerous oil companies had strong economic incentives to answer such questions. These companies stood ready to explore the risks and rewards of offshore operations, in part because of the scarcity of good leases onshore, where large oil companies had locked up giant acreage at low costs in the depressed 1930s. Seismic surveys in the 1930s had revealed numerous promising salt domes in the Gulf of Mexico. It made good geological sense that the large oil fields discovered in the early twentieth century along the Texas-Louisiana coasts did not stop at the water's edge. In the late 1940s, several major oil companies eagerly extended their ongoing quest for oil out into the Gulf. A handful of smaller companies looked out in the same direction seeking "break-through" discoveries that could vault them up the ranks of the independent oil producers. These companies faced an uphill battle offshore. If they could not develop a dependable technological system capable of getting offshore oil to markets onshore at a competitive price, they could not sustain operations in the Gulf of Mexico.

History was kind to the pioneers of the offshore industry in the Gulf of Mexico. They arrived at the right shore at the right time. The Gulf sloped very gently out, stretching for a hundred miles in places along the continental shelf before reaching water depths of three hundred feet. Companies thus could walk gradually, step-by-step, into deeper waters as they developed new technologies. As they moved out, they could draw on the workforces and expertise of clusters of oil-related manufacturing and service companies that had grown previously to meet the needs of a booming onshore industry in the region. Best of all, significant discoveries in the Gulf quickly rewarded their initial efforts, encouraging them to make larger investments.

In developing new fields, the offshore industry could draw on previous experiences gained near the shore in California and in a variety of inland waters around the world. Before the 1930s, oil had been developed off the Southern California coast near Summerland through the use of a system of trestles that reached out into the edge of the Pacific Ocean to tap oil fields that extended from known onshore deposits. But this region lacked the threat of the extreme weather produced by hurricanes. Extensive development of oil in the protected waters of Caddo Lake in Louisiana, Venezuela's Lake Maracaibo, and the Caspian Sea generated knowledge useful in everyday operations offshore. Finally, work in the marshy areas of "inshore" Louisiana in the 1930s helped prepare the way for operations in nearby areas offshore. None of these previous projects, however, had to be designed to stand up to hurricanes in the open sea.[2]

Griff Lee, a design engineer for Humble Oil and then for offshore construction giant McDermott, aptly summarized the situation facing the industry in 1945: "There had been no construction of open frame structures in open water before." Designers could look at data on the wave and wind forces exerted on seawalls or on ships at sea, but such data could not predict the forces that would come to bear during a hurricane on structures permanently fixed to the ocean's floor.[3] Given this void of knowledge about conditions offshore, those eager to explore for oil in the Gulf of Mexico would have to take risks amid considerable uncertainty as they learned by doing. This was not unusual in the oil industry exploration business, which lived by the oft-repeated adage, "Fortune favors the bold."

With great confidence born of past technical successes and fed by the profits promised to the first movers into the Gulf, the oil industry used very rough "best estimates" of wind and wave forces in the initial design and construction of offshore facilities. When severe storms exposed problems in offshore designs, industry engineers solved them "on the run." In practice, this could mean working on facilities out in the ocean rather than on models in wave tanks. Meanwhile research went forward to generate the data needed to improve the best estimates. The offshore industry faced the challenge of sustaining innovation through time, because the movement into deeper water raised new technical issues. In addition, going farther out into the ocean generally made offshore work more expensive, increasing the industry's economic incentives to protect itself against costly repairs and disruptions of production.

From the earliest days in the Gulf, an offshore fraternity grew. All involved understood that no single company could long prosper alone; the success of each company ultimately depended on the expansion of a network of technical expertise and specialized services and equipment. The oil companies typically had extensive research and development efforts to support offshore production, but they could also count on the technical expertise of specialized drilling and service companies. Consulting firms grew to provide a variety of services, including weather forecasting. Major universities along the Gulf Coast took the lead in academic research on conditions offshore. As new opportunities arose, new companies came in and out of this evolving network of innovative organizations. For example, the adaptation in the 1950s of the helicopter for offshore transportation services dramatically altered planning for evacuations. The role of different companies changed as the offshore fraternity grew and the technical challenges became more complex, with greater specialization over time.

An aggressive, risk-taking culture emerged in the Gulf of Mexico, focused not so much on individual entrepreneurial risks as on the collective will of people involved in various parts of the industry and employed by numerous companies to push on out into deeper waters.[4] The exuberance of successful expansion bound the fraternity together, while also at times blinding them to the risks of such environmental factors as hurricanes.

If fortune favored the bold in the formative years of offshore development, unusually good weather favored the foolhardy. Until 1964, no major hurricanes swept through areas with high concentrations of offshore operations. Thus for almost twenty years the offshore industry amassed the data and the experience needed to improve the design of its equipment in the relative calm before three major storms moved through "offshore alley," the area of concentrated facilities off the coasts of Louisiana and Texas. Hilda (1964), Betsy (1965), and Camille (1969) severely tested the technical system that had evolved in the Gulf of Mexico.[5] The industry received a gentleman's "C" on these tests. Although showing remarkable creativity in managing calculated risks and making engineering adjustments on the run, the offshore fraternity had gravely underestimated the risks presented by major hurricanes.

EARLY FORAYS INTO THE GULF

The oil industry first stuck its toe into the Gulf of Mexico to test the waters before World War II, and the results of these early forays identified several key problems presented by storms. In the late 1930s, Humble Oil (then a Houston-based, majority-owned subsidiary of Standard Oil of New Jersey) constructed one of the first drilling sites in the Gulf at McFadden Beach, south of the giant refineries at Port Arthur, Texas. Borrowing from the approach that had proved successful in Southern California, the company extended a trestle more than a mile out from shore, with drilling rigs at the end of the line supported by men and materials brought out on a train track over the trestles. The drillers struck no commercial deposits of oil, and after a small hurricane in August of 1938 ripped apart the entire facility, Humble abandoned this venture. The industry subsequently ratified Humble's decision: trestles could not be built high enough or strong enough to withstand hurricane-driven waves in the Gulf.[6]

The first real test of offshore construction came up the coast about fifty miles near Cameron, Louisiana, a small coastal town near the Texas-Louisiana border. In 1937 and 1938, Pure Oil and Superior Oil, two large independent oil companies, together built a large wooden platform about a mile offshore in approximately fourteen feet of water. This Creole field became the first producing property in the Gulf. It proved that profits could be made offshore while also revealing the severe challenges posed by hurricanes and the limitations of applying onshore technology in an offshore environment.

The companies constructed a giant platform measuring 320 feet by 180 feet from which to drill the exploratory well and then to produce any oil found. The primary task was to drive some three hundred treated yellow pine piles 14 feet into the sandy bottom using pile drivers mounted on barges. This "stick-building" approach sought safety and strength through the clustering of many wooden piles; it sought stability against wave forces by driving the piles as far as possible into the sand. It sought protection from hurricane winds by using design criteria developed for onshore buildings to construct a structure that could survive winds of up to 150 miles per hour.[7]

Hurricane-driven waves were another thing entirely. With no available data on wave heights or wave forces, I. W. Alcorn, the designing engineer from Pure Oil, chose to build the deck fifteen feet above the water. He figured that such height would provide sufficient protection from normal high waves. He could not calculate the strength and height needed to survive a major hurricane; nor did he have the capacity to build such a structure with existing tools. So he struck upon a reasonable compromise. He designed the deck so that it would be swept off the piles by very high waves, thus limiting the damage done by a severe hurricane to the extensive system of piles. The wooden deck could then be replaced after the storm.[8]

The Creole platform produced its first oil in March 1938. Once production began, the problems of transportation and communication became more pronounced, foreshadowing similar problems in the post–World War II offshore industry. Workers lived in houseboats at Cameron, the closest town. But the platform itself was some ten miles along the coast from Cameron, meaning that all men and supplies came to the platform via a long and often rough ride out in shrimp boats leased for this purpose. A one-way ride might take up to an hour and a half. Without communication between the supply point, the boats, and the platforms, the shrimp boats often arrived at the platform only to find seas at the site too rough to allow workers to transfer from the boat to the platform. Rope ladders hanging from the platform could be lowered down to the deck of the shrimp boats in relatively calm waters, but not in rough seas. In the thick fog that often hovered over the platform, boat captains would at times simply cut their engines and listen for noise from the platform in order to find this man-made island. From the start, it was understood that in the event of a major storm, the men would be evacuated after the equipment on the deck had been secured.

The Creole platform proved quite successful in finding and producing oil. Using directional drilling to tap the field at several surrounding locations, it produced over four million barrels of oil over the next thirty years, during which time it was constantly upgraded as the offshore industry became more experienced at construction. Alcorn proved farsighted on one key point. In 1940 a small hurricane moved through the region, sweeping the deck into the ocean and badly damaging the piles. Crews drove some new piles and quickly rebuilt the deck, and the platform returned to production, the first offshore structure in the Gulf to survive a hurricane.[9]

World War II halted development in the Gulf. Workers on several small platforms being built offshore in 1942 remember scanning the horizon nervously in search of the periscopes of German submarines. But the war set in motion several processes that proved quite helpful to the offshore industry when peace returned. First and foremost was the work of the U.S. Army's oceanography and weather service, which created a corps of well-trained specialists who forecast wind, wave, and soil conditions for use in the amphibious landings in northern Africa, Normandy, and the Pacific. These "weather officers" accumulated data on the behavior of waves and soils in different storm conditions. From such information they sought to predict whether conditions at a specified place and time might be appropriate for an amphibious landing. After the war, several of the weather officers led the industry's efforts to collect and interpret better data on winds, waves, and soil in the Gulf of Mexico. They developed the creative technique of "hindcasting," which used observations drawn from the histories of past hurricanes to help forecast the path and impact of current and future hurricanes.[10]

The war paved the way for postwar developments in many other ways. Much-improved communications at sea could be adapted for use offshore. War-surplus vessels produced in great numbers to support amphibious landings could be purchased and converted for offshore uses at bargain basement prices after the war. Perhaps the most important impact of the war, however, was on attitudes, not equipment. Veterans who had postponed their lives for four or five years returned eager to get back to normal work and family lives. They came back with a sense of urgency and a sense of adventure, two characteristics required of those who pushed out into the Gulf in search of oil after World War II.

The race offshore was on in the late 1940s. Despite uncertainties between coastal states and the federal government over the ownership of offshore lands, despite economic uncertainties, despite technical uncertainties, numerous oil companies headed out into the Gulf in search of big, virgin fields. Economics shaped their technical choices. One young Shell engineer recalled asking an old hand at Brown & Root (one of the two dominant offshore construction companies in these early years), "In just how deep of water do you think Brown & Root could build an offshore platform?" The simple answer was, "First, young man, you will have to tell me how much money Shell is prepared to spend on such a platform."[11]

In these formative years, two basic approaches to offshore exploration and production emerged. The first was the Creole approach writ large. Humble, Superior (a large independent), and Magnolia (a Dallas-based majority-owned subsidiary of Standard of New York) chose to build permanent platforms to find and develop oil in the Gulf. These platforms could hold crews of up to fifty workers, as well as all needed equipment and supplies. They were sturdy enough to last the life of the field and to survive harsh weather. They were also expensive to build and fixed in place once constructed, attributes that greatly magnified the risk of building them for use in drilling wildcat wells.[12]

A smaller company, Kerr-McGee, developed the less risky, more flexible small platform with tender approach. By using refurbished war-surplus landing ships, tanks (LST) to house men and supplies, this approach required the construction of a relatively small platform to support the drilling rig needed to find and produce oil. The LSTs were more than three hundred feet long; once most of their insides, including their engines, had been gutted, they could be converted into a sort of giant, mobile facility for living quarters and storage space. Kerr-McGee recognized the obvious economic attractions of using the LSTs, at least while war-surplus vessels remained plentiful and inexpensive. In the event of a dry hole, the tender, unlike the large fixed platforms being built to explore for oil by other companies, could be towed to a new location and at least a portion of the cost of the small platform could be salvaged.[13]

Severe weather had implications for both systems. Large platforms could be designed and built to withstand hurricane-level storms much more easily than the small platforms with tenders. High decks and safe procedures for transferring workers could be incorporated in their designs. The first generation of fixed platforms constructed from 1946 through 1948 placed decks from twenty to forty feet above the mean level of the Gulf, reflecting the broad range of opinion on what was the most likely wave height in a severe hurricane.[14] In contrast, the tenders posed serious problems in high wind and waves. These heavy vessels were not self-propelled, and in rough seas they could become floating sledgehammers posing serious dangers to the small platform.

After the success of Kerr-McGee's small platform with tender, Humble invested millions of dollars in buying surplus LSTs and converting them for use as tenders. It developed a mooring system using chain two inches in diameter to hold these large vessels alongside small platforms. Company engineers designed the ship's anchoring system to withstand one-hundred-mile-per-hour winds. To accommodate the height of the tender, decks on the small platforms were often placed forty feet above the ocean. Men and equipment moved from the tender to the platform over a bridge that could be raised from the vessel to the deck. So difficult was passage over this bridge in rough seas that workers came to call it "the widow maker." If a hurricane seemed likely to affect a tender operation, the company would move the tender away from the platform so that it could ride out the storm at anchor while posing less danger of pulling off of its moorings and smashing into the platform. Until the coming of helicopters, Humble maintained large vessels near its offshore locations to evacuate workers in the event of severe weather.[15]

198 Problems with the tenders in rough weather did not, however, outweigh the economic advantages the small platform with tender had over the large fixed platforms. The huge downside of permanent platforms remained: a dry hole meant that literally "sunk costs" could not be recovered. Until the development of dependable, cost-effective mobile drilling rigs that could stand up to rough conditions in the open sea, the "semi-mobile" small platform with tender remained the dominant approach to offshore exploration and production.

Oil companies active in the Gulf went forward using both approaches until the late 1940s, when the "tidelands" controversy temporarily halted leasing while the federal government and state governments turned to Congress and the courts to resolve questions of ownership of offshore lands. This controversy became quite heated, particularly in the 1952 presidential campaign. But the pause in leasing gave the offshore industry a short breathing space in which to reexamine assumptions about design criteria for offshore structures and to begin a generation of basic research about waves and soil conditions in the Gulf.

This research proceeded on a number of loosely coordinated fronts. The major oil companies created their own research groups, which worked closely with leading research institutes such as Scripps and the University of California, Berkeley. Consultants also provided much input into the studies of basic conditions. In the 1950s, the American Petroleum Institute (API), the industry's primary trade association, became more active in the collection of improved data about waves and soil. In these formative years for the offshore industry, a pattern of cooperation emerged in the quest for reliable data on the forces exerted on structures by wind and waves and on the load-bearing capacity of the soft soil in the Gulf.

The leading authorities on soil conditions were the founders of McClelland Engineers, a New Orleans–based consulting firm that extended the work of the weather officers into the Gulf of Mexico. Bramlette McClelland, John Focht, and Robert Perkins pioneered the applications of soil mechanics to the problems of the offshore industry. To do this, they had to have information on conditions in the Gulf. With industry funding and cooperation, in 1947 they began boring soil samples offshore, building a database for use in offshore construction. At times they studied conditions for contractors preparing to install a structure at a specific location; at other times, they investigated general conditions in areas likely to be explored in the future. Their analysis of the results of oil-company-sponsored tests also led the way in applied research on the load-bearing capacity of the piles used to support offshore platforms.[16]

The API took the lead in the collection of other sorts of data on the soil in and along the Gulf. In 1951 the Institute launched what came to be known as Project 51, which spent four years undertaking basic work on conditions in the Gulf, using core drillings, serial mapping, and seismic surveys. This work, as well as that of McClelland Engineers, provided fundamental information vital to the safe construction of offshore structures. It did not, however, directly address a question that was later revealed as important: What would be the reaction of soil in various parts of the Gulf to the extreme conditions generated by severe hurricanes?

Other research studied the force of waves on offshore structures, both in normal times and in times of extreme weather. Here the oceanography department at Texas A & M University led the way. C. L. Bretschneider and Robert Reid, two more former weather officers, cooperated with several major oil companies to conduct field measurements to determine the wave forces exerted on vertical cylinders placed in the ocean. J. R. Morison later added considerations of inertial components to this work.[17]

Other primary research was much more directly tied to hurricanes. From 1947 into the 1970s, extreme wave heights generated by severe storms remained a critical question on the minds of offshore engineers. This question was attacked from two directions. The first sought to develop better means to track storms and to predict where they would hit; the second sought better information about the maximum height of waves that could be expected in different parts of the Gulf. Weather forecasting in general had advanced steadily over the decades before World War II, but the offshore industry needed more detailed and more frequent forecasts than the U.S. Weather Service could make available to them. To meet this demand, A. H. Glenn, a former weather officer with graduate training at the Scripps Institute of Oceanography and UCLA, mustered out of the Air Force and created Glenn and Associates, a New Orleans–based weather forecasting agency designed to meet the special needs of operators of offshore facilities. Glenn and others made great strides in using historical data about past hurricanes to "hindcast" the path and the intensity of future hurricanes. By analyzing all available information about past hurricanes with sophisticated theoretical models of the behavior of winds and waves, Glenn and a growing group of hindcasters gave platform designers a much-improved understanding of potential wave forces while beginning the process of categorizing hurricanes according to their intensity.[18]

But forecasting storms was not quite the same as forecasting maximum wave heights; the origins, path, size, location, and power of a hurricane could influence wave heights, particularly in localized areas near the eye wall. How could a designer improve his estimate of the maximum wave height and wave force that might challenge the structural integrity of a platform over its life in a specific place in the ocean? With no trustworthy measured data on extreme wave heights, different companies placed their bets using the best guesses of dueling consultants, many with connections to prestigious universities or research institutes. Highly publicized reports by two such consultants, retired naval officers

F. R. Harris and H. G. Knox, stated authoritatively that "in 100 feet of water waves will probably seldom, if ever, exceed 20 feet in height." Decks thus should be placed "20 feet above the still water line."[19]

Other leading experts on waves disagreed. W. H. Munk, another former weather officer, who had forecast conditions for the Normandy invasion, came in with a higher figure. After analyzing existing data with theoretical models of wave formation and behavior, Munk settled on a maximum wave height of about twenty-five feet and a recommended deck height of thirty-two feet above the water. With a wide range of "expert" opinions from which to choose, companies designed their platforms based on their willingness to take risks and their sense of the odds against a twenty-five-year storm hitting their particular location during the life of their particular field. The safe consensus in these early years hovered around a maximum wave height of about twenty-nine feet in the shallow waters of the Gulf, with a frequency of perhaps once every forty to fifty years.

A series of relatively weak, small hurricanes in 1947–1952 quickly called this consensus into question. After a small but intense hurricane off Freeport, Texas (south of Houston), in October 1949 severely damaged a platform, the post-mortem suggested that waves as high as forty feet had buffeted the platform. The observed wave damage to several platforms in these years led to estimates of waves in the twenty-two- to twenty-nine-foot range in each case. Once every fifty years, indeed. Observations also showed more clearly than had been previously understood that the key problem was to keep these mammoth waves from cresting on the deck. During the Freeport storm, a platform with a deck twenty-six feet above the ocean suffered damages that cost its owner more than $200,000 in losses while a nearby platform with a thirty-three-foot deck showed no damage.[20] The owner of the damaged platform came away convinced that a relatively small investment to build a slightly higher deck would have been justified to avoid the very high costs of cleaning up a damaged platform and the loss of production and revenues from shutdown time when oil could not be produced.

Here was a strategy for managing risks under uncertainty that held a special appeal for the largest and most conservative companies. The California Company (Calco, a subsidiary of Standard Oil of California) had a particularly dangerous encounter with the first hurricane of this era, and its leaders responded by greatly improving safety standards. In early September 1948, a hurricane rose quickly off Louisiana, without sufficient warning for the evacuation of all offshore workers. The hurricane hit Calco's large operations off Grand Isle, Louisiana, placing more than fifty men in harm's way. Twenty-five of them huddled aboard a converted LST tender that had been placed under tow to try to reach safe harbor. Unable to make much headway, the captain of the tug decided to cut his lines and take his tug to safety, leaving the LST adrift in the hurricane. Meanwhile a derrick barge with thirty men aboard also bounced about in the rough seas after a rescue boat sent for it ran aground. Hours later tugs finally managed to control both vessels and bring them to safety. The men aboard came ashore "wet, but unhurt," but company officials knew that only good luck had prevented a disastrous loss of life. Those involved in this incident came away determined to make changes to avoid risks to workers and to minimize the damages that the hurricane had done to Calco's platforms.[21]

With such concerns in mind, Calco went back to the drawing board, applying significantly higher estimates of maximum wave heights and forces in its designs. In the words of Paul Besse, one of the engineers at Calco who took the lead in redesigning its offshore facilities, "That [storm] certainly elevated every platform that Chevron put in from that day forward." The company also elevated the decks of two platforms already installed in the Gulf, staking claim to leadership in the offshore industry in moving decks up higher to avoid wave damage in severe storms. Seeking better information to use in designing platforms, Besse found little, because "there had never been a time when anyone was crazy enough to try to build a platform in the open ocean and place men and equipment on it. . . . We had to go on theory, and the hurricane . . . caused Chevron to start thinking about placing wave measuring equipment on a platform offshore."[22]

Others agreed that it was time to obtain better measurements of wave heights. After Chevron installed three separate pilings in the Gulf with devices to measure wave heights in 1954, Humble Oil helped analyze the data obtained. The companies then calculated new design criteria for severe hurricanes in Texas and Louisiana. A. H. Glenn used these calculations along with wind and wave measurements from onshore and from ships to generate for the industry a new estimate of projected hundred-year storm conditions in the Gulf and other locations around the world. Calco and Humble, later joined by Shell, became the offshore industry's leading advocates for using such data to adopt higher, safer standards for platform construction and deck placement. Humble's leading offshore engineer, Arthur Guy, expressed the philosophy behind this new attitude with a simple sentence: "Error [on the side of greater safety] is cheap." Many of the largest offshore companies agreed that the costs of potential for damage and even loss of life far outweighed the relatively small costs of building safer platforms. Better safe than sorry—and less expensive in the long run.[23]

THE PROBLEMS OF HURRICANES "SOLVED" IN THE 1950S

The election of Dwight Eisenhower and the end of the stalemate in offshore leasing in 1953 unleashed a burst of activity in the Gulf. At that time, there were already approximately seventy separate platforms in water depths up to seventy

202 feet in the Gulf.[24] Both numbers increased dramatically from 1953 until the economic downturn in the Gulf in the late 1950s. In this building boom, the offshore industry created a fully developed "Gulf of Mexico system" for exploring and producing oil.

At the heart of this approach was the development of mobile drilling rigs that could explore for oil in different locations, leaving production of oil for permanent platforms. This dynamic new offshore industry evolved quickly and became an important part of the cluster of innovative organizations that constituted the offshore fraternity. It developed in several directions at once, as entrepreneurs organized companies to create various technologies for drilling at sea. Submersible rigs, jack-up rigs, drilling ships, and semi-submersible drilling rigs evolved side-by-side in the 1950s and 1960s. Each type of rig had characteristics that made it attractive for certain water depths and locations, and all were used to find oil in the Gulf and in other regions from the 1950s forward. These drilling rigs had one common characteristic that made them vulnerable to severe storms: they were designed to drill oil wells, not to move gracefully through the ocean. Awkward to control and use in the open sea, the drilling rigs used to find oil and natural gas proved much more vulnerable to high winds and rough seas than the fixed platforms built to produce oil and natural gas.[25]

Deadly and costly accidents involving drilling rigs highlighted a key problem facing offshore operators: uncertainty over insurance. Hedging risks with insurance made good business sense, but underwriters shied away given the "perils beyond their [the offshore operators] reasonable control and not heretofore encountered in their land operations." Yet after deciding that risky offshore work might not yet be insurable, insurance companies examined more closely their existing policies and found that they were already liable for hundreds of millions of dollars under policies covering such things as damage to vessels, explosions, and injuries to workers. The lull in activity during the tidelands controversy afforded these companies the opportunity to begin to sort out the key questions facing them. Were mobile drilling rigs vessels or drilling rigs? Should their workers be considered seamen or drillers? Was the blow-out of an oil well in the ocean the same as an explosion at sea? Providing legally binding answers to such questions was the first step in providing adequate coverage for offshore operations.[26]

In comparison to mobile drilling rigs, underwriters had less trouble in insuring the permanent platforms most companies built to provide a safe, sturdy foundation for long-term development. By the mid-1950s, these platforms were much-improved versions of those first built by Magnolia, Superior, and Humble in the late 1940s. The Gulf of Mexico system of this era came to be dominated by "piled jackets," large metal structures constructed in specialized

fabrication yards onshore, transported by purpose-built barges, installed with
specialized equipment, and then pinned by piles driven down through the
jacket into the ocean floor. Once the piles had been driven, prefabricated decks
could be welded onto the jacket. Fabrication onshore produced a stronger,
more uniformly built frame; the time spent on construction in the rough,
unpredictable conditions out in the open sea could be minimized. The completed structure was self-contained, including quarters for work crews.[27]

Transportation and communication improvements allowed these platforms
to be supplied more easily, while also ensuring that the crews could be evacuated in the event of a storm. Fleets of purpose-built supply boats owned and
operated by emerging firms such as ODECO quickly replaced the shrimp
boats and war-surplus boats that had provided much offshore transportation in
the earliest years in the Gulf. These boats were faster, stronger, and more comfortable, and they were equipped with modern communications. But they still
required long hours in the water to ferry men and supplies back and forth from
platform to shore.[28]

For safety and convenience, it was only a matter of time before local entrepreneurs developed helicopter service out to the rigs. By the early 1950s, Humble had contracted with a local company to lease helicopter service to platforms
far out in the Gulf. The first entrant into this new business was PHI (Petroleum
Helicopters Incorporated), which grew quickly in the 1950s and operated a fleet
of thirty-three helicopters as of 1958. Once oil companies made the investment
in helicopter landing pads out on the platforms and drilling rigs, the industry
had a greatly improved capacity to respond to emergencies. This innovation in
transportation quickly became the offshore industry's first line of defense
against hurricanes. When a hurricane threatened, the skies filled with helicopters ferrying workers to safety onshore. By sharply reducing the risks of loss
of life in storms, helicopters allowed the offshore industry to focus on the calculation of the potential risks of property damage in deciding on the best design
criteria to protect its facilities from major storms.[29]

Effective evacuations, however, required more accurate and up-to-date
weather forecasting. To monitor the path of hurricanes, many offshore companies subscribed to a well-developed private forecasting service that kept in touch
with their offshore facilities via advanced communications equipment. Before
the creation of a well-developed satellite network in the 1960s, the U.S. Weather
Service simply did not have the resources to deliver the quality of forecast information available through New Orleans–based Glenn and Associates, which provided frequent detailed reports on wind, weather, and waves in areas of the Gulf
containing offshore operations. This private weather service supplemented government data with its own long-range radar system and with the four daily observations submitted from the rigs of subscriber companies. The companies could

have personal consultations with meteorologists if in doubt about storms. In this era before satellite observations, the offshore industry had far superior information about storms than was available to others; its special needs gradually led to the improvement of forecasting in general.[30]

An overview of the response of this system of operations when faced with a hurricane comes from an article in 1956 in *The Humble Way*, the employee magazine of Humble Oil. In this case, a private forecasting service warned the company of a gathering storm that might ultimately pass over one of its major facilities. Careful monitoring of the storm convinced management to prepare for the worst. Workers then cleared the decks of the small platform in use at the site, storing some materials in the tender vessel, which was then battened down and moved away from the platform using winches on the mooring system. After anchoring the tender, workers evacuated in ships. Once the storm had passed with little damage, the workers returned and the platform was back in production the next day.[31]

Humble was a major company with well-built platforms and well-developed safety procedures. The storm that threatened its facility was relatively small and did not score a direct hit. Later in 1956 and 1957, Humble and the rest of the companies in the Gulf had a more demanding test, as two fairly large hurricanes skirted areas with numerous offshore platforms.

The first was Hurricane Flossie, which moved through clusters of facilities offshore near the western edge of Louisiana in September of 1956. Labeled the "first real hurricane test" for offshore operators since drilling activity began in 1947, Flossie unleashed 110-mile-per-hour winds and fifteen- to twenty-foot waves that caused the shutdown of several hundred offshore producing wells and many drilling rigs for two to three days. Although costs from downtime exceeded actual damages, this minimal hurricane did teach operators several valuable lessons.

The first lesson reflected the attitudes produced by a decade of relatively mild weather. Again, as in 1948, nearly fifty men "rode out" the storm on tenders and other vessels. After a Calco tender vessel had been torn from its anchor, twenty-five crewmen fighting to survive in the high seas floated serenely in the eye of the storm for a while before one-hundred-mile-per-hour winds returned from the opposite direction and their struggle began anew. The companies and the men involved took a calculated risk that they would be safe. After noting that Flossie was only half as forceful as hurricanes that could hit the area, one trade journal, *World Oil*, echoed the arguments of operators who "say more attention should be given to complete evacuation, doing away entirely with the calculated risk." The industry took justifiable pride in its lack of fatalities in hurricanes, a record not exactly guaranteed by asking workers to ride out storms in clumsy converted LSTs.[32]

Numerous tenders broke their mooring chains or moved off their anchors during Flossie. One of Humble's tenders suffered breaks in six of eight mooring chains and swung around into the adjoining platform, causing some $200,000 in damage. Other companies reported problems with damaged risers, the conduits for the pipe from the platform to the ocean bottom. Yet despite such problems, all in all, the reports on Flossie stressed the effectiveness of existing designs and safety procedures, with the oft-repeated caveat that this was not a major storm. One respected trade magazine writer gave an optimistic interpretation of the lesson of Flossie: "The greatest fears of the offshore oil operators have been dispelled by the arrival of Hurricane Flossie." This "full-blown hurricane" had shown conclusively that the industry's "engineering estimates were correct."[33]

Nine months later, Hurricane Audrey, the first major hurricane to skirt Louisiana's "offshore alley," inflicted expensive damage, reminding the industry that it still had not experienced the effects of the direct hit of a major storm. In June of 1957, this storm arose quickly in the Bay of Campeche, took a straight path up toward the Texas-Louisiana state line, and slammed ashore at Cameron. The storm surge and high winds of this category 4 hurricane killed four hundred to five hundred people. As of July 2008, it remains the seventh deadliest hurricane in U.S. history. Yet damage offshore was relatively minor. One mobile drilling rig sank in the storm and four tenders suffered damage when they pulled loose from their moorings and ran aground. Estimated damage to all offshore facilities reached about $16 million.[34]

What registered most clearly in the harsh aftermath of the storm was that the offshore industry had fared dramatically better than the communities along the coast. After helping clean up the carnage in Cameron, the industry proudly concluded that "forethought minimized hurricane damage to offshore installations." On the key issue, the industry's record remained spotless: not a single life was lost offshore during Audrey. Two offshore workers reportedly died, but only after they had been evacuated from a platform to an interior location and then chose to return to Cameron to try to protect their homes. In its overview of the "scars" left by Audrey, one of the major offshore trade journals concluded that "the industry has scored an overwhelming though costly victory."[35] This "lesson" of the industry's triumph over a major hurricane encouraged a sense of complacency that paved the way for later troubles.

The industry could not be quite so optimistic concerning the performance of mobile drilling rigs. In quick succession in 1956 and 1957, five mobile rigs capsized—four in the Gulf of Mexico and one off Qatar in the Middle East. Some were in rough waters; one was at dock being readied for sea. These five disasters caused more than $7 million in damages, with thirteen fatalities in the four accidents in the Gulf of Mexico. The first imperative of mobile drilling

206   rig design was the effective drilling of oil wells once on locations, but all had to be seaworthy enough to be towed in calm conditions. Although these "ungainly monsters of the sea" had been designed "to float within a reasonable degree of safety," they continued to experience difficulties from rough seas and high winds.[36]

In September of 1957 still another hurricane, Bertha, moved inland near Cameron, sinking one drilling tender and driving another aground. The industry had been put on notice by nature, not once, but three times in a twelve-month period. It responded by raising new questions about the origins and properties of hurricanes. The focal point of investigations was a newly formed API committee, the Advisory Committee on Fundamental Research on Weather Forecasting. Staffed by industry experts who had the resources to fund research by academics and consultants, this new committee tackled fundamental issues that had long eluded explanation. It brought to bear the best knowledge available in the offshore fraternity on a critical question: What caused hurricanes to form, and could their paths and intensity be forecast with greater certainty?

To address such issues, the API committee engaged the services of Herbert Riehl, a professor of meteorology at the University of Chicago, to prepare a "think piece" on what was known about hurricanes and what sorts of research were needed to advance knowledge. In the years from 1956 through 1962, the committee explored these issues with the best available theoretical ideas about hurricane formation and motion and the creative use of data supplied by A. H. Glenn on past hurricanes and potential hurricanes that did not develop. The committee, like the oil industry as a whole in these years, made use of rudimentary computers. Computer analysis helped the committee improve the art and science of hindcasting, giving the designers of offshore equipment useful information on which to base design criteria. In 1962 the API decided to sponsor no more research on hurricanes and the committee went out of existence. Its last publication reminded the reader of the great economic value of research that could predict the path of hurricanes, but apparently those who funded the work of the API could not see concrete results coming from the work of this advisory committee.[37]

MAJOR HURRICANES BRING INSTITUTIONAL INNOVATIONS IN THE 1960S

The offshore industry had its hands full with many issues other than hurricanes in the 1960s. The push out to produce oil in the deeper waters of the Gulf reached the 100-foot mark in 1957 and then quickly moved on out to 225 feet in 1965 and more than 300 feet in 1969. More than a thousand platforms had been built in the Gulf by the mid-1960s. The technology of exploration and production, as well as that of deep water pipelines, moved forward by leaps and bounds, enabling the industry to increase offshore production in the Gulf of Mexico to more than one million barrels a day by the late 1960s. At the same time that the Gulf of Mexico system was being improved to operate effectively in deeper water in the Gulf, it was also being adapted for work offshore in the Middle East, in earthquake-prone California, and in the powerful ice floes of the Cook Inlet in Alaska.[38] As the offshore industry tackled this array of challenging technical problems, there was a sense that the hurricane problem had been contained, if not solved, by research, measurements, and experience.

In these heady years, the stakes grew higher for those working offshore, because the costs of development tended to rise sharply upward as water depths increased. Yet despite this growing economic incentive to build sturdier platforms, many companies refused to depart from traditional practices. Despite a growing consensus on the basic oceanographic issues—wave, wind, and soil mechanics—the "design criteria used by various major oil companies differed by more than 200 percent for the same wave height considerations."[39] On the key issue of deck height, common practices ranged from the use of the 1950s standard of twenty-nine to thirty-two feet above mean Gulf level all the way up past the fifty-foot range by safety-conscious companies such as Calco. The offshore fraternity had generated improved knowledge about the dangers of hurricanes, but absent strong leadership from an industry organization such as the API or government regulation, each company interpreted this data according to its own calculation of acceptable risks. Higher meant safer and more expensive, and each company placed its own bets on the right combination of safety and cost for the particular location and water depth of each particular project.

In 1964 through 1969, a series of devastating hurricanes called these bets. Hilda (October 1964) and Betsy (1965) both measured as "one-hundred-year" storms; then four years later in August 1969, Hurricane Camille, labeled a "four-hundred-year storm," roared through the western Gulf. These three major storms in rapid succession showed conclusively that hindcasters had underestimated the potential frequency and power of severe storms, and that design criteria adopted by much of the offshore industry simply could not withstand the power of major hurricanes.

Hilda was not the largest hurricane to hit the Gulf of Mexico in the postwar years, but it did more damage to the offshore industry than any previous storm. In late September of 1964 Hilda spun into the Gulf and grew into a very scary storm, with winds estimated as high as 150 miles per hour before the storm lost power while moving over cooler waters near shore before making landfall

208   in central Louisiana. Before coming ashore, however, Hilda moved slowly through offshore facilities valued at more than $350 million. In the words of one executive from a company that suffered severe damage, "Instead of spreading out over a big area . . . , she seemed to gather her energy into one tight mass and moved in and really tore things up."[40] When the sun came out after the storm, cleanup crews returning to the evacuated platforms found stunning devastation. Losses reached more $100 million, with thirteen platforms destroyed and five more damaged beyond repair. By scoring a direct hit on an area rich in offshore facilities, this relatively small hurricane delivered a jolt of reality to an industry grown complacent about its capacity to withstand the power of major hurricanes.[41]

One response was a meeting of concerned offshore operators at the Roosevelt Hotel in New Orleans in November of 1965. Sixty-four people attended, including representatives of most of the major oil companies active in the Gulf, the major contractors, gas transmission companies with pipelines in the Gulf, oceanographic consultants, and several university researchers. No organization called the conference; Hilda simply frightened individuals and companies into collective action. Those who had previously been satisfied to go it alone in designing offshore platforms now looked about for help in understanding what had happened and what needed to be done to avoid future catastrophes. Griff Lee, who had been active in offshore design and construction with a major oil company (Humble) and a major contractor (McDermott) since World War II, described the meeting as "a turning point for the industry. Before then, it had almost been every man for himself. This put together a cooperative spirit."[42]

The meeting began with a somewhat apologetic speech by A. H. Glenn, one of the leading weather forecasters and hindcasters employed by the offshore industry. After reviewing the history of Hilda's development, Glenn addressed a question on everyone's mind: What was the practical meaning of the phrase "twenty-five-year storm?" Hilda, labeled a "one-hundred-year storm," differed from previous postwar hurricanes more because of its path and its slow lateral speed than because of the force of its winds or waves. As Glenn lectured the audience about the problems of defining a twenty-five-year or a one-hundred-year wave and the distinctions between a one-hundred-year storm and a one-hundred-year wave, many in the room must have wondered why they had paid so much for so long for forecasts and hindcasts and why they had ever been so confident that hurricane conditions could be accurately predicted.[43]

When Glenn sat down, group discussion began. Representatives of individual companies summarized the amount of damage they had suffered and then described in great engineering detail how the damage had affected the various parts of their platforms. These reports had a somber tone, as those who had ordered platforms and those who had built them traded notes about how Hilda had mangled their handiwork.

Near the end of the meeting Griff Lee took the floor to review "the complete failure" of a major platform that his company, McDermott, had recently built for Union Oil. Lee included a pointed reminder that McDermott had used A. H. Glenn's predictions of the forces generated by a twenty-five-year storm in designing the platform. An examination of the wreckage made it clear that Glenn's estimates had been much too low. Working from severely flawed design data, the company had produced a severely flawed design with a lower deck that, at least in retrospect, had no realistic chance of surviving the fury of Hilda's waves.

The retrospective analysis of the problems with the design of this destroyed platform had a hard practical edge, because its twin had been loaded on a barge awaiting installation at a nearby site when Hilda hit. Lee gave the audience a classic account of engineering on the run, relating how McDermott had carefully studied the destroyed platform to make "some reasonable modifications of the (twin) structure," which it then installed. This was the ultimate wave tank test, using a real hurricane in the real Gulf of Mexico to test design assumptions. With strengthening near the ocean floor, stronger deck legs, and a higher deck, the one-time twin took its place as an only child out in the Gulf, near where the destroyed platform had once stood.[44]

After summarizing the overall destruction of Hilda, Lee concluded with an impassioned call for the industry to change its ways. He noted that all but one of the platforms destroyed by Hilda had been designed to meet the projected forces of a twenty-five-year storm. This meant, in effect, that they had been "designed with the owner accepting a risk." The prevailing attitude was "that the 25-year storm was only going to occur once in the whole Gulf of Mexico every 25 years, and if I'm lucky it will be over by your platform, not by mine."[45] In a speech he subsequently repeated many times at industry gatherings, Lee admonished the group to cut through the uncertainty about wind and wave forces by moving toward design criteria based on the forces generated by a one-hundred-year storm. This meant strengthening platforms, with emphasis on raising the decks, given that Hilda had provided striking evidence of the dangers to platforms when crashing waves "get into the decks." Two practical incentives pushed those present to heed Lee's call for action. The first was economic; the costs of cleanups and repairs were quite high compared to the incremental costs of building stronger platforms. The second was a matter of engineering pride. Good engineers did not like waste and inefficiency, and the images of platforms crumpled over into the Gulf were not ones they cared to see again.[46]

210   Unfortunately, they saw many more less than a year later in September of
1965, when Hurricane Betsy emerged in the Atlantic, crossed Florida, and
moved through an area off the eastern coast of Louisiana containing more than
$2 billion in offshore investments. The storm destroyed eight platforms and
damaged many others. In the massive damage caused by Betsy, one event came
to symbolize the dangers of hurricanes. "Maverick," a state-of-the art jack-up
drilling rig owned by George H. W. Bush's Zapata Corporation and at work on
a project for Calco when Betsy struck, simply disappeared. The future president later received a check for $5.7 million from a New Orleans underwriter
who had placed the insurance for the rig with Lloyd's of London. The offshore
industry as a whole received another unmistakable warning that it had not cor-
rectly understood the risks posed by major hurricanes.[47]

Insurance could ease the financial pain only if insurers continued to accept
the extreme risks of providing coverage for mobile drilling rigs. "Maverick's"
destruction was only the latest in a line of accidents involving such rigs, and
underwriters had begun to revisit the question of whether this segment of the
offshore industry might be uninsurable. A representative of John L. Wortham
& Son, a major Houston-based insurance company, acknowledged that the
"tremendous risks" required "extra efforts" from insurers. Others in the underwriting business continued to debate the basic issue of whether a mobile
drilling rig should be insured as a vessel or as a drilling rig, its workers as "land-
lubbers or seamen." The compromise gradually struck was to make the rigs
safer as they were towed to the drilling site by having inspections of them by
experienced naval architects while they were under construction and then having qualified naval engineers aboard while they were under tow. This compromise satisfied Lloyd's and others, and an insurance crisis was avoided.[48]

Insurance could be used to reduce the risks from losses due to hurricane,
but the prevention of damage through better design and construction was obvi-
ously cheaper and more efficient. Upon further review after the devastation of
Hilda and Betsy, the offshore industry discarded its previous optimism and
reevaluated its traditional approach to the threats posed by hurricanes. Greater
cooperation was needed to define better design standards. More systematic
effort by the offshore fraternity was needed to effectively address problems that
quite clearly had not yet been solved.

The conference after Hurricane Hilda was followed by a similar conference
after Hurricane Betsy, which had dramatically reinforced the calls of Griff Lee
and others for change. At Houston's Rice Hotel in November of 1966, representatives of the offshore industry met to create what became the API's Offshore Committee. Under the auspices of the industry's major trade association,
this committee gradually became a permanent focal point of efforts to define
uniform standards that would limit future damage from hurricanes.[49] From the

wreckage and uncertainty caused by Betsy and Hilda came the institutional response long needed to foster the spread of best practices and the systematic discussion of hurricane-related issues. Here finally was the organizational structure long needed by the offshore fraternity to manage a systematic, continuing effort to define and spread best practices for the industry.

Basic research and measurement of wind, waves, and soil continued, at times in cooperative efforts and at times within individual companies. Shell Oil led the way in the gathering of data on wave heights with a project that placed sophisticated measuring devices on a string of large platforms in the Gulf. These devices could provide real measures to confirm the existing theoretical models of maximum wave heights during severe storms.

Or, as it happened, they could show finally and conclusively that the maximum waves from hurricanes had been consistently and grossly underestimated. During Hurricane Camille in August of 1969 Shell measured waves seventy to seventy-five feet high. This figure stunned offshore veterans who remembered early "expert" predictions that waves in the Gulf would "seldom, if ever, exceed 20 feet." Of course, twenty years of experience and the movement into deeper water had replaced such early guesses with higher and higher figures. But seventy feet made a mockery of the common wisdom about wave heights.

Before Camille ripped apart the region around Biloxi, Mississippi, this monstrous category 5 hurricane passed through a heavily developed offshore region south of New Orleans. Initial estimates of $100 million in property damages raised questions about what the toll might have been had the storm taken a track one hundred miles to the west through the heart of offshore alley. But the "quality," as well as the quantity, of damage drew as much attention as the astonishing reality of a seventy-foot wave in the Gulf. Included among the platforms destroyed were three modern ones installed by Shell, the generally acknowledged leader in offshore design. One of these was only five months old and was at the time the world's tallest fixed deepwater platform.[50]

Suddenly, more than thirty years after the first successful offshore venture in the Gulf of Mexico, Camille had washed up a new design problem. The giant new platform lost by Shell had been designed to withstand one-hundred-year waves, but a mudslide caused by the storm, not wave forces alone, had toppled the structure, which had come to rest on its side some one hundred feet away from its original site. Before 1969, shifting ocean sediments caused by earthquakes had been known to break telephone cables on the ocean floor, and as early as 1950, oceanographic consultants had studied the possibility that unburied offshore pipelines might move during hurricanes. But before Camille, platform designers had not anticipated that, under extreme conditions, mudslides could pose catastrophic threats to platforms. The soil analysis routinely conducted for platform construction simply had not examined this possibility.[51]

Shell's failed platform was in three hundred feet of water in "South Block 70," located offshore from the mouth of the Mississippi River. In retrospect, it was not surprising that the ocean bottom in a region covered by sediments deposited by a large river would be soft and relatively unsettled. Under extreme hurricane conditions—Camille had two-hundred-mile-per-hour wind gusts to go with its seventy-foot waves—such sands could behave almost like a liquid. Shell's studies of the failed platform's site revealed a phenomenon not previously observed by the offshore industry. Camille had dramatically altered the contours of the Gulf of Mexico in South Block 70, lowering the ocean floor and, in effect, moving platforms into deeper water.[52]

This stunning development spurred a race to understand mudslides and to find ways to design platforms to withstand them. The process of change was similar to earlier efforts to generate better estimates of maximum wave heights while at the same time developing practical designs using the best available calculations of maximum wave forces. First came the careful postmortems of the platform that had been swept away in Camille and another one nearby that had been displaced. The information from these studies was placed in the context of the scant existing scientific literature on the frequency and intensity of mudslides. From this starting point, research was undertaken to fill in the wide gaps in information about mudslides. As this research moved forward, preliminary engineering analysis of the forces exerted by mudslides could begin. Design criteria gradually emerged from this analysis, as did the realization that in extreme hurricanes some areas of the Gulf simply might not support platforms built with existing technology.

By 1970 the process of adaptation to hurricanes had reached a turning point. The offshore industry had pushed ahead for a quarter of a century, solving engineering problems on the run when necessary by using the best available estimates of hurricane-generated forces and then adapting these standards after they were called into question by additional research or by damage caused by hurricanes. The three major hurricanes in the 1960s removed much of the uncertainty about the power of severe storms in the Gulf, and the offshore industry responded by taking a hard, collective look at its traditional assumptions.

They did so within two important new venues for cooperation among oil companies, construction companies, academic specialists, and consultants. After its establishment in 1966, the API's Offshore Committee quickly grew into an effective instrument for defining, publicizing, and modifying the best possible standards for offshore operations. The definition of industry standards had been an important part of the work of the API, which was ideally suited to bring together experts from various areas of the industry to share information about best practices. The Offshore Committee simply extended this tradition to matters concerning standards of safety and design offshore. The sharing of basic research on various aspects of offshore operations went forward after 1969 at the Offshore Technology Conference (OTC), which hosted an annual meeting at which industry specialists gathered to present papers about their research. The combination of the API Offshore Committee and the OTC fundamentally altered the generation and exchange of information about managing the risks of hurricanes by creating central new organizations that could reach across all sectors of the offshore fraternity.

Peter Marshall, a Shell engineer who entered the offshore industry in 1962, had a front-row seat from which to observe such changes. He had designed a platform installed in 1965 in 283 feet of water, earning the record for water depth. Two days after its installation, almost before he could brag about his efforts, the platform suffered severe damage during Hurricane Betsy. Examination of the platform revealed pieces of the "Bluewater 1." This semi-submersible drilling vessel had been an epoch-defining technological breakthrough in offshore drilling when built by Shell in the early 1960s. Hurricane Flossie had capsized the vessel in 1964. As a new owner readied it to return to work the next year, Hurricane Betsy displayed a stormy sense of irony by sending the "Bluewater 1" careening into its former company's record-holding platform.[53] Almost forty years later, Marshall could recall this episode with a laugh.

As Marshall continued his work for Shell, he witnessed sharp changes in the approach of the offshore industry in its formative years and in the years after the 1960s. With the coming of computer-assisted design, "Intuitive design and an entrepreneurial spirit gave way to computers and an era of no surprises." Marshall summarized the key change in attitude with the simple declaration that "we were less afraid of failure then." He lamented the passing of the days when offshore engineers had been given greater latitude to do their jobs more creatively while accepting greater risks.

These comments point to two critical changes that gradually occurred in the 1940s through the 1960s. "Intuitive design" was unavoidable in an era when great uncertainties remained about the design forces generated by hurricanes. With more experience and the accumulation of better information about wave, wind, and soil, more precise calculations could be made. From a design perspective, the "era of no surprises" was a definite advance; risks could be calculated with much greater certainty once big surprises had been eliminated. As noted by Marshall, computer-assisted design completed this transformation in the calculation and management of the risks of hurricanes. In all phases of offshore design and construction, computers allowed engineers to make better, faster calculations using the growing body of data available to them. Of course, even the growing power of computers could not match the power of major hurricanes, which revealed with awesome efficiency weaknesses in offshore structures.

POSTSCRIPT: HURRICANES KATRINA AND RITA

Events in recent years reinforce the lessons of the past while placing them in a new historical context. Indeed, the response to the devastation offshore from Hurricanes Ivan, Katrina, and Rita in 2004 and 2005 recalls the hand-wringing after the three devastating hurricanes of the late 1960s. Like the severe storms of the 1960s, this recent wave of storms came after an extended period of relative quiet in the primary producing regions offshore in the Gulf. In eerie similarities between past and present, Katrina closely followed the route inland of Camille and the storm surge from Rita once again flattened Cameron, Louisiana, as had Audrey in 1957. Katrina and Rita serve as the most recent reminders of the severe challenge of sustaining innovation to continue to meet the force of hurricanes as the offshore industry evolves.

Although the underlying process of technical adaptation remains the same, much has changed in the years since Camille. The offshore industry in the Gulf of Mexico has grown dramatically along the Louisiana and Texas coast, with almost four thousand platforms operating in the Gulf compared to about one-fourth that number in the 1960s. In that decade, platforms in three hundred feet of water were viewed as "deepwater projects," whereas in 2005 the deepwater platforms damaged by hurricanes were in six thousand feet of water. To move that far out into the Gulf required far-reaching changes in design, including the substitution of semisubmersible platforms tethered or moored to the bottom of the ocean for those literally pinned to the bottom with piles. Much of the operation of the new deepwater platforms is done by subsea systems using robotics, while basic production facilities are placed on the decks of traditional fixed platforms. All such technological changes, as well as the location of the newest deepwater platforms far from shore in more than two miles of water, have raised new sets of issues on the potential impact of hurricanes.

Questions about the hurricane-worthiness of the newest generation of deepwater platforms could not be avoided after the publication of a series of stunning photographs in July 2005 of BP's massive Thunder Horse platform, at the time the largest semisubmersible production platform ever constructed. Photos on the front page of newspapers around the nation captured the impact of a direct hit by Hurricane Dennis, a relatively small storm. This much-publicized, billion-dollar platform listed dangerously, bringing to mind Shell's state-of-the-art platform that disappeared from sight in 1969 during mudslides caused by Camille. While awaiting repairs after being righted, the giant platform rode out the high winds and waves of Katrina and Rita, two major hurricanes which swept through the Gulf of Mexico later in 2005. As with any major project using relatively new technology, Thunder Horse had been tested in its new operating environment. In this case, an unforeseen vulnerability had been revealed, forcing the first production of its much-needed 250,000 barrels per day of oil to be pushed back by three years. Initial production finally began from Thunder Horse in June of 2008, with full production expected by the end of the year.

The postponement of Thunder Horse's production points to a fundamental change in the oil industry since the 1960s, one that greatly affects the attitude toward hurricane risks. Although the total production of oil from the Gulf of Mexico had increased to the 1.5-million-barrel-per-day mark since the 1960s, the growing importance of offshore production in a very tight energy market has made the nation much more vulnerable to the shutdown of offshore producing properties. Because of their size and trajectories, the one-two punch of Katrina and Rita, which plowed through offshore alley in rapid succession, affected approximately three-fourths of the existing platforms in the Gulf. The shutting down of production to prepare for the hurricanes followed by extensive damage to platforms, drilling rigs, and pipelines led to the shutting in of perhaps 1.2 million barrels of oil production a day for an extended period. This came at a time of significant oil shortages and sharply rising oil prices. Repair work was made more difficult by the widespread damage onshore to support industry and refineries. As in the 1960s, the severe storms of 2005 emphasized the oil industry's vulnerability to the risks of hurricane damage. But now the offshore industry has to manage not only the technical risks but also the economic and political risks posed by disruptions of production in a time of energy scarcity.

Looking back on the management of hurricane-related risks from the perspective of fifty years of work on offshore structures, Griff Lee offered a sobering appraisal that suggests how little the industry knew as it plunged into the Gulf of Mexico: "In light of today's data, the early load estimates were off (too low) by a factor of ten." The offshore oil industry of the post–World War II era nonetheless survived and prospered, overcoming its initial ignorance of the force of hurricanes through a combination of unusually good weather, extraordinary technical innovations, and the practical experience of able engineers working in a variety of oil-related organizations. By observing the impact of hurricanes over several decades, the offshore industry gradually developed both the organizations and the technical expertise needed to understand and then to better manage the risks posed by hurricanes. By regularly adapting to new technical challenges revealed by hurricane damage, this evolving fraternity of offshore specialists continued to find and produce much-needed domestic oil

216   and natural gas reserves from the Gulf of Mexico.[54] Armed with new technologies but facing intense public pressure in an era of severe energy shortages, today's generation of offshore specialists continues to venture deeper out into the Gulf of Mexico in spite of the ever-present dangers of one of the most powerful and unpredictable forces of nature, the hurricane.

NOTES

1. For an overview of the history of the offshore oil industry, see Hans Veldman and George Lagers, *50 Years Offshore* (Delft, Holl.: Foundation for Offshore Studies, 1997). See also Tyler Priest, *The Offshore Imperative: Shell Oil's Search for Petroleum in Postwar America* (College Station, Texas: Texas A&M University Press, 2007); and F. Jay Schempf, *Pioneering Offshore: The Early Years* (Houston: Offshore Energy Center, 2007). Joseph A. Pratt, Tyler Priest, and Christopher Castaneda discuss this history from the perspective of one of the largest offshore construction companies in *Offshore Pioneers: Brown & Root and the History of Offshore Oil and Gas* (Houston: Gulf, 1997).

2. For an overview of these experiences, see Veldman and Lagers, *50 Years Offshore*, 13–25.

3. Griff Lee, interview by Joseph A. Pratt, June 13, 1996, Offshore Energy Center Collection, Houston, Texas, 1–5 (hereafter cited as OECC).

4. This attitude of "can do" engineering is illustrated throughout Pratt, Priest, and Castaneda, *Offshore Pioneers*. Government regulation of offshore activity before the 1970s came from a variety of agencies, none of which exercised strong control. Both the state and federal government had authority to lease offshore lands. The Army Corps of Engineers held the power to issue construction permits for projects in navigable waters, and it required offshore companies to clearly mark their platforms and to dismantle them once they were no longer in use. The Coast Guard had authority over safety and limited powers over oil pollution.

5. Lawrence S. Tait, "Hurricanes . . . Different Faces in Different Places," paper presented at the 17th Annual National Hurricane Conference, Atlantic City, New Jersey, Apr. 11–14, 1995.

6. Henrietta Larson and Kenneth Wiggins Porter, *History of Humble Oil & Refining Company* (New York: Harper & Brothers, 1959), 422, 433. Also, Henrietta Larson and Kenneth Wiggins Porter, "Drilling Wells Off Shore in Texas Bays and Inlets," *Oil & Gas Journal* (Apr. 14, 1938): 113.

7. I. W. Alcorn, "Derrick Structures for Water Locations," *Petroleum Engineer* (Mar. 1938): 33–37.

8. I. W. Alcorn, "Marine Drilling on the Gulf Coast," paper presented at API-Southwestern District, Drilling Division, Drilling and Production Practices, Fort Worth, Mar. 24–25, 1938.

9. "First Well in Gulf of Mexico Was Drilled Just 25 Years Ago," *Offshore* (Oct. 1963), 17–19.

10. For profiles of several of these weather officers, see Schempf, *Pioneering Offshore: The Early Years*, 139–40. See also Robert Reid, interview by Malcolm Sharples,

Houston, Oct. 17, 1998, OECC; Curtis Crooke, interview by Tyler Priest, Houston, Oct. 6, 2001, OECC; John A. Focht, interview by Tyler Priest, Houston, Oct. 6, 2001, OECC; and Bramlette McClelland, interview by Tyler Priest, Houston, October 6, 2001, OECC.

11. C. H. Siebenhausen Jr., "Outline of Notes for a Shell History," document in OEC Archives, Houston, Texas.

12. Dean A. McGee, "Exploration Progress in the Gulf of Mexico," *Drilling* (May 1949): 50–53, 117–20. See also Dean A. McGee, "Magnolia Testing Offshore Formations in the Gulf," *World Petroleum*, (Mar. 1947): 60–61; and Dean A. McGee, "Giant in the Gulf," *The Humble Way* (Jan.-Feb. 1948): 15–17.

13. Pratt, Priest, and Castaneda, *Offshore Pioneers*, 21–30; and "LSTs Help Drill for Oil," *The Humble Way* (July-Aug. 1948): 6–7.

14. "Gulf of Mexico Oil Play," *Shell News* (Oct. 1949), 4–9.

15. "LSTs Help Drill for Oil," 6–7; C. E. Kolodzey, technical paper, Humble Oil, Apr. 23, 1954, OECC.

16. OEC, *Offshore Pioneers*, 34–36.

17. R. O. Reid, interview by Malcolm Sharples, Houston, Oct. 17, 1998, OECC.

18. E. G. Ward, interview by Tyler Priest, Houston, Oct. 17, 1998, A. H. Glenn folder, OECC.

19. F. R. Harris and H. G. Knox, "Marine Construction: Some Important Considerations," *Oil & Gas Journal* (Oct. 18, 1947): 131.

20. R. C. Farley, "Hurricane Damage to Drilling Platform," *World Oil* (Mar. 1950): 85–92; and M. B. Willey, "Structures in the Sea," *Petroleum Engineer* (Nov. 1953): B-38–47.

21. C. Paul Besse, interview by Joseph A. Pratt, New Orleans, Sep. 2000, OECC, 9–11; and "Drilling Program Moves Ahead," *The Calco News* (Sept. 1948): 1.

22. C. Paul Besse, interview by Pratt, 12.

23. Pat Dunn, interview by Joseph A. Pratt, Columbus, Texas, July 1, 1996, 18.

24. Jack Toler, "Offshore Petroleum Installations," *Proceedings of American Society of Civil Engineers* 79 (Sep. 1953): 289–95.

25. There is a well-developed historical literature on mobile drilling vessels. See Veldman and Lagers, *50 Years Offshore*, 49–58.

26. J. E. Pike, "An Underwriter Looks at Insurance for Offshore Drilling," *Drilling* (May 1949): 49, 108–9.

27. M. B. Willey, "Structures in the Sea," B-43–47.

28. Alden "Doc" LaBorde, interview by Joseph A. Pratt, Houston, Oct. 17, 1998, OECC.

29. John Persinos, "I Am Not a Compromising Woman," *Rotor & Wing* (Feb. 1999): 39; John Persinos, "Offshore Airlift," *The Humble Way* (Mar.-Apr. 1957): 14–21; and John Persinos, "Offshore Operators Gear for Decade of Steady Growth," *Petroleum Week* (Feb. 19, 1960).

30. "Offshore Weather Forecasting," *The Calco News* (Oct. 1949): 3–4.

31. "Storm at Grand Isle," *The Humble Way* (Jan.-Feb. 1956): 8–21.

32. "A 'Sea Story' of the LST S-24," *The Calco News* (Dec. 1956): 3; and Don Lambert, "Offshore Operators Look at Flossie's Damage," *World Oil* (Nov. 1956): 73–75.

218   33. R. F. Bailey, "Progress Report on the Serviceability of Used ST Mooring Chain Cables," Humble Oil Report, Feb. 1958, OECC; and James Calvert, "Gulf Offshore Activity Booming," *World Petroleum* (June 1957): 48–51.

34. "Forethought Minimized Hurricane Damage to Offshore Installations," *Offshore Drilling* (Aug. 1957): 15–18, 25; and "Third Offshore Evacuation," *Offshore* (Oct. 1957): 21.

35. "Forethought," 25; and "Weather Compounds Offshore Risks," *The Humble Way* (July-Aug. 1957): 8–9.

36. James Calvert, "The Mobile Rig Disasters," *World Petroleum* (June 1957): 30–33.

37. Herbert Riehl, "The Hurricane," *Drilling* (Aug. 1957): 65–69; and Mercer Parks and Herbert Riehl, "Hurricane Formation in the Gulf of Mexico," Southern District, API Division of Production, Mar. 1963.

38. Pratt, Priest, and Castaneda, *Offshore Pioneers*, 95–179.

39. Griff Lee, "Offshore Platform Construction to 400-foot Water Depths," *Journal of Petroleum Technology* (Apr. 1963): 384.

40. Transcript, Hurricane Hilda Damage Conference, New Orleans, Nov. 23–24, 1964, 3–4. Copy provided by Griff Lee.

41. "Betsy's Damage Will Surpass Hilda's," *Offshore* (Oct. 1965): 26–28.

42. Hurricane Hilda Damage Conference, 75–78; Transcript, Hurricane Andrew Structural Performance Information Exchange, API Meeting, Washington, D.C., Oct. 29, 1992, 5–7. Copy provided by Griff Lee.

43. Hurricane Hilda Damage Conference, 5–8.

44. Hurricane Hilda Damage Conference, 75.

45. Hurricane Andrew Structural Performance Information Exchange, 6.

46. Hurricane Hilda Damage Conference, 76.

47. "The Future of the Offshore Drilling Industry," *Drilling* (Nov. 1965): 46–48.

48. Herbert Kuhlmann, "Insurance Problems Mount with Offshore Operations," *Drilling* (Aug. 1956): 74–75; Herbert Kuhlmann, "Insurance: A King-Sized Marine Drilling Problem," *Drilling* (Aug. 1957); and L. K. Griffin, "Insurance Savings: Offshore Experience May Hold the Key," *Drilling* (Oct. 1959): 57, 131.

49. Griff Lee, interview by Pratt, 27–29.

50. "Camille Knocks Out 300,000 b/d and Costs Industry $100,000," *Offshore* (Sep. 1969): 33–35.

51. Robert O. Reid, "Oceanographic Considerations in Marine Pipe Line Construction," *Gas Age* (Apr. 26, 1951): 1–6; R. G. Bea, "How Sea-floor Slides Affect Offshore Structures," *Oil & Gas Journal* (Nov. 29, 1971): 88–91; and John Focht, interview by Tyler Priest, Houston, Oct. 6, 2001, OECC, 10–11.

52. R. G. Bea, "How Sea-floor Slides Affect Offshore Structures," 89.

53. Peter Marshall, interview by Joseph A. Pratt, Houston, Sep. 2002, OECC.

54. Overall, the offshore industry had more serious safety problems in such areas as the development of deep-water diving and blow-outs of offshore wells, especially in the early years, when mobile drilling rigs also presented problems in rough seas.

7

MEMENTOS: LOOKING BACKWARDS AT THE HONDA MOTORCYCLE CASE, 2003–1973

Christopher McKenna

POST-SCRIPT

The nouveau-noir film *Memento* opened to rave reviews in 2001.[1] *Memento*, adapted and directed by British filmmaker Christopher Nolan, told the story of a man searching for his wife's killer. Unfortunately, the husband had been injured in the attack. His injury was extraordinary and specific—a blow to the head had destroyed his ability to form new memories. Although the husband remembered precisely what had happened prior to the attack, everything that he witnessed after the attack soon vanished from his short-term memory. The protagonist's solution to his constant memory loss was to carry a Polaroid camera to capture those images that he thought were important; to use a notebook to jot down potential reminders; and, not least of all, to draw a series of tattoos over his body to record his quest—the eponymous "mementos" in the film's title. If this plot was not jarring enough, Nolan's narrative structure would make the film an instant favorite of cinephiles; for instead of following the usual chronological timeline, viewers were forced to follow the story *backwards*. The movie moved in reverse, through fifteen-minute increments, trailing the distraught husband from an opening scene where he brutally executes a man we presume to be his wife's killer through successive narrative sweeps, each one ending where the previous segment finished. The result was an unsettling narrative divorced from a traditional timeline yet uniquely dependent on the underlying historical chronology.[2]

Although it may not be immediately obvious, Christopher Nolan's message in *Memento* is particularly relevant to scholars of corporate innovation—that merely knowing the end of the story does not guarantee that one understands

220   the underlying causation. There was also a second message in the movie, however, of which academics should be equally cognizant—that a single document or "memento" is a precarious basis upon which to make conclusions and from which to judge ultimate causality. As thoughtful scholars are only too aware, many accumulated small truths do not necessarily aggregate to one big truth. Unfortunately, as mementos accumulate, so too does the certainty that only one explanation is right and, as a result, looking beyond these immediate conclusions becomes ever-more difficult. In the end, the fragments become the story itself, misdirecting subsequent attempts to analyze the original event. What was once truly novel and important is eventually lost under the weight of accumulated half-truths.

This chapter takes its narrative structure from the film *Memento*, moving backwards to strip away the accumulated mementos of a particularly overdetermined business case study of corporate innovation: The Boston Consulting Group's *Strategy Alternatives for the British Motorcycle Industry*, published in 1975.[3] Over time, as we shall see, the management consulting firm's report, and the subsequent mementos, became the basis for literally a dozen different scholarly reinterpretations of the competitive and corporate strategies pursued by the Japanese motorcycle manufacturer Honda in its conquest of the American market. My purpose here is not simply to recontextualize the analysis—for each series of mementos has done just that—but to peel away the accumulated layers of analysis in order to return to the original problem. In doing so, I seek to explain how the original problem came to be buried under a series of didactic mementos created by each succeeding generation and why this example has become, in the words of one prominent academic, "the perfect juxtaposition of two versions of the same story."[4]

CONCLUSION

That this chapter should be included within a volume concerned with the challenge of remaining innovative is apt given that the history of Honda's conquest of the American motorcycle market is a particularly revealing case study of business-government relations, the disciplinary behavior of professionals, and the creation of new international markets. The rise of Honda from a regional producer of underpowered scooters to an international automotive giant remains a powerful illustration of a dynamic, innovative industrial firm. Moreover, in an era when advocates of globalization are often focused on the development of new markets outside of the United States, Honda's exploitation of a novel "niche" market for its smaller motorcycles is a reminder that not all new markets need be within developing countries.

This chapter, however, is not simply about the innovations that the executives within Honda pursued but also intended to remind others of the potential for innovation in writing scholarly narratives. To be more direct, the reverse chronological structure of this chapter is not an accident but rather a self-conscious reaction to the formulaic structure of most business history scholarship. So common is the aversion to this style that best-selling books such as *Liar's Poker* and popular films such as *Being John Malkovich* have parodied traditional business history as "clumsy fascist propaganda."[5] Business history, however, need not be formulaic in its literary style. Even the late Alfred Chandler, who was never a self-conscious stylist, employed an intensely hierarchical structure in his academic writing that provided a self-conscious parallelism to Chandler's broader theoretical argument about the importance of managerial structures in achieving strategic goals.[6] In this chapter, running the narrative backwards highlights the discontinuous nature of scholars' episodic arguments about the Honda Motorcycle case; thus my theoretical arguments about the underlying sources of innovation are intimately connected with the chapter's novel form. And innovative narrative structures have value even beyond the information conveyed: as *The New York Times* reviewer noted, perhaps the best reason to watch the movie *Memento* "is for the disorienting pleasure of its unusual narrative technique." There are many narrative structures that writers can choose to adopt, and one of the intended conclusions of this chapter is that academics should consider employing unusual narrative structures to better illustrate their particular theoretical arguments.

At the theoretical level, this chapter argues that a succession of "mementos" has plagued the academic analysis of Honda's competitive strategy and corporate innovations for more than thirty years. The six written "mementos" that resulted from the five distinct temporal periods are Andrew Mair, "Learning from Honda" (1999); Henry Mintzberg, Richard Pascale, Michael Goold, and Richard Rumelt, "The 'Honda Effect' Revisited" (1996), which also reprinted Henry Mintzberg, "Learning 1, Planning 0" (1991); Richard Pascale, "Perspectives on Strategy: The Real Story Behind Honda's Success" (1984); Dev Purkayastha, "Note on the Motorcycle Industry—1975" (1978); and The Boston Consulting Group, *Strategy Alternatives for the British Motorcycle Industry* (1975).[7] As each successive generation has used "mementos" from earlier eras to reinterpret Honda in light of their contemporary perspective, the successive accumulation of artifacts has made each subsequent round of analysis ever more dependent on previous interpretations. Looking backwards, over the past thirty years, it is clear that successive generations of scholars have grafted the imperatives of postmodernism, disciplinary debates, Japan's presumed economic superiority, case method teaching to M.B.A.s, and the crisis of national industries onto their historical analyses of innovations in motorcycle production. Because these mementos have accumulated like successive sedimentary layers of silt, the only way to reach solid bedrock is to dig through the layers into the past. Although the most recent set of "answers" may be initially appealing, they are also based on the greatest number of accumulated half-truths that, in turn, has meant that they are built on the shakiest of theoretical foundations. To see just how shaky the scholarship has ultimately become requires us first to consider the final section of the debate—where it really all begins.

STRATEGY AS NARRATIVE: 1997–2003

In 1997, only a year after the *California Management Review* published a special issue reviewing the debates over the Honda motorcycle case, the conflicting arguments put forward in its retrospective were already serving as a prime example for an emerging group of postmodern scholars of management.[8] Ever since the academic study of management emerged in the late nineteenth century through the writings of Frederick Taylor and Henri Fayol, business school academics, like their predecessors in schools of commerce and industrial relations, had often described management as a "science," even if others continued to regard it otherwise.[9] For example, leading disciplinary journals such as *Organizational Science* and *Administrative Science Quarterly* still require authors to present their "findings" complete with their preliminary "hypotheses" and "methodology," in order to permit their "results" to be systematically checked. By 1997, however, postmodern deconstruction had crept into the "science" of management, and authors such as Derrida and Foucault, more familiar to literary theorists than business school professors, were beginning to storm the last bastion of the academy still untouched by linguistic analysis: corporate strategy. For the first time, academic strategists were describing their professional field as a form of narrative, dependent not only on the ideological constraints of national culture but also on the status of power-laden disciplinary disputes within the profession.[10]

By 1997, after two decades of debate over the source of Honda's competitive success, the facts seemed clear and the standard interpretation was divided into two opposing, if friendly, camps. But the postmodern strategists had a solution to this apparently irresolvable division—they bypassed the ossified analysis of their predecessors by recasting the Honda story as a tale of the disciplinary strategies of the individual academics who had constructed the competing interpretations. Thus their reinterpretation of the case was to describe the debate over Honda as a "story." In this new interpretation, Honda became not only the best-documented example of an innovative corporate strategy but also an introduction for students to the nature of academic debates within the discipline.[11] For example, the Honda case became so important to teachers that it was the very first case study included in Bob De Wit and Ron Meyer's textbook from 1998, *Strategy: Process, Content, Context—An International Perspective.*[12] That De Wit and Meyer would use Honda as a case study was not unusual, for Honda had been the source of several iterations of business school case studies ever since Harvard Business School's original "Note on the Motorcycle Industry—1975," written by Dev Purkayastha under the direction of Professor Robert Buzzell in 1978.[13] From 1997 onward, however, the difference was that case writers no longer focused exclusively on Honda but began to build their narrative around the competing interpretations of the case offered by different strategists—the De Wit and Meyer case, written by Andrew Mair, was titled: "Reconciling Managerial Dichotomies at Honda Motors."[14] Similarly, even though Clayton Christensen, a faculty member at the Harvard Business School, only wanted to use the specific example of Honda's expansion to the United States to illustrate his larger point that Honda's small motorcycles represented an innovative "disruptive technology," Christensen felt the need to set out the two competing histories in his best-selling book from 1997, *The Innovator's Dilemma.*[15] Like the postmodern anthropologist in James Hynes's satirical novel *Publish and Perish,* who revolutionized his discipline by arguing that the proper subject of anthropology was not the study of primitive peoples but the study of anthropologists, so too had the study of competitive strategy moved from an analysis of business strategies to the study of business strategists.[16]

Once strategists had seized the story of the dispute as the central lesson from the Honda case, any attempt at reconciliation or a solution to the underlying dispute appeared pointless. The primary lesson, after all, was the academic debate. As Andrew Mair argued, in the case of Honda, "reductionist one-sided theories . . . only deepened when strategy thinkers debate 'the meaning of Honda.'"[17] That the form of the narrative could overwhelm the underlying story was not unusual, but in this example retaining both sides of the narrative became not only the literary but also the literal solution for these postmodern strategists. Thus Andrew Mair's answer to the complicated narrative of Honda was not to eliminate the apparent duality—to argue that one side was wrong, that the debate was meaningless, or even that Mair had conclusive proof that one side was right—but instead to argue that the management within Honda "uniquely" embodied "a dichotomy-reconciling approach to management thinking in which both poles are retained."[18] The postmodern solution to the Gordian knot was not to cut the rope but to bronze its ornamental design. Honda, it seems, was no longer an exemplar of corporate success because of either "learning" or "planning" but because of its innovative ability to manage intellectual dichotomies. Thus Honda had created, in Mair's new formulation, "a dichotomy reconciling strategic capability."[19] Those of us who are sensitive to "undergraduate grey"—the tendency of undergraduates to try to solve complex academic debates by taking the middle ground—would have immediately perceived Andrew Mair's solution as a novel form of "postmodern grey."

This memento from 1999 is more symbolic than definitive—Andrew Mair's article has not yet become an artifact to rank alongside the previous mementos of Honda—yet Mair, like the authors of the various mementos before him, caught the tenor of the times. Despite the widespread perception among academics that the original explanation of Honda's strategy is clearly wrong, the Honda case study remains a best-seller at the Harvard Business School Press. There is, however, no longer just one case to assign, but two—Honda (A) and Honda (B)—both of which present radically different interpretations.[20] Moreover, should a business school faculty member be tempted to resolve the conflict—to try to solve the puzzle—the accompanying teaching note explicitly cautions them against discussing "which is right," because

> [t]he cases were not meant to offer either view as the right interpretation. Rather, they were meant to lead students to a more sophisticated understanding of the value for a manager to have both perspectives when setting out to implement strategic change."[21]

Unlike most Harvard case studies in which the teaching note includes overheads that highlight financial or statistical data, the only slide in the accompanying teaching note for Honda is an optical illusion (exhibiting "rival-schematic ambiguity") in which one can alternately see the picture of an old woman or a young lady—a visual metaphor in which, as the teaching note explains, "like the Honda cases, both are valid, both coexist, and each has separate value."[22] The most recent memento from Honda is ultimately a postmodern illusion that gives Richard Rumelt's description of "the many faces of Honda" an unintended *double entendre* for the next generation of academics trying to untangle the ongoing riddle.[23] The true meaning of Honda, in its current iteration, is a Rorschach test in which any reader's interpretation is necessarily the correct answer. Perhaps, however, more traditional academics might be unhappy with this postmodern solution; in which case, let's skip back in time to 1990.

PICK A POSITION, BECOME A STRATEGY GURU, 1990–1996

The early 1990s were a very good period for what Richard Pascale dubbed the "strategy" industry.[24] The revenues of the elite strategy consulting firms grew at a double-digit rate for more than a decade while consultancies such as McKinsey,

A. T. Kearney, Monitor, Bain, and BCG struggled to hire new M.B.A.s to meet the surging demand for their services.[25] Similarly, the leading business schools in America and in Europe found it increasingly difficult to attract and retain faculty as the lure of consulting fees, speaking engagements, and lucrative book contracts became ever harder to resist.[26] For academics perennially taunted by the question, "If you're so smart, why aren't you rich?" the criticism that strategists simplified their theories in order to repackage them for a wider audience seemed less damning than rational—at $25,000 a day, who needed academic credibility?[27] The answer, of course, was that the top academic strategists struggled both to retain academic respectability and to serve as successful consultants. At Harvard, Michael Porter and Rosabeth Moss Kanter both sought to enhance their academic reputations even as they maintained active ties with their Boston-based consulting firms.[28] Similarly, John Kay, then teaching economics at the London Business School, also served as chairman of his consulting firm, London Economics.[29] Yet these academics were only too aware of the apparent conflict of interest often voiced by their academic colleagues— for as academics were prone to wonder: If strategists were so rich, could they really be that smart?

For Henry Mintzberg, a professor of strategy at McGill University in Canada and at INSEAD in France, the answer was clear—being smart was more important than being rich.[30] As Mintzberg explained in the introduction to his 1994 book *The Rise and Fall of Strategic Planning*, "a certain tone of cynicism pervades much of this book," and that cynicism was aimed at his colleagues who valued easy answers over messy realities.[31] In particular, being smart was always a game for Mintzberg within the academic discipline of strategy, for Mintzberg loved to demolish existing managerial myths, to point out that managers did not behave like the hyper-rational strategists described by Alfred Chandler and Igor Ansoff. In particular, he insisted that corporate strategy did not necessarily proceed down an orderly path crafted largely through logic as the rival strategists at Harvard had long argued.[32] Strategies, in Mintzberg's formulation, were as often "emergent" as they were "planned"—a stream of actions that appeared to have a pattern only in retrospect. At the heart of Mintzberg's distinction was his insistence that a vast number of apparently brilliant business strategies were not truly "intended" and that "we often fool ourselves, as well as others, by denying our subconscious motives."[33]

Although Mintzberg first introduced his heretical view of strategy in 1987, it wasn't until 1990 that his position became deeply polarized, when Mintzberg and Igor Ansoff, one of the acknowledged founders of the discipline, debated Mintzberg's redefinition of strategy in the leading journal, and Mintzberg claimed victory almost before the fight had begun.[34] Perhaps not surprisingly, Mintzberg's

226   biggest gun in this war for intellectual leadership of strategy would be the example of Honda. As Mintzberg argued first in 1991, and then again in 1996:

> I would like to introduce just one fact here. In one sense, it is the only real fact I know in all of the literature of strategic management. . . . It is Richard Pascale's account by several Honda executives about how they developed on site the strategy that captured two-thirds of the American motorcycle market. What is especially fascinating about this messy account is that it stands in sharp contrast to the brilliantly rational strategy imputed to these executives by BCG consultants who apparently never bothered to ask.[35]

To claim his *coup de grace* against Ansoff, Henry Mintzberg used only three references: Richard Pascale's 1984 article on Honda, BCG's report on the British motorcycle industry from 1975, and a passage from Ansoff's classic text, *Corporate Strategy*, from 1965—thus his "single data point" was supported by just three citations.[36] Mintzberg was so swayed by the narrative power of Pascale's original account that he organized, in 1996, a special issue of the *California Management Review* to revisit the "Honda Effect." This special issue would become a memento of Mintzberg's celebration of Honda as the ultimate example of emergent strategy—that single data point upon which Mintzberg concentrated his considerable energies.

The problem, of course, for Mintzberg was that in order to better illustrate his larger point, he allowed Pascale's description of Honda to stand in for his carefully researched database of emergent strategies within Canadian companies. Moreover, once Mintzberg had accepted this metonymy and was intent on "scoring points" from Igor Ansoff, the common elements between planned and emergent strategies disappeared as Mintzberg emphasized the more obvious differences. Unfortunately, Mintzberg had never done any historical research on Honda and it showed—in one glaring *faux pas*, Mintzberg expressed surprise that the academic strategist Michael Goold had identified himself as one of the three authors of the BCG report despite Goold's name being prominently listed among the authors on the title page of the original report that Mintzberg had himself frequently cited.[37] The problem, however, was not only that Mintzberg was not familiar with the specific details of the Honda case, but also that his own legendary sensitivity to more subtle analysis began to break down when he sought to buttress his theory of emergent strategy at the expense of a fuller explanation of Honda's management.[38] As Andrew Mair later argued, "product differentiation leads to polar opposite 'strategy-bites.'"[39] Indeed, in one of the most telling exchanges, Mintzberg failed to accept the olive branch offered by Goold and Rumelt, who sought to create a middle ground linking the two opposing explanations. It would have been possible, for example, for Mintzberg to argue that Honda's middle managers in America pursued a haphazard strategy that was not directly supervised by Japan (and hence "emerged" from the Tokyo executive suite), but was reasonably coherent and rational at the business-unit level in Los Angeles (and thus planned by the regional managers). If this were true, Mintzberg and Ansoff would have both been right. Such a solution is not unreasonable given that, in 1995, Ikujiro Nonaka and Hirotaka Takeuchi argued in their highly influential book *The Knowledge Creating Company* that Honda thrived precisely because of its innovative ability to encourage middle managers to "serve as a bridge between the visionary ideals of the top and the often chaotic reality of those on the front line of business."[40]

Henry Mintzberg's decision to transform Honda into a one-dimensional example in order to buttress his argument may well have earned him the admiration of management guru Tom Peters; however, the lasting result of his obstinacy would be a stylized memento that caricatured Richard Pascale's revisionist account without adding anything new to what was already known about competition in the motorcycle industry during the 1960s and 1970s.[41] At its core, Mintzberg's account was still based exclusively on Richard Pascale's 1984 article, and after twelve years of extensive debate, he had not added anything substantial to the original revision. Mintzberg had simply transformed Pascale's account into a polarized clash between "learning and planning."

EXPLAINING JAPAN, BEMOANING STRATEGY, 1984–1989

Richard Pascale's article on Honda in 1984 was a product of his broader research into the cultural explanations for the Japanese postwar industrial miracle—a very topical question during the 1980s.[42] In 1981, Pascale, with coauthor Anthony Athos, published *The Art of Japanese Management: Applications for American Executives*, which, alongside Tom Peters and Robert Waterman's book *In Search of Excellence* and Terence Deal and Allan Kennedy's influential *Corporate Cultures: The Rites and Rituals of Corporate Life*, became the three studies that popularized the concept of corporate culture.[43] There was, however, an unrecognized irony in the fact that Richard Pascale, who would later bemoan the influential role of Western strategy consultants, had himself been supported by the consulting firm of McKinsey & Company during his research on Japanese management from which he took his subsequent analysis.[44] While we can't infer intentionality from the outcome, the end result was that Pascale used research funded by McKinsey to attack the best-known strategy report of McKinsey's archrival, The Boston Consulting Group.[45] For The Boston Consulting Group, Pascale's critique came at a particularly difficult time. BCG was already under attack in the press, from analysts who were increasingly critical of the firm's "tired" theoretical models; from old and new competitors following the defection of BCG

228   partner William Bain to found his own consulting firm, Bain & Company; and from within the firm as some of the partners in BCG grew resentful of the "excessive commercialism" of the new managing director, Alan Zakon, who took over in 1980 following the retirement of the firm's founder, Bruce Henderson.[46] Pascale's critique of the power of the elite strategy consulting firms came at the very moment when BCG was at its weakest. Indeed, the fact that Pascale's attack followed devastating coverage of The Boston Consulting Group in *Fortune* magazine meant that the partners at BCG saw little reason to respond; the first semi-official explanation would come from former BCG consultant Michael Goold eight years later.[47]

If Pascale's critique was motivated by two underlying concerns—an intense desire to explain the Japanese miracle and a growing sense that the "strategy" industry (including consultants, academics, and the business press) had it wrong—he appeared genuinely surprised by the alternative version that he uncovered. As Pascale recounted in his 1984 article in the *California Management Review*, while talking with the Honda executives in Japan who had "masterminded" the export of small Japanese motorcycles to the United States in the early 1960s, he complimented them on their brilliant strategic planning. When the executives replied that the actual process had been riddled with false starts, strategic blunders, managerial oversights, and serendipitous outcomes, Pascale soon realized that the case study in corporate innovation that he had been teaching his students in the business school at Stanford was simply not true. As Pascale explained:

> History has it that Honda "redefined" the U.S. motorcycle industry. In the view of American Honda's start-up team, this was an innovation they backed into—and reluctantly. It was certainly not the strategy that they embarked on in 1959.[48]

It turned out that Honda's infamous advertising campaign, "You Meet the Nicest People on a Honda," had been a source of some disagreement before it was adopted, "thus, in 1963, through an inadvertent sequence of events, Honda came to adopt a strategy that directly identified and targeted a large, untapped segment of the marketplace that has since become inseparable from the Honda legend."[49] Pascale would also describe, in great detail, how the Honda executives had first tried to sell large American-style motorcycles until, short of cash and unable to fix the faulty larger bikes, they reluctantly sold their small "super cub" motorcycles and surprisingly discovered a new market for these smaller motorcycles in America.[50]

It is telling that Richard Pascale quoted at length from the Japanese executives and also at equal length from the Harvard Business School case, but only sparingly from the original BCG report.[51] Indeed, although Pascale attacked the "inflexibility" of "ideas such as the experience curve or portfolio theory," he clearly felt most let down by the business school cases and teaching notes from which he had first learned the presumed details of Honda's strategy. Although Harvard Business School cases always contained the caveat that the case was written only "as the basis for class discussion rather than to illustrate either effective or ineffective handling of an administrative situation," such warnings, like the boilerplate disclaimers on a financial prospectus that "past performance is no indication of future results," have little real meaning.[52] Investors invariably buy stocks based on their past results just as business school students and teachers ultimately presume that the cases that they read accurately reflect the profiled companies. In short, Richard Pascale was deeply shocked and dismayed to learn that the Harvard Case Study that he had been teaching to his introductory classes in strategy, and the teaching note written by his colleague, Richard Rumelt, were both wrong.[53] Like the proverbial American high school students when first told that George Washington did not really throw a silver dollar across the Potomac River (but that the story was good for their moral development), Richard Pascale asked rhetorically, "then what good is Western business education?" His dismay with corporate strategy, with consultants, and with the Western conception of theoretical models would serve as a touchstone for a generation of disillusioned students of strategy.

It is striking, therefore, to realize that the central motivation for Richard Pascale's devastating critique of The Boston Consulting Group's analysis of the strategies adopted by Honda was not an inherent distrust of consultants—given that Pascale himself had accepted research support from McKinsey and would continue to serve as a strategy consultant during most of his career. Instead Pascale's memento was a result of the apparent deception of the business school case writers who had made him feel like a fool for putting his faith in their work. The blame, it seems, for Boston Consulting Group's troubles lay not in their faulty analysis of strategy at Honda, but in the mysterious alchemy involved in translating a BCG report into a Harvard Business School case study.

TEACHING STRATEGY TO M.B.A.s, 1977–1983

In 1977, strategic management, or "business policy" as it was better known then, was still an emerging academic discipline, just making the transition from the empirical work of Kenneth Andrews and Alfred Chandler (both at the Harvard Business School) and Igor Ansoff (best known for his association with Carnegie Mellon) to the new stylized economic models being put forward by Michael Porter (who was then a young faculty member at Harvard).[54] What these teachers of strategy lacked, however, was not new ideas or potential colleagues but good examples in the public domain.[55] The academics at the Harvard Business School were superb in writing decision-oriented cases in a highly stylized

230   form ("Mike Brady looked out of his office window and wondered what steps his start-up company, Mirada, should take next . . . "), but there were almost no publicly available documents that provided detailed, real-life examples of strategic management in action.[56] It is therefore not surprising that the faculty at the business school at Harvard (and Virginia and UCLA) would quickly convert BCG's published report on the strategic actions available to the British motorcycle manufacturers into a standard HBS case study, albeit a case originally intended for strategic marketing.[57] For the Harvard case writers, BCG's report was a perfect illustration of a real-life problem (revitalizing a national industry beset by international competition), placed within the context of a useful strategic tool (the experience curve), and containing detailed data that were available in the public domain. The fact that it also detailed the innovative strategic decisions of a major Japanese manufacturer made it all the more attractive in the classroom.

    If the writers and teachers of the industry note were happy, their business school students were particularly pleased because here, after all, was a real-life example of a top management consulting firm at work. In the 1970s, as the market for the traditional consulting assignments involving administrative reorganization at firms like McKinsey and Booz Allen declined, strategy consulting took off, and no strategy consulting firm was considered more influential than The Boston Consulting Group.[58] By the late 1970s, BCG was winning its annual competition with McKinsey & Company to hire the greatest number of Baker Scholars (the top 5 percent of the graduating class at the Harvard Business School) as the HBS students increasingly saw BCG as an equally prestigious and much more "cutting-edge" consultancy than McKinsey.[59] To be taught an HBS strategy case was one thing, but to learn how the consultants at BCG had developed their analysis of a complicated strategic problem was like getting an inside view of the business elite. This act of quasi-voyeurism would prove immensely popular in the classroom and, as a result, the BCG study was a central case for a generation of M.B.A.s. By 1983, the year before Richard Pascale's article appeared, the "Note on the Motorcycle Industry" had become a "minor classic," as the eighth best-selling case for Harvard Business School Case Services with sales of nearly seventeen thousand copies in the preceding two years.[60]

    In translating BCG's report, the first volume of which alone was 120 pages long, to a case study, the Harvard case writers, supervised by Professor Robert Buzzell, had to change the tone from a carefully worded, apolitical document to an HBS-style case study. As Andrew Mair has described the resulting shift, the Harvard Business School case study "added a wealth of detail, as well as a bullish business school 'success story' tone not apparent in the original BCG study."[61] Because Buzzell had intended that teachers use the case to teach mar-

keting, the HBS case would place greater emphasis on the innovative market   231
segmentation of Honda motorcycles, describing the "strategic" transformation
of the existing market from a bad-boy image best remembered in Marlon
Brando's *The Wild One*, from 1953, to the more wholesome images conjured
up in Honda's 1963 advertising campaign, "You Meet the Nicest People on a
Honda."[62] This accessible style, of course, was the model for all Harvard Business School cases; the idea was to lure the student into a sense of narrative complacency before the instructor dissected his or her misunderstanding of the case
in a Socratic dialogue before an audience of eager classmates.[63]

The most troubling part of the case, however, was not its distillation of the
original report but the questionable section on "historical development" added
to the analysis by the Harvard case writers.[64] This history, which Richard Pascale later quoted at length in his 1984 critique (in Pascale's words "history has
it . . . "), was the worst type of business history—for in order to give their history of the motorcycle industry an active voice (unlike the British writers of the
BCG case, who depended on their characteristic passive-voice sentence structure to avoid assigning causality), the Harvard case writers imputed strategic
intentionality from the historical outcomes.[65] The result was an immensely
readable, logical story of the historical development of the Japanese motorcycle
industry that wove into its narrative the presumed strategic decisions taken
along the way. In other words, almost all that Richard Pascale would later claim
was wrong with the Harvard case study originated in the attempt by the Harvard case writers (and those who wrote the teaching notes) to insert their under-
researched history of the motorcycle industry into the statistical and theoretical
strategy framework developed by The Boston Consulting Group.[66] The mis-
application of business history, in retrospect, was the root of all of the subsequent debates.

NEW MARKETS, OLD MODELS (FOR BCG), 1973–1976

One of the reasons that the rapid decline in the British motorcycle industry was
so shocking—and was worthy of public debate in Parliament—was that the
industry had, only twenty years earlier, experienced a boom in sales in the United
States such that by the mid 1950s, "nearly half of the motorcycles sold in America were British."[67] In the early 1960s, as other British industries struggled to
secure export markets, British motorcycle manufacturers sold nearly 80 percent
of their industry output in America—the sudden collapse of the British motorcycle market was almost impossible to foresee in the 1950s. British success in the
1950s and early 1960s, however, would soon be reversed as the Japanese manufacturers, led by Honda, began selling larger, more technically sophisticated, and

232   reliable models first in the United States and later in Britain itself.[68] As the consultants from The Boston Consulting Group reported, the British manufacturers responded by retreating from those competitive markets in which they were being directly challenged — giving up first on small and then medium-size machines — until they couldn't retreat any further.[69] By the early 1970s, the British motorcycle industry, one of the few stars of the British postwar economy, was on its deathbed as exports to the United States dried up and Japanese rivals relentlessly gained ground on the British manufacturers.

In 1973, as BSA motorcycles of Britain was collapsing, the British Government orchestrated the merger of the final two remaining motorcycle companies in Britain into a new enterprise known as Norton Villiers Triumph (NVT). In 1973, the British motorcycle manufacturers produced only five different models, their three largest motorcycles accounting for 19 percent of the U.S. market.[70] In its first twenty months of operation, however, the newly combined NVT lost £7.4 million on total equity of only £9 million. Despite the immense union pressures applied to the British Government, for political reasons, executives resisted a proposal to consolidate manufacturing into two factories from three. The consultants from The Boston Consulting Group estimated that even the least ambitious strategy alternative they could propose would result in a deficit of £15 million and take more than twenty-five years to achieve final payback.[71] The more ambitious alternatives considered by the consultants would have incurred deficits nearly three-and-a-half times as great, even if they were lucky enough to be successful. As the consultants from BCG recognized, this report was not an ordinary business plan but was instead the basis upon which the British Parliament would debate the future of Britain's industrial economy.

Management consulting firms such as BCG were generally reluctant to take on public assignments like an analysis of the British motorcycle industry because the resulting publicity rarely proved beneficial. Although this assignment was a sign of how important BCG's model of the experience curve had become — for at the heart of BCG's analysis was its model of how large production volumes drove down costs, thus making market share a crucial component of strategy for industrial manufacturers — the assignment was also a mark of its own desperation in the British market.[72] Bruce Henderson, the founder of BCG, may have been the leading light of strategy in America in 1975, but in Britain this was not the case. Even though they were struggling through the economic downturn, the consultants at McKinsey & Company continued to dominate the British market for consulting services, because they had worked for at least one-quarter of the hundred largest industrial companies prior to 1973.[73] Indeed, as newspaper coverage would demonstrate, the control that McKinsey exerted within Britain was extraordinary — the consult-

ants at McKinsey had previously worked for such prominent state-owned enterprises as the BBC, British Rail, the Post Office, and, most famously, the Bank
of England. To the consultants at BCG, struggling through the downturn in
the economy that began with the OPEC crisis, the study of the British motorcycle industry might have had obvious political pitfalls, but the potential to
publicly trump the consultants from McKinsey was simply irresistible.[74]

The irony, of course, was that BCG managed to produce a report that was so
widely accepted within Parliament as an accurate overview of the strategic (and
political) alternatives available to the British secretary of state for industry that
Parliament decided to have an excerpted copy printed for distribution to the
wider public. The printed report of BCG's findings, in turn, was such an unusual
documentary source that it was quickly seized upon by the leading business strategists on both sides of the Atlantic. What the BCG consultants had feared was a
document that could be used against them for its policy implications about the
British economy, particularly in the politically charged atmosphere of the 1970s,
was instead used against them for its historical inaccuracy about the Japanese
economy a decade later. The consultants at BCG could hardly have imagined
that outcome when they initially sat down to draft the report.

INTRODUCTION

In her work on the development of the social sciences, historian Dorothy Ross
argues that in the period from the late nineteenth to the early twentieth century, academic disciplines such as economics, political science, psychology, and
sociology moved from a primarily historicist to an ahistoricist understanding of
their subjects.[75] In contrast, from their origins in scientific management during
the early twentieth century until now, as this case suggests, academics in business schools have become increasingly more historicist. Although neither the
consultants at BCG nor the case writers at Harvard ever imagined that getting
the history of Honda right would have mattered, more than twenty-five years of
subsequent debates over the "true story" of Honda's innovative strategy suggests
just how historically sensitive management academics have unwittingly become.

One of the crucial implications of the Honda case, therefore, is not only
that history matters—for every good scholar will readily agree to that easy tru-
ism—but that history matters increasingly to those business school academics
who only a generation ago might have dismissed questions of intentionality,
causation, and context as secondary to financial and market data to identifying
the underlying sources of innovation. As it turns out, the analysis of rival theories in strategic management relies as much upon intentions as outcomes,
and business strategists are increasingly finding the answers to those crucial

234 questions in business history, not simply in their analysis of the present. Had the case writers at Harvard or the consultants at BCG felt more comfortable placing organizational change and corporate innovation within the context of historical change, Richard Pascale might not have found the rival explanations of Honda's strategic success and innovative processes quite so divergent or quite so shocking.

PROLOGUE (AS PAST)

The tag phrase used in the advertisement for the movie *Memento* was "some memories are best forgotten." Our memories, however, are not as easily forgotten, as Richard Pascale's innocent discussion with the executives at Honda subsequently revealed. Business, like all social institutions, is a deeply political activity. In the past, at least, business school academics and business historians have too often forgotten that apolitical statistics of market share, theories of strategic planning, and innovative advertising campaigns often mask deeper political decisions occurring behind closed doors. If we let the ahistorical mementos, the surviving artifacts of the deeper political process, guide our interpretations, then we had better be certain that we understand the context from which these mementos emerged and the cumulative impact that they have on our interpretation of original events. To infer causality from a series of mementos is not only historically dubious but may well undercut otherwise valuable quantitative data and theoretical arguments when all is revealed to be wrong. For, as the protagonist in *Memento* eventually learns, it is one thing to find the final answer, but you must also remember the original question.

EPIGRAPH

"Time flies like an arrow; fruit flies like a banana."—Groucho Marx

NOTES

I would like to thank Ted Buswick of The Boston Consulting Group, Professor Richard Rosenbloom of the Harvard Business School, and Barry Hedley of Gonville and Caius College, University of Cambridge (one of the three original authors of *Strategy Alternatives for the British Motorcycle Industry*) for their comments on this chapter. All opinions and mistakes are mine alone.

1. *Memento,* directed by Christopher Nolan (Culver City, Calif.: Columbia TriStar, 2001). For a representative review, see A. O. Scott, "Backward Reel the Grisly Memories," *New York Times,* Mar. 16, 2001, E14.

2. That such a reverse narrative is particularly dependent upon a very tight chronology is something that Christopher Nolan has himself emphasized by pointing out that "it's totally linear, it's just reversed." See also "Interview of Christopher Nolan," *Memento*, DVD, by Elvis Michel (Burbank, Calif.: Columbia TriStar Home Entertainment, 2001).

3. The Boston Consulting Group, Limited (hereafter BCG), *Strategy Alternatives for the British Motorcycle Industry: A Report Prepared for the Secretary of State for Industry* (London: Her Majesty's Stationery Office, 1975).

4. Henry Mintzberg, "Introduction," in "The 'Honda Effect' Revisited," Henry Mintzberg, Richard Pascale, Michael Goold, and Richard Rumelt, *California Management Review* 38 (Summer 1996): 78.

5. *Being John Malkovich*, directed by Spike Jonze (Universal City, Calif.: USA Films, 1999); Michael Lewis, *Liar's Poker: Rising Through the Wreckage on Wall Street* (New York: W. W. Norton, 1989), 159.

6. Alfred D. Chandler Jr., *Strategy and Structure: Chapters in the History of the American Industrial Enterprise* (Cambridge, Mass.: MIT Press, 1962); Alfred D. Chandler Jr., *The Visible Hand: The Managerial Revolution in American Business* (Cambridge, Mass.: Harvard University Press, 1977).

7. Andrew Mair, "Learning from Honda," *Journal of Management Studies* 36 (Jan. 1999): 25–44; Mintzberg et al., "The 'Honda Effect' Revisited," 78–117; Henry Mintzberg, "Learning 1, Planning 0: Reply to Igor Ansoff," *Strategic Management Journal* 12 (Sep. 1991): 463–66; Richard T. Pascale, "Perspectives on Strategy: The Real Story Behind Honda's Success," *California Management Review* 26 (Spring 1984): 47–72; Dev Purkayastha, "Note on the Motorcycle Industry—1975," case# 9-578-210 (Boston: Harvard Business School Press, 1978); BCG, *Strategy Alternatives*.

8. Mair, "Learning from Honda," 25–44; Mintzberg et al., "The 'Honda Effect' Revisited," 78–117.

9. Frederick Winslow Taylor, *The Principles of Scientific Management* (New York: W. W. Norton, 1967). The description of management as a science has waxed and waned over the intervening years with the rise of operations research after World War II on one extreme and the analysis of corporate culture during the 1980s on the other extreme.

10. David Barry and Michael Elmes, "Strategy Retold: Toward a Narrative View of Strategic Discourse," *Academy of Management Review* 22 (Apr. 1997): 429–52.

11. Richard Whittington, *What Is Strategy—and Does It Matter?* 2nd ed. (London: Thompson Learning, 2001), 33.

12. Andrew Mair, "Reconciling Managerial Dichotomies at Honda Motors," in *Strategy: Process, Content, Context—An International Perspective*, eds. Bob De Wit and Ron Meyer, (London: International Thompson Business Press, 1998), 893–911.

13. Purkayastha, "Note on the Motorcycle Industry—1975"; James Brian Quinn, "The Honda Motor Company," in *The Strategy Process*, eds. Henry Mintzberg, James Brian Quinn, and Sumantra Ghoshal (London: Prentice-Hall, 1998), 293–314.

14. Mair, "Reconciling Managerial Dichotomies at Honda Motors," 893.

15. Clayton M. Christensen, *The Innovator's Dilemma: When New Technologies Cause Great Firms to Fail* (Boston: Harvard Business School Press, 1997), 153–56.

236   16. Pankaj Ghemawat, "Competition and Business Strategy in Historical Perspective," *Business History Review* 76 (Spring 2002): 37–74; and James Hynes, *Publish and Perish: Three Tales of Tenure and Terror* (New York: Picador, 1997).

17. Mair, "Learning from Honda," 25, 38.

18. Mair, "Learning from Honda," 39.

19. Mair, "Learning from Honda," 26.

20. E. Tatum Christiansen and Richard T. Pascale, "Honda (A)," case # 9-384-049 (Boston: Harvard Business School Press, 1989); E. Tatum Christiansen and Richard T. Pascale, "Honda (B)," case # 9-384-050 (Boston, Harvard Business School Press, 1989).

21. E. Tatum Christiansen, "Honda (A) and (B) Teaching Note," (Boston: Harvard Business School Press, 1989), 1.

22. Christiansen, "Honda (A) and (B) Teaching Note," 4.

23. Richard P. Rumelt, "The Many Faces of Honda," *California Management Review* 38 (Summer 1996): 103–11.

24. Pascale, "Perspectives on Strategy," 47–72.

25. A. T. Kearney & Co., *Worldwide Management Consulting Firm Reports 15th Consecutive Year of Double-Digit Growth*, report, February 10, 1999; John A. Byrne, "The Craze for Consultants," *Business Week*, July 25, 1994, 60; James O'Shea and Charles Madigan, *Dangerous Company: The Consulting Powerhouses and the Businesses They Save and Ruin* (New York: Times Books, 1997).

26. John Micklethwait and Adrian Wooldridge, *The Witch Doctors: Making Sense of the Management Gurus* (New York: Times Books, 1996), 43–60.

27. Donald N. McCloskey, *If You're So Smart: The Narrative of Economic Expertise* (Chicago: University of Chicago Press, 1990). Jim Collins, coauthor (with Jerry I. Porras) of the best-selling business book *Built to Last: Successful Habits of Visionary Companies* (New York: HarperBusiness, 1994), would brag that his financial success due to royalties and consulting allowed him to resign his position at the Stanford Business School to become "a self-employed professor who endowed his own chair and granted himself tenure."

28. Micklethwait and Wooldridge, *The Witch Doctors*, 51.

29. John Kay, "The Structure of Strategy," *Business Strategy Review* 4 (Summer 1993): 17–37.

30. J. I. Moore, *Writers on Strategy and Strategic Management* (London: Penguin, 2001), 49–63.

31. Henry Mintzberg, *The Rise and Fall of Strategic Planning* (New York: The Free Press, 1994), 4.

32. Henry Mintzberg, "The Manager's Job: Folklore and Fact," *Harvard Business Review* 53 (Jul.-Aug. 1975): 49–61.

33. Henry Mintzberg, "Crafting Strategy," *Harvard Business Review* 65 (July-Aug. 1987): 67.

34. H. Igor Ansoff, "Critique of Henry Mintzberg's 'The Design School: Reconsidering the Basic Premises of Strategic Management,'" *Strategic Management Journal* 12 (Sep. 1991): 449–51; Henry Mintzberg, "The Design School: Reconsidering the Basic Premises of Strategic Management," *Strategic Management Journal* 11 (Mar.-Apr. 1990): 171–95; Mintzberg, "Learning 1, Planning 0: Reply to Igor Ansoff," 463–66.

35. Henry Mintzberg, "Learning 1, Planning 0," *California Management Review* 38 (Summer 1996): 92.

237

36. Pascale, "Perspectives on Strategy"; BCG, *Strategy Alternatives*; H. Igor Ansoff, *Corporate Strategy: An Analytic Approach to Business Policy for Growth and Expansion* (New York: McGraw-Hill, 1965).

37. BCG, *Strategy Alternatives*, v; Michael Goold, "Design, Learning and Planning: A Further Observation on the Design School Debate," *California Management Review* 38 (Summer 1996): 94; Henry Mintzberg, "Reply to Michael Goold," *California Management Review* 38 (Summer 1996): 96.

38. Bruce Ahlstrand, Joseph Lampel, and Henry Mintzberg, *Strategy Safari: A Guided Tour Through the Wilds of Strategic Management* (New York: The Free Press, 1998), 230.

39. Andrew Mair, "Learning from Japan? Interpretations of Honda Motors by Strategic Management Theorists," Nissan Occasional Paper Series, Number 2, Nissan Institute of Japanese Studies, University of Oxford, 1999, 45.

40. Ikujiro Nonaka and Hirotaka Takeuchi, *The Knowledge Creating Company: How Japanese Companies Create the Dynamics of Innovation* (New York: Oxford University Press, 1995), 15.

41. Moore, *Writers on Strategy and Strategic Management*, 49.

42. For one of the most influential of these critiques, see William J. Abernathy and Robert H. Hayes, "Managing Our Way to Economic Decline," *Harvard Business Review* 58 (July-Aug. 1980): 67–77.

43. Anthony Athos and Richard Pascale, *The Art of Japanese Management: Applications for American Executives* (New York: Simon & Schuster, 1981); Terence E. Deal and Allan A. Kennedy, *Corporate Cultures. The Rites and Rituals of Corporate Life* (Reading, Mass.: Addison-Wesley, 1982); Thomas J. Peters and Robert H. Waterman Jr., *In Search of Excellence: Lessons from America's Best-Run Companies* (New York: Harper & Row, 1982). Mauro Guillén, among others, describes these three books as the centerpieces of the corporate culture movement. Mauro F. Guillén, *Models of Management: Work, Authority, and Organization in a Comparative Perspective* (Chicago: University of Chicago Press, 1994), 289–90.

44. On the links between McKinsey & Company and Richard Pascale's early work on Japanese corporate culture, see Christopher D. McKenna, *The World's Newest Profession: Management Consulting in the Twentieth Century* (New York: Cambridge University Press, 2006), 192–94.

45. It is worth noting as well that Richard Pascale's chosen title, "Perspectives on Strategy," would have appeared, at least to those in the know, to be a dig at BCG's long running internal publication simply titled "Perspectives."

46. Thomas C. Hayes, "McKinsey & Co., Problem Solvers," *New York Times*, Aug. 19, 1979, F1; McKinsey & Company, "Competitor Analysis Project," Montsoult, France, March 18–20, 1981, 3–4, McKinsey & Company Corporate Archives, New York; James Abegglen, April 27, 2001, tape 4, The Boston Consulting Group Internal Archives, Boston.

47. Goold, "Design, Learning, and Planning," 169–72; Walter Kiechel, III, "The Decline of the Experience Curve," *Fortune*, October 5, 1981; Walter Kiechel, III, "Corporate Strategists Under Fire," *Fortune*, December 27, 1982.

238

48. Pascale, "Perspectives on Strategy," 63.

49. Pascale, "Perspectives on Strategy," 64.

50. With more than twenty-six million motorcycles produced to date, the Honda Super Cub would become the single best-selling motorcycle in history, the motorcycle equivalent of the Ford Model T, Volkswagen Beetle, or Toyota Corolla. See Steve Koerner and Jun Otahara, "The Honda Motor Co. and the American Motorcycle Market," paper presented at the Business History Conference, Wilmington, Delaware, Apr. 20, 2002, 4.

51. Pascale, "Perspectives on Strategy," 49–76.

52. Purkayastha, "Note on the Motorcycle Industry—1975," 1.

53. Pascale quotes from Rumelt's teaching note in "Perspectives on Strategy," 53.

54. Moore, *Writers on Strategy and Strategic Management*, 1. As a point of reference, the main disciplinary association for strategists, the Strategic Management Society, was founded in 1980.

55. Robert D. Buzzell, "Note on the Motorcycle Industry—1975: Teaching Note" (Boston: Harvard Business School Press, 1985), 1.

56. In contrast to BCG's study of the British motorcycle industry, McKinsey & Company, *A Study of Western Electric's Performance* (New York: American Telephone and Telegraph, 1969) was largely a celebratory account and one that offered little guidance to managers operating in unregulated markets, because AT&T was justifying its performance as a regulated monopoly.

57. Pascale, "Perspectives on Strategy," 59.

58. George W. Stalk Jr. and Carl W. Stern, *Perspectives on Strategy from The Boston Consulting Group* (New York: John Wiley & Sons, 1998), xiii–xiv.

59. John Thackray, "Winning the Game with a Hot Theory: Companies Seek Advice of Boston Consulting Group," *New York Times*, Apr. 15, 1979, sec. 3, 4.

60. Buzzell, "Note on the Motorcycle Industry—1975: Teaching Note," 1.

61. Mair, "Learning from Honda," 29.

62. Purkayastha, "Note on the Motorcycle Industry—1975," 5; *The Wild One*, directed by László Benedek (Culver City, Calif.: Columbia Pictures, 1953).

63. Louis B. Barnes, Abby J. Hansen, and C. Roland Christensen, *Teaching and the Case Method: Text, Cases, and Readings* (Boston: Harvard Business School Press, 1994).

64. Although Professor Richard Rosenbloom, the former associate dean for research and course development at the Harvard Business School who approved the budgets and staffing for all cases written between 1976 and 1980, including the 1978 "Note on the Motorcycle Industry," argues that there "was no single model for HBS case writing in the 1970s," Rosenbloom also acknowledges that the writers of *Industry Notes* did follow established patterns, including a section, often historical, that listed the principal competitors in the industry. Richard S. Rosenbloom, written comments to the author, Nov. 2, 2002.

65. BCG, *Strategy Alternatives*; Pascale, "Perspectives on Strategy," 61.

66. Pascale, "Perspectives on Strategy," 61; Purkayastha, "Note on the Motorcycle Industry—1975," 5–6.

67. Koerner and Otahara, "The Honda Motor Co. and the American Motorcycle Market," 6.

68. Steve Koerner, "The Japanese Motor Cycle Industry, 1945 to 1960," paper presented at the Association of Business Historians Conference, Chapel Hill, North Carolina, Sep. 2, 1999, 8.

69. BCG, *Strategy Alternatives*, ix.

70. BCG, *Strategy Alternatives*, x.

71. BCG, *Strategy Alternatives*, xxiv.

72. Ghemawat, "Competition and Business Strategy in Historical Perspective," 45–6.

73. Derek F. Channon, *The Strategy and Structure of British Enterprise* (Boston:
Harvard Business School, 1973), 239; Michael C. Jensen, "McKinsey & Co.: Big Brother
to Big Business," *New York Times*, May 30, 1971, sec. 3.

74. Stephen Aris, "Supermanagers," *The Sunday Times*, Sept. 1, 1968.

75. Dorothy Ross, *The Origins of American Social Science* (New York: Cambridge
University Press, 1991).

INTRODUCTION TO PART III

Sally H. Clarke, Naomi R. Lamoreaux,
and Steven W. Usselman

Up to this point, the chapters in this volume have focused primarily on private institutions. Perhaps to a greater degree than any other modern industrial society, the United States has persistently placed its fortunes in the hands of private parties operating in competitive markets. It would be a tremendous mistake, however, to conclude that government played little role in efforts to sustain innovation across the twentieth century. Although previous chapters did not always treat politics explicitly, they pointed to several ways in which government policies and actions impinged upon the pursuit of innovation. The patent system, so important to inventors and corporate laboratories alike, was of course an arm of government. AT&T was a regulated monopoly for most of the twentieth century, and this status clearly influenced both its accounting techniques and the conduct of its research. The substantial support for research that flowed from government to private industry (both directly and through universities) from World War II onward shaped both Stanford and its surrounding business community, as did a variety of policies intended to favor small business. In these and many other ways, government had a hand in business activity.

The chapters in Part III draw out the role of the state in much sharper relief. Each examines how firms interacted with government agencies in ways that restructured their corporate boundaries, altered market structures, and profoundly influenced their pursuit of innovation. In one case, the boundaries were those among competitors rushing to innovate in the dynamic, emergent sector of digital computing. Here the instrument shaping relations among firms was the long-standing (though evolving) one of antitrust action pursued by the Department of Justice. In the second case, the crucial boundaries involved

244 those between buyers and sellers, or consumers and managers, in the mature automobile industry. Here the state moderated market relations through the newer, less familiar mechanism of policies administered by the Federal Reserve that affected access to consumer credit.

Antitrust has been a persistent, recurrent feature of the American economy since the turn into the twentieth century. Few large firms have escaped its influence. Both AT&T and Corning, as discussed in Part I, altered their innovation strategies in response to changing antitrust doctrine. Among the many corporations subjected to such scrutiny, few have stood out more prominently than International Business Machines (IBM). As IBM transformed itself from a leading supplier of complex mechanical accounting systems into the dominant firm in the booming postwar electronic computing industry, it faced a stream of antitrust suits from competitors and from the Department of Justice. The most famous of them stretched for more than a decade; by the time the case ended in 1982, it had helped fuel a profound transformation in thinking about competitive behavior among firms pursuing innovation.

In "Unbundling IBM: Antitrust and the Incentives to Innovation in American Computing," historian Steven Usselman traces IBM's long involvement with antitrust and detects in it a persistent concern among regulators, competitors, and other interested observers that large firms potentially impeded technical change. Antitrust advocates worried that large firms such as IBM might retard or distort the course of innovation by unduly leveraging their control of patented technologies. Early antitrust statutes expressly banned "ties" that compelled consumers of one proprietary device to acquire other products or services from the same supplier. New Dealers such as Thurman Arnold, who assumed control of the Department of Justice's antitrust division shortly before World War II and launched a vigorous round of antitrust prosecutions, pressed for mandatory licensing of corporate patents. Through such measures Arnold and others hoped to "unbundle" technical innovations from the larger systems they served and to identify distinct technical accomplishments, which they hoped would diffuse widely among competitive firms (as occurred in the case of the transistor, which AT&T agreed to license under terms of a consent decree signed in the 1950s).[1] Antitrust thus worked to redraw corporate boundaries in ways its enthusiasts thought would push industries dominated by large firms toward a looser organizational structure, not unlike that we have come to associate with Silicon Valley. In modern parlance, they sought to promote processes of open innovation, with multiple players competing to contribute pieces of comprehensive technical systems.[2]

As Usselman explains, this goal initially ran headlong into a core set of business practices IBM had developed under the leadership of its famed CEO,

Thomas J. Watson. IBM leased comprehensive accounting systems, with all
components, services, and parts (including the critical punched cards) bundled in a single price unique to each customer. Watson resisted the Department of Justice at every turn. Over time, however, as Watson's son Tom sought
to orchestrate transitions to digital computing and solid state electronics, IBM
assumed a more conciliatory attitude. It did so at least in part because the
younger Watson and a new cohort of managers came to worry about whether
highly integrated systems might retard incentives to innovate. Watson and his
successors were looking to redraw boundaries in ways that forged new alliances
within IBM and opened these reconstituted units to market pressures, much
as we observed in the cases of AT&T and Corning. In many respects, government intervention merely compelled IBM to reorganize its routines and practices along lines the firm itself was gravitating toward under the press of
competition. In the late 1990s, government and IBM actually ended up as allies
promoting the benefits of open systems in a suit against Microsoft.[3]

Whereas Usselman treats a case in which the state shaped relations among
competitors on the supply side of the market, Sally Clarke examines how the
state altered relations between firms and their customers on the demand side.
Clarke's case study, "Credit and the Mature Market for Automobiles," concerns
consumers' use of installment credit to purchase automobiles during the years
after World War II. From $9 billion in 1919 (as reported in inflation-adjusted
2005 dollars), total installment credit in the United States grew to $63 billion
in 1939 and then ballooned to $521 billion by 1969.[4] Credit specifically for autos
followed a similar pattern. Few observers would doubt the centrality of credit
to enabling Americans to purchase goods that have substantially improved their
standard of living during the twentieth century. As Clarke cautions, however,
the significance of credit as a financial innovation should not lead us to assume
that all Americans have had ready access to credit. Nor does it mean that the
terms of credit have always been conducive to its use.

Clarke reminds us that during the 1920s the terms for automobile loans typically required buyers to make a down payment of one-third or more of the car's
purchase price and to pay off the loan within twelve months.[5] During the late
1930s, creditors liberalized terms to a small degree, but then credit sales dried
up during World War II, only to recommence after the war. By the 1960s, car
loans were much more generous: buyers often made down payments of less
than a third the purchase price and repaid their loans in more than twenty-five
months for used cars or thirty-one months for new cars.[6]

At first blush, one might assume that the terms changed as a result of private
decisions by car dealers who interacted directly with consumers, or perhaps
by managers at financial institutions such as the General Motors Acceptance

246 Corporation (better known as GMAC). Clarke tells a different story. She finds that during the 1950s dealers indeed wanted to liberalize credit terms but were frustrated by government policy. In particular, the Board of Governors of the Federal Reserve System (the Fed) had put in place credit restrictions in their effort to combat inflation. Undaunted, dealers wrote numerous letters to Fed officials demanding changes, and the Fed relented. Dealers achieved this success by invoking claims that the existing credit regulations were discriminatory: they limited the ability of farm or working-class families or others with limited incomes or financial resources to purchase new cars.

Although dealers in the 1950s lobbied to relax obstacles based on class or income distinctions, they did not address other types of discrimination, such as that based on sex, marital status, age, or color. During the 1970s, victims of such discrimination—especially women, who routinely had been denied access to credit by lenders—pressed for further change. Congress responded with the Equal Credit Opportunity Act in 1974 (and amended it in 1976). Under its Regulation B, the Fed now took responsibility for establishing fair lending standards and monitoring creditors' lending practices. Clarke thus finds that the state played a key role not only in broadening access to credit and expanding the automobile market but also in confronting creditors' cultural prejudices about who was trustworthy for loans. After the passage of the ECOA, a much broader array of Americans was able to access credit markets and purchase goods that raised their living standards.[7] Although it is beyond Clarke's study, the implication of the ECOA was that these consumers may also have been able to voice their wishes and needs more strongly and thus potentially shape the direction of innovation.

NOTES

1. See Ellis W. Hawley, *The New Deal and the Problem of Monopoly: A Study in Economic Ambivalence* (Princeton: Princeton University Press, 1966), Part III; and Chapter 4, by Kenneth Lipartito, in Part I of this volume.

2. As Margaret Graham makes clear in her chapter on Corning in Part I (Chapter 3), antitrust action could work at cross-purposes by interfering with attempts by firms to enter alliances that might facilitate the emergence of innovative networks. Congress attempted to remedy this situation in the 1990s by revising antitrust law in ways more sympathetic to inter-firm research alliances. On these reforms, see Thomas M. Jorde and David J. Teece, eds., *Antitrust, Innovation, and Competitiveness* (New York: Oxford University Press, 1992).

3. On IBM's embrace of open innovation, see Henry W. Chesbrough, *Open Innovation: The New Imperative for Creating and Profiting from Technology* (Boston: Harvard Business School Press, 2003).

4. Sally H. Clarke, *Trust and Power: Consumers, the Modern Corporation, and the Making of the United States Automobile Market* (New York: Cambridge University Press, 2007), 254.

5. Clarke, *Trust and Power*, 255–56. See also Geoffrey H. Moore and Philip A. Klein, *The Quality of Consumer Instalment Credit* (New York: Columbia University Press, 1967), 9 and 12. See also Martha L. Olney, *Buy Now, Pay Later: Advertising, Credit, and Consumer Durables in the 1920s* (Chapel Hill: University of North Carolina Press, 1991), 109, 113.

6. Clarke, *Trust and Power*, 255–56. See also Moore and Klein, *The Quality of Consumer Instalment Credit*, 9 and 12. On the development of installment credit prior to World War II, see also Olney, *Buy Now, Pay Later*, 86–134.

7. Louis Hyman also addressed the importance of changes in credit markets during the 1970s in *Debtor Nation: Changing Credit Practices in 20th Century America* (Princeton: Princeton University Press, forthcoming), chapter 7. On the ECOA, readers should consult the National Consumer Law Center, *Credit Discrimination*, 3rd ed. (Boston: National Consumer Law Center, 2002).

8

Unbundling IBM

Antitrust and the Incentives to Innovation
in American Computing

Steven W. Usselman

Few American corporations of the twentieth century achieved greater notoriety than International Business Machines (IBM). Under the sustained leadership of Thomas J. Watson and his son Tom, IBM rose from a modest-sized supplier of punched-card accounting equipment in the 1910s to become the world's dominant supplier of electronic computers, the glamour product of the American Century. When the younger Watson relinquished the helm in 1971, IBM had the highest market capitalization of any American company.[1] Foreign governments strained to create their own champions capable of matching the American giant, while vanquished competitors such as General Electric and RCA left the computer market to Big Blue.[2] As further advances in hardware and programming drove computing down in price and size and brought it into a much wider realm of applications, IBM retained a powerful presence in virtually every segment of the broadening market. Its revenue and stock values soared ever higher for another two decades, until the early 1990s, when competitors from Asia and the West Coast seized the initiative and plunged IBM into the sort of crisis that had periodically plagued companies such as Ford.[3]

As IBM navigated a course through the startling technological transformations that have characterized modern information technology, it persistently faced a challenge that did not loom quite so large at Ford and some of the other firms studied in this volume: the threat of antitrust prosecution. In 1936, just months after IBM secured a major contract from the new Social Security Administration that would help make its punched-card equipment a ubiquitous feature of private and public bureaucracies, the Department of Justice won an antitrust suit against the company. As IBM vaulted to leadership in electronic computing during the early 1950s, Justice launched another investigation, which culminated in 1956 with a comprehensive consent decree.[4] A decade later, with IBM still in the throes of executing the massive System/360 project that replaced its entire product line with machines built from solid state components manufactured in its own plants, Justice intervened once more. The resultant suit, one of the most significant in the annals of antitrust, lasted from early 1969 to 1982.[5] By the time a judge dismissed the suit as "without merit," IBM had successfully launched its widely heralded PC. Even when the firm's fortunes later turned sharply downward, antitrust remained a significant element in its strategic thinking. During the mid-1990s, IBM provided vital testimony on behalf of the Justice Department in its case against Microsoft, the company that had displaced Big Blue atop the industry. Meanwhile, IBM renegotiated its own agreement with government, which agreed to remove key provisions from the consent decree that had governed IBM's behavior since 1956.

The ongoing engagement between IBM and the Department of Justice bore many marks of an epic struggle between rival combatants. Antitrust is an inherently adversarial process. The elder Watson had once gone to jail for antitrust violations while employed at NCR. He brooked no compromise with the Department of Justice and castigated his son for negotiating a consent decree. The younger Watson implied in his memoirs that the stress of countering the suit launched in 1969 and several accompanying private antitrust actions contributed to the heart attack that forced his early retirement as CEO.[6] IBM spent a small fortune defending itself against these claims, as legions of lawyers and economists on both sides devoted years of their professional lives to the cases.

For all the persistent combativeness, however, relations between IBM and the Department of Justice never devolved into a simple doctrinaire feud between habitual adversaries. Government and corporation did not wage some sort of grand battle over the virtues of *laissez-faire*. The disputes and their resolutions turned upon more nuanced considerations. At root, they involved complex questions regarding the ways in which market structure and firm organization shaped incentives and promoted or impeded technological change. The antitrust cases involving IBM essentially provided a forum in which lawyers, economists, and other experts actively worked through the issues that animate this book. The exercise was, of course, far from academic. The immediate economic stakes were large. Yet the cases also served as an extended learning experience—a sustained laboratory experiment, in a sense—through which a large community of participants and observers came to understand better the nature of innovation and the role of the large firm in the innovative process.

At issue, from the first dispute of the 1930s through the Microsoft case, was a fundamental set of business practices that the elder Watson had instilled upon his arrival at IBM in 1914.[7] Those practices were aimed at providing comprehensive services obtained from tightly integrated systems. At the heart of these systems stood a proprietary technology, known as the accounting machine in the electromechanical age and later as the central processing unit (or CPU) during the computer era. Attached to it were an array of peripheral devices, including readers to input stored data and printers to present results in various forms. The separate components in an installation were linked through distinctive means—by punched cards of unique format in the electromechanical era, and later by exclusive input-output channels and software programs known as operating systems—over which IBM also retained close proprietary control. The precise mix of devices varied from customer to customer, as IBM representatives in the field worked to tailor each system to the needs of the particular client. IBM technicians visited these sites regularly, in some cases maintaining a virtually constant presence, in order to keep the equipment running and to devise further uses for it. This package of equipment and services came at a single comprehensive price, unique to each installation. For many decades, such prices were expressed strictly in terms of a monthly rental charge, as IBM in all cases retained ownership of the equipment and leased it to customers. The practice remained common even after the 1956 consent decree required IBM to sell as well as lease its products.

Antitrust proceedings against IBM aimed at breaking apart this integrated approach to conducting its business. Both government and private litigators sought to compel IBM to separate the various components of its bundled products and services, to make each available independently from the others, and to set distinct prices for each of them. Though the specific remedies varied from case to case, as technology changed and antitrust doctrine evolved, this was the persistent objective. IBM's adversaries in the antitrust arena looked to "unbundle" the components and services that made up the integrated systems. They sought to establish clearly defined targets of modest scope at which competitors could take aim. Their objectives paralleled precisely those pursued more recently by government prosecutors, and by IBM itself, when they attempted to force software giant Microsoft to draw clear boundaries between its operating systems and applications software.[8]

In pursuing this course, government grappled with problems that were quite familiar to IBM itself. Managers such as the younger Watson worried incessantly about how to establish accountability and to foster incentives to innovate within sprawling yet highly integrated operations. To this end, they reorganized IBM with astounding frequency, redrawing its internal boundaries in ways that realigned incentives. The wisdom of bundling various products and services recurrently came under close scrutiny. Managers, like government officials, struggled continually to assess trade-offs between standardization and product

252 differentiation. Both groups worried about whether research should be coupled tightly to specific product development efforts or whether researchers should be given broad leeway to pursue innovation and then license their results widely.

It would be a gross distortion, of course, to characterize IBM management and antitrust officials as working in partnership. IBM certainly would have preferred to operate without such oversight, to be left free to redraw its boundaries as it saw fit. The company fought hard to retain such autonomy. Yet at many points in the process, government and the firm actually moved in symphony. At times, government pushed IBM in directions its management hoped to go. The company's enduring record of success resulted in no small measure from its ability to recognize the commonalities and to engage the process constructively.

ANTITRUST DOCTRINE THROUGH MID-CENTURY

IBM began its long engagement with antitrust in the mid-1930s, at a time when its fortunes otherwise ran high. Though rental income dipped slightly following the Crash of 1929, business demand for punched-card equipment soon stabilized, as firms seeking paths to recovery frequently revamped their accounting procedures. New government programs mandating standard reporting of data further stimulated demand. A federal contract to manage information for the new Social Security Administration, secured in 1935, soon accounted for a significant portion of IBM's rising revenues. Watson's profit-sharing agreement made him the nation's highest paid business executive.[9] Such conspicuousness attracted scrutiny. The Department of Justice took IBM to court in May 1936. Sixteen days later, a panel of justices returned a resounding verdict against the company.[10]

The central issue in this case involved IBM's practice of requiring customers who leased its machines to purchase punched cards from IBM. Government prosecutors complained that the provision constituted an illegal "tie" that enabled IBM to earn inordinately large returns from sales of cards to customers who were effectively held captive by the lease agreements. To support their claim, prosecutors noted that the federal government itself, when contracting with IBM for services associated with Social Security and other programs, had secured a special agreement under which it could manufacture its own cards in the standard IBM format. Lawyers for IBM countered that the company must maintain control over cards in order to ensure quality, thereby keeping its leased machines in good working order and preserving its reputation for reliable performance. The government production facility, defense attorneys suggested, would utilize manufacturing equipment supplied by IBM and operate under the watchful eye of its technicians. Paper for the cards would come from suppliers approved by IBM. A skeptical court, noting that government had agreed to pay substantially higher rental charges in order to gain the privilege of producing its own cards, shrugged away the arguments about quality control. IBM could achieve the same end, judges asserted, by publishing technical standards for cards. The court ordered IBM to drop the card purchase requirement and compelled the firm to assist alternative suppliers of cards in starting production facilities that would compete with IBM's.[11]

In taking these steps, the court drew upon economic reasoning that had long informed antitrust law. Advocates of antitrust activity had persistently tried to restrict firms from leveraging their monopoly control of patented technologies by tying additional products and services to those patented techniques. Concerns about such practices extended back to the very origins of the Sherman Act in the late nineteenth century, when public resentment toward patent monopolists such as the Bell telephone interests ran strong.[12] When Congress revised the antitrust statutes in 1914 with passage of the Clayton Act, Section 3 expressly condemned tying contracts requiring consumers to acquire one product or service with the purchase or lease of another.[13] Conspicuous cases involving the motion picture industry and the United Shoe Machinery Company gave the measure teeth.[14] A supplier such as United Shoe might legally assemble a portfolio of patented machines, the Supreme Court ruled, but it could not prevent lessees of one machine from utilizing others supplied by competitors. Each product must stand on its own merits. Judges cited the case liberally in reaching their quick judgment against IBM. Beyond its immediate potential effects on revenues, which in some recent lean years had come largely from cards, the ruling thus put IBM on guard about its essential business strategy of marketing integrated systems.[15]

Although the judgment against IBM drew upon established precedent, it came at a time when questions about patent monopolies had acquired new prominence. Corporate research facilities at firms such as Du Pont, General Electric, and AT&T had assembled large patent portfolios during the 1920s and early 1930s. In some instances, these firms had entered into controversial patent pooling agreements, through which they shared rights to patents covering such conspicuous consumer technologies as paints, gasoline additives, electric light, and radio. Always suspect, these pooling arrangements came under increasing scrutiny during the long Depression decade. Early in the Depression, most discussion of technology focused on the possibility that investment in industrial technology had generated oversupply and unemployment. Such thinking ultimately fed into Keynesian economics, with its emphasis on macroeconomics and stimulation of demand through fiscal policies. By the mid-1930s, however, a substantial contingent of economic observers had come to consider the possibility

254 that the real failure of capitalism lay with its inability to generate sufficient innovation and foster new areas of enterprise. The corporate patent pools came under attack for allegedly suppressing innovation by clogging the paths of independent inventors who might pursue more radical change.

As the Depression persisted, this line of reasoning gained influential footholds within Congress and the Roosevelt administration. Pressured by Congress, the Federal Communications Commission (FCC) conducted a prominent inquiry into the pooling of patents pertaining to radio. At the Department of Justice, FDR appointee Thurman Arnold deployed the tools of antitrust against the patent pools. Arnold lent his support to the radio investigations, spurring an ultimately successful private antitrust action by upstart Zenith, and pursued a highly visible case against petroleum and chemical interests who had pooled patents to form the Ethyl Corporation.[16] In December 1938, Arnold secured another platform when he joined the Temporary National Economic Committee (TNEC). Convened jointly by the Roosevelt administration and Congress and commonly referred to as the Monopoly Committee, this group immediately cast its investigative eye on the patent pools, which FDR had expressly identified as one of three principal areas of concern.[17]

Patent pools drew such ire in part because they seemingly provided a ready subterfuge for cartel behavior. By artificially adjusting prices for shared intellectual property rights, skeptics argued, parties to these agreements effectively fixed prices for final goods and services. Such suspicions enjoyed widespread currency in the economics profession of the 1930s, much of which fully embraced antitrust as an appropriate remedy for horizontal alliances aimed at pricefixing.[18] Those looking to limit such opportunities for subterfuge often pushed for reforms that would curb the numbers of patents granted. Growing numbers of observers also embraced the idea of requiring patent holders to license their patents to anyone willing to pay a reasonable fee. Mandatory licensing, which various parties had proposed since the late nineteenth century, became a central objective of Arnold and his reinvigorated antitrust initiative.[19]

Calls for patent reform and mandatory licensing met with intense resistance from many quarters. Corporate research directors such as Frank Jewett of Bell Labs fought mandatory licensing at every turn. Jewett and many others defended strong patent rights as an essential component of the dynamic, innovation-based competition that economists such as Joseph Schumpeter heralded as the essence of modern economic life. In his *Capitalism, Socialism and Democracy*, Schumpeter dismissed antitrust as irrelevant, because it focused not on emergent new industries but on static competition in highly developed markets for stable goods. In tempering the effects of price competition in established markets, antitrust simply forestalled the day when firms faced the crises

that might stimulate transformative innovative activity. Companies would take   255
such bold steps, moreover, only if assured the prospect of recouping monopoly profits for some time after expending their effort and investment on innovation. Growth occurred through the ongoing struggle to derive monopoly
returns from the creation of distinctive assets providing at least temporary barriers to entry. Patents offered an essential means of establishing such barriers.[20]

Even many advocates of dynamic competition, however, harbored doubts
about whether the patent system functioned in quite such benign fashion in a
world increasingly populated by highly organized research institutions. The
brilliant young economist Alfred Kahn, who would later become a prominent
figure in the campaign for government deregulation, wrote an influential essay
in 1940 portraying the corporate patent pools as necessary evils forced upon
innovative corporations by an anachronistic patent system. In assigning rights
so indiscriminately to every little improvement, the patent system clogged the
modern mechanisms of technical advance, which relied upon a broad-based
cooperative effort conducted by organized research laboratories operated by
firms, universities, and government. Companies looking to commercialize particular improvements derived from the collective effort simply must pool their
patents, Kahn argued, or risk having their creations blocked by others claiming patent infringement. Convinced that the barriers erected by the patent system thus worked against innovation, Kahn called for sweeping reform of the
patent system. Failing that, he embraced antitrust as a means of ferreting out
those who abused patents and impeded technical change.[21]

Similar strains of ambivalence came from no less a figure than Vannevar
Bush, the former dean of engineering at MIT, who came to Washington in
1939 to head the Carnegie Institution and direct FDR's wartime science policy.[22] Like Kahn, Bush stressed that technological innovation had necessarily
taken on an increasingly collaborative character. Corporate laboratories had
emerged in response to the changing nature of the innovative process, and they
were now indispensable to it. But Bush, who had seen some of his own early
inventions absorbed into the corporate morass, also insisted that small firms
and entrepreneurs must always remain vital to the process as well. So too must
the large cohort of independent scientists and engineers working in American
universities.

As Bush orchestrated the federal government's massive wartime investment
in research and development and contemplated the shape federal support for
science and technology should take after the war, the issue of who should take
ownership of the fruits of collaborative research grew ever more prominent. A
1945 report issued by Roosevelt's National Patent Planning Commission, which
included famed research director Charles Kettering of General Motors and

256   president Owen Young of General Electric, managed to squelch the TNEC proposal for mandatory licensing but failed to suppress the idea entirely.[23] When Bush pressed Congress to create a permanent National Science Foundation, skeptics wondered who would own rights to patents generated through its activities. Firmly committed to private enterprise, whether large or small, Bush successfully resisted proposals that government itself should retain rights to such patents.[24] In securing this triumph, however, he may well have lent further momentum to the cause of compulsory licensing of corporate patents, because the laboratories generating those patents might now be perceived as having benefited substantially from public funds. Massive spending by Cold War defense agencies, much of which also flowed to corporate research programs, only reinforced the perception.

A decade after the first antitrust ruling against IBM, the idea of using antitrust law to curb patent rights thus steadily gained momentum. As efforts to reform the patent system languished, a spate of developments in antitrust law filled the void. In a major case launched during the war, the Supreme Court in 1947 embraced DOJ reasoning that effectively established a clear *per se* rule banning ties involving patented technologies.[25] Other cases sought to fix universal principles establishing that dominant firms necessarily abused their market power when they deployed certain types of leveraging tactics.[26] A new suit against United Shoe, ultimately resolved in the DOJ's favor in 1953, banned the firm from bundling free service with leases for its patented equipment.[27] In 1950, Congress weighed in with the Celler-Kefauver Act, which severely constrained the freedom of dominant firms to acquire others.

Bolstered by what observers came to refer to as the New Sherman Era, Thurman Arnold and his allies at the Justice Department began to negotiate consent decrees requiring some of the most prominent firms in the United States to license their patents. One of the first and perhaps the most famous of these actions involved AT&T and Bell Laboratories. Launched in January 1949 and culminating with a consent decree signed seven years later, this suit marked a fateful turn in the ongoing controversies that had swirled around the telephone monopolist since its founding three quarters of a century before. The Labs would now function as something like a public resource, providing the fruits of science and technology at minimal cost in exchange for AT&T continuing to enjoy a virtual monopoly in telephone services.[28] Serendipitously, the onset of the investigation coincided almost precisely with Bell Labs's announcement of its revolutionary new electronic component, the transistor. As a good faith gesture, AT&T agreed to make the transistor and other solid state technologies available to anyone willing to pay a $25,000 license fee, years before the final consent decree imposed mandatory licensing of all patents emerging from the Labs.[29]

THE TRANSITION TO ELECTRONIC COMPUTING

This stream of developments in antitrust law had a profound impact on IBM. Like so many other institutions in the United States, IBM had been transformed by World War II. Demand for its traditional products soared, as the military draft, weapons procurement contracts, and a greatly expanded income tax system all generated record-keeping requirements of staggering proportions.[30] Annual net income grew nearly threefold over the course of the war, topping $36 million in 1945, when the value of IBM's installed base of rental equipment reached more than $115 million.[31] Confident that customers had grown accustomed to IBM equipment and would soon find additional uses for these systems, Watson authorized design of a new generation of accounting devices utilizing a punched card of larger format.

The war and its aftermath left IBM in healthy financial condition, but they also presented the firm with a new set of challenges. Wartime projects had rapidly pushed back the frontiers of advanced calculating among scientists and engineers. Several projects utilized state-of-the-art electronic and magnetic devices to store and manipulate arithmetical information in novel ways. Such developments held obvious implications for IBM. Anyone doubting this needed only to follow the well-publicized exploits of J. Prespert Eckert and John Mauchly, the University of Pennsylvania professors who had headed the effort to create the Electronic Numerical Integrator and Calculator (known as ENIAC) used to calculate ordnance tables. Shortly after the war, the two resigned their university posts and boldly announced that they would sell a revised version of ENIAC for use in the business world. They targeted major banks and insurance companies, long some of IBM's most prestigious and lucrative accounts.[32]

The handful of electronics engineers employed by IBM, most of whom had temporarily left the firm for wartime defense assignments, operated in the shadows of a research and manufacturing culture in which a mechanical engineering ethos held sway. They returned to IBM humbled by their limited understanding of electronics and daunted by its vast potential. Their wonder and discomfort only grew when Bell Labs announced in 1947 its discovery of the solid state transistor, a device which promised to substitute for the expensive and troublesome vacuum tubes that constituted the backbone of most electronic circuits.[33]

Although electronics certainly posed a formidable challenge that IBM could not long ignore, the rapidly emerging technology did not plunge the company into immediate crisis. IBM was not facing the sort of emergency that confronted the Ford Motor Company during its postwar reconversion to civilian production or the oil shocks of the 1970s.[34] Advanced electronic calculating

258   presented an opportunity for growth in an emergent market of scientists and engineers.[35] Outside this niche, in realms where performance was judged across a broader range of criteria than mere calculating speed, electronics still had a great deal to prove. Eckert and Mauchly would need to achieve considerable economies in order to make a machine developed under the largesse of a military contract pay dividends in the commercial marketplace. Neither Eckert and Mauchly nor anyone else engaged in the field of advanced calculation, except IBM, possessed the established capabilities in manufacturing and service necessary to accomplish that sort of refinement. Those competitors, moreover, faced just as daunting a challenge as IBM in trying to stay abreast of further developments in electronics.[36]

As IBM plotted its course into the world of electronics, the firm thus found itself in a curious position. Seizing the new opportunity within the scientific and engineering market called for highly entrepreneurial activities of a sort often associated with small startup organizations. Thomas Watson turned the task over to his eldest son, an unseasoned executive who had just returned from military service and was looking to gain additional managerial experience.[37] Yet IBM was not a small entrepreneurial firm. In its established realm of punched-card data processing, IBM operated as a dominant firm whose presence had grown ever more prominent, despite efforts to tame it through antitrust.

As much as the younger Watson might have wished to pursue electronics and scientific calculation free of antitrust concerns, in open competition with startup firms such as Eckert-Mauchly and Engineering Research Associates, such a course was effectively foreclosed to him. IBM's increasingly dominant position in the commercial sector made it highly vulnerable to the new antitrust regime. The company soon discovered, moreover, that it could not so readily insulate its entrepreneurial activities from its established commercial activities. Over the long run, both old and new equipment were destined to share a common technical underpinning in electronics. Lessons learned on the entrepreneurial side might well make their way into the established commercial line. That certainly was young Tom Watson's fervent hope. Along the way, those entrepreneurial activities in turn drew considerable strength from business practices developed on the commercial side. Developed routines in areas such as manufacturing, customer service, and field engineering provided vital assistance that often gave IBM a crucial advantage over its rivals in emergent markets.[38]

This convergence might have unfolded quite gradually, over the course of a decade or longer, if not for the bold aggressiveness of Eckert and Mauchly. In late 1947, the upstarts secured a design contract from Prudential Insurance, one of IBM's largest customers and also its sole lender. Two years later, Prudential contracted with Eckert-Mauchly for an electronic computer utilizing data input from magnetic tape. Use of tape for data storage threatened to displace the bays of punched cards that had long kept customers such as Prudential so dependent upon IBM.[39]

As a galvanized IBM scrambled to plot a response, its management received a sharp lesson about how antitrust would shape its freedom of action. Eckert-Mauchly, despite securing support from Prudential, still found itself woefully short of funds necessary to carry out its contracts. Its founders actively sought a more experienced suitor with ready access to capital. Tom Watson met with the inventors to discuss a buyout, but he soon abandoned the idea after consulting with lawyers at IBM and the Justice Department.[40] At the time, Congress was about to pass the Celler-Kefauver Antimerger Act, which severely curbed acquisitions by firms with large market shares.[41] Rebuffed by Watson, Eckert-Mauchly turned instead to Remington Rand, IBM's longtime chief competitor in the commercial data processing sector. The Justice Department gave its imprimatur to that merger and also permitted Remington Rand to acquire Engineering Research Associates, a startup company that had provided IBM with some of its stiffest competition in the market for scientific and engineering computing.[42]

These steps by its major competitor and by the Department of Justice sent clear notice to IBM that electronic calculating and computing fell within the broader orbit of its established business activities. Yet in so doing, they ironically reinforced the very message Tom Watson himself was struggling to promulgate within IBM. Frustrated with slow progress in electronics, Watson in 1949 had pledged to the IBM salesforce that within a decade all IBM products would be built exclusively from electronic components. In cutting IBM off from the option of acquiring expertise in electronics in wholesale fashion through merger, while allowing its principal competitor to exercise that option at will, Justice also lent further urgency to the younger Watson's insistence that IBM must reorient its own internal research and development efforts around the new technology of digital electronics. At the dawn of the electronics age, government pushed IBM in directions in which it was itself straining to move.

As the younger Watson orchestrated IBM's transition into the new era under the ground rules established by Justice, he and his colleagues found they could draw on a variety of outside sources of expertise through means other than outright merger. At the earliest opportunity, for instance, IBM sent personnel to seminars in solid state electronics offered at minimal charge by AT&T under terms of its consent decree.[43] It hired skilled personnel from RCA and other electronics companies who also benefited from knowledge promulgated by Bell Labs.[44] Later in the decade, as new startups gained the upper hand in semiconductor

260 production, IBM entered a critically important joint agreement with Texas
Instruments. By guaranteeing TI large purchases of electronic components and
providing it with technical assistance in circuit assembly technology, IBM
obtained expertise in silicon devices that proved vital to the System/360 line
of computers introduced in the mid-1960s.[45]

Another avenue open to IBM was defense contracting. Watson turned
aggressively in that direction during the early 1950s, as defense spending surged
with the escalating arms race and the onset of conventional military action in
Korea. Spillovers from these projects occurred more slowly than Watson had
imagined, however, and in some cases left IBM embroiled in expensive patent
disputes with fellow contractors.[46] More modest development projects, funded
internally and aimed at established markets, often generated surprisingly robust
returns. These successes drew extensively upon skills and business practices
that antedated the rise of electronics. The ability to mobilize and coordinate a
broad range of ongoing activities, including systems assembly, maintenance,
and custom programming at the site of customers, loomed large. Such com-
petencies, perfectly suited to the vaunted salesforce the elder Watson had
always considered the heart of IBM, were useful even in the rarified realm of
military contracting. IBM secured such contracts in large measure because it
knew how to close a deal and keep a customer satisfied.[47]

As with the Social Security contract in the 1930s, success in winning government business apparently came at a price in the antitrust arena. The advantages
IBM derived from its established business practices generated complaints from
its competitors and scrutiny from the Department of Justice. In January 1952, the
guillotine fell. The Department of Justice filed suit. While publicly vowing to
fight the claim, the younger Watson negotiated with Justice, over the strenuous
objections of his father.[48] When the Supreme Court ruled against United Shoe
in 1953, in a case which again had close parallels to that against IBM, government clearly held the upper hand. In 1956, the same year his father died, Tom
Watson entered into a formal consent decree with the Department of Justice.[49]

GOING IT ALONE: THE 1956 CONSENT DECREE
AND THE BIRTH OF SYSTEM/360

The 1956 decree bore marks from virtually every arrow in the antitrust quiver.
Revisiting the principal matter from the 1936 case, it called for IBM to take
concrete steps to ensure that competitive card manufacturers entered the field,
so that IBM would have less than 50 percent of the market for cards. New
clauses took direct aim at IBM's core business practices. IBM agreed to sell as
well as lease its products and to set distinct prices for the various devices and
routine services, such as regular maintenance, that constituted its integrated

systems installations. All of this paralleled the latest ruling against United Shoe. 261
In an added wrinkle, IBM would be required to set up an independent service
bureau, not for the purpose of maintaining its machines but to develop and
market data processing applications using IBM equipment. The clause establishing the service bureau marked an early effort to separate hardware from
software, before the latter had come fully into the consciousness of either IBM
or its customers.[50] Last but not least, IBM also agreed to license its proprietary
technologies for reasonable fees, while paying rival Remington Rand substantial licensing fees for its patents.

Although many of its details were new, the 1956 consent decree reflected the
persistent concern that had informed discussions of antitrust, patents, and innovation during the previous half century or more. Every element of the decree
sought to isolate the various technical components that made up an accounting or data processing system, to force IBM to make each stand on its own merits and to give competitors clear targets at which to focus their own efforts. The
clause regarding licensing of technology pushed this vision further up the innovative pathway. Competition would not occur merely at the machine level; it
would at least to some degree involve individual components and processes that
went into the machines themselves. Stipulations regarding maintenance charges
and the service bureau would impose the same philosophy but work in the
opposite direction, toward applications in the field at customer installations.
The unifying goal of government was to draw clear boundaries among the various machines and activities that together composed an integrated data processing operation.

In placing such emphasis upon boundary making, antitrust activity once
again mirrored to considerable degree concerns that already occupied Tom
Watson himself. At the time he consented to the decree, Watson was struggling
to reorganize IBM's internal operations. In 1956, he broke the firm into divisions responsible for various products. Three years later, he dramatically reshuf-
fled those divisions, added a new one responsible for developing advanced
systems applications, and created a new level of "group executive" responsible
for certain clusters of products.[51] In the interim, IBM announced the creation
of its first central research facility.[52] Early in 1961, the firm added a Components Division. Distinct from Research, it would supply basic electronic circuits for all IBM machines produced by the various product divisions.[53] Those
product divisions, throughout all the reshuffling, shared a common sales organization that stretched across the entire firm.

This incessant organizational churning reflected IBM's ongoing concern
about how best to stimulate technological learning and product innovation.
Watson and others within IBM worried that a project-based approach, in which
contractors essentially funded learning and innovation through purchase of

state-of-the-art systems, had not yielded the widespread benefits they desired. Three years into the initial reorganization, the division responsible for these high-end projects had absorbed some 70 percent of IBM's engineering resources, while generating just 30 percent of its revenue.[54] Those projects, moreover, contributed little innovation of obvious utility to the divisions that earned the lion's share of revenue by providing serviceable computer systems to the broad business and educational markets. Neither branch of IBM, moreover, appeared willing to undertake dramatic forays into advanced solid state technology that Watson and many other observers considered essential to sustaining leadership in computing.[55]

At root, these matters involved nagging questions about incentives and accountability. Stand-alone product development projects, aimed at a particular segment of the commercial market, offered a high degree of accountability. But by targeting limited markets, such projects generated little incentive (or resources) to take bold steps at innovation. High-end projects focused on enhanced technical performance, in contrast, tended to lack economic discipline and ultimate accountability. Susceptible to large cost overruns, they rode off of the faith that they would generate useful spinoffs.

One solution to this dilemma was to create a centralized research function, independent of any particular product development effort, with a clear mandate to generate bold but useful innovation that divisions could draw upon when developing new products. But how, under such an arrangement, could IBM maintain accountability within either the centralized research facility or the product programs that drew upon it? What would keep some product programs from acting as free riders, collecting the fruits of the centralized effort without shouldering their fair share of the cost? And what would keep centralized research from going off in directions that favored one division over another, or even worse, none at all?

Such questions tormented Watson and IBM during the late 1950s and early 1960s. Announcing the 1959 reorganization, Watson described the prevailing practice as "management according to who shouts the loudest."[56] He hoped the new organizational arrangements would create a less capricious mechanism. Within a year, however, such issues again reached a crescendo as IBM contemplated creating its Components Division.[57] Some managers complained vigorously that requiring divisions to obtain essential electronic building blocks from a common internal source would undermine their autonomy and with it their accountability. For a time, top management seriously entertained the idea of having the new Components Division sell its products in the open market. Outside sales, in addition to building volumes and spreading capital costs, would provide a check on the division's performance and establish a value for its offerings.

At this point, antitrust considerations appear to have substantially influenced Watson's course of action. If IBM chose to sell components on the open market, Watson and his management team realized, it must under terms of the consent decree agree to license the techniques involved in component production to outside parties. Components would, like peripheral devices, become susceptible to direct competition from outside providers. Such licensing ran directly against the purposes of IBM's backward integration into components production, through which IBM aimed to recapture the initiative from Texas Instruments and other electronics firms as technological change forced a convergence between components production and systems design. If IBM wished to integrate backward, then, it must do so definitively and develop close ties between its new capabilities in components and its production of "boxes" such as central processors. It could not go halfway. In the end, Watson authorized the Components Division without fully resolving the issues of accountability.[58]

The tension over components was a particularly dramatic example of a more general phenomenon. In creating divisions charged with producing systems for different segments of the market for computing, IBM ran considerable risk of duplicating its engineering efforts. This possibility existed not only with regard to basic electronic components but also in other areas such as input-output devices and applications programming. In these realms, too, IBM felt mounting incentives to concentrate its resources upon common products that might be used across a range of systems. In addition to reducing development and manufacturing costs, standardization in these areas might generate significant economies in sales and service.

The lure of standardization ultimately prompted IBM to embark on development of a comprehensive new product line. Eventually known as System/360, it featured a series of central processing units built from common components. These processors would possess "compatibility," by which IBM meant that they could run common applications programs, with higher-performance machines executing the routines faster. To every extent possible, these systems would utilize common input-output devices such as card readers, disc and tape drives, terminals, and printers. Though they were initially slated to appear over the course of some eighteen to twenty-four months, IBM ultimately introduced most of the line at a single blockbuster announcement in April 1964.[59]

THE COMPLEX LEGACY OF SYSTEM/360

System/360 has been widely heralded as one of the boldest and most successful strategic initiatives in the annals of American business.[60] Many observers have echoed a famous contemporary article from *Fortune* magazine, which dubbed it a "$5 billion gamble" because IBM had "bet the company" on a single product.[61] With System/360 leading the way, IBM's revenue and profits jumped to new heights. In 1969, worldwide revenues topped $7 billion and generated net income of nearly $2 billion, double the levels of 1965.[62] In its wake, foreign governments in Europe and Asia scrambled to create firms that could imitate IBM's success.[63]

In the eyes of some influential observers within IBM itself, however, System/360 was hardly an unmitigated triumph. The announcement plunged the firm into a state of crisis. Delays at facilities responsible for component production and circuit assembly created a massive scheduling problem with ramifications across the entire line. Programmers missed deadlines and accumulated alarming cost overruns as they struggled to achieve the promised compatibility while also scrambling to accommodate new applications such as time-sharing, which the salesforce had promised to dissatisfied customers. By early 1966, Watson had placed the entire program in the hands of an emergency management team. Headed by future CEOs Frank Cary and John Opel, this small group of executives imposed order over existing operations while also contemplating the lessons for the future. Its basic recommendation was neatly captured in the mantra, "Never Again."[64]

Criticism of System/360 went beyond the immediate difficulties of delivering products to market. Skeptics such as Cary and Opel also questioned the strategy System/360 embodied. Architects of that strategy had tried to address multiple concerns in one fell swoop. By relentlessly pursuing modularity in hardware and compatibility in software, they sought to gain the economies of Fordist mass production, without turning the computer industry into strictly a commodity business. IBM would continue to provide distinctive systems to each customer, only now, those systems would be assembled from common building blocks and would run common programs. By imposing this change across the entire line and building large volumes, IBM would be able to concentrate its resources and take a leap forward technologically, while also establishing a new hegemonic design. Customers would make a wholesale transition to the new system in order to gain the substantial advantages of the technological leap. Once having done so, customers would remain captive to System/360, as its modularity and compatibility enabled them to add computing capability more readily and cheaply than through any other alternative.[65]

Holes in the strategy came to light almost as soon as IBM announced the new line.[66] Customers at the lower end of the performance spectrum looked for incremental improvement rather than the clean break IBM had desired. They pressured IBM to offer an emulator program that would make System/360 compatible with previous generations of equipment, including machines recently introduced by major competitors. Meanwhile, users at the higher end

complained that System/360 failed to meet their needs. Rival Control Data
Corporation (CDC) won prestigious contracts for a new top-of-line computer
for use in advanced scientific applications.[67] Another gap in the System/360
line opened when rivals began offering popular time-sharing systems, which
allowed several operators to utilize a single central processing unit at the same
time.[68] Much to Watson's chagrin, IBM could not meet either challenge without substantially modifying System/360 hardware or software.

The sudden rise of time-sharing also exposed the risks inherent to modularity. Time-sharing systems effectively shifted computing expenditures away
from central processors and toward peripheral devices such as terminals, storage systems, and printers. Under terms of the 1956 consent decree, IBM had
to permit customers to obtain these stand-alone, "plug-compatible boxes" from
rival manufacturers. The modular design of System/360 had made it easier than
ever for potential rivals to exploit such opportunities and drive down the cost
of individual boxes. Any attempt to deviate from easy interconnectivity, say by
designing a time-sharing system that utilized distinctive programming channels housed in each of the attached boxes, would undoubtedly be seen as an
effort to retard the growth of plug-compatible manufacturers and would almost
certainly arouse further scrutiny under antitrust law.[69]

And, sure enough, antitrust action soon followed. CDC sued, claiming IBM
had prematurely announced what amounted to a phantom machine intended
to stave off competition at the high end. Its lawyers quietly collected a book of
case histories documenting how the IBM salesforce had strained to retain valued accounts through dubious practices, including exaggerated claims and provision of free services. Other firms launched private suits over time-sharing and
peripherals. Most important of all, the Department of Justice itself again began
investigating IBM. By late 1967, Watson once more found himself considering
how to avoid a major government antitrust suit. Only this time, to his shock
and disgust, Watson failed. In January 1969, on the last day of the Johnson
administration, government brought suit. Amazingly, the suit would stretch
across more than a dozen years. It ended about the same time John Opel succeeded Frank Cary as CEO. During his nearly decade-long tenure, Cary would
never know a day without the suit.[70]

The government antitrust action, though vigorously contested by Cary and
ultimately dismissed as without merit, must be counted as one more black mark
against System/360. The case gained strength from the successful action of Control Data, which quickly secured a highly favorable settlement from IBM. As
part of the agreement, IBM took possession of the case histories assembled by
CDC's lawyers and immediately destroyed them. This act, which by Watson's
own admission deprived government of evidence that would have significantly

266   enhanced its case, further fueled the resolve of government and helps account
for the inordinate length and vituperation of the case.[71]

Even before government launched the suit, moreover, the Justice Department extracted further concessions from IBM regarding its fundamental business practices. As before, government directed much of its concern toward tying
or bundling. It objected in particular to IBM's practice of offering many services, including much of its programming and software, without explicit charge.
Government asserted that IBM won key accounts from its competitors by lav-
ishing customers with software and programming services for which it charged
no published price. Such bundling undercut the objective of establishing clear
price and performance targets for computer products. IBM counsel Burke Marshall, who in his former capacity as assistant attorney general had overseen government antitrust litigation, characterized the practice to Watson as "a classic
tie" and advised him to adjust the policy if he wished to avoid a lawsuit.[72] Following months of internal study embracing every branch of the corporation,
IBM announced on December 6, 1968, that it would begin charging set prices
for much of its programming at the start of the following July.[73] The move
seemed so significant within IBM that managers preparing for the anticipated
competitive environment dubbed their work "New World."[74]

WHAT WROUGHT ANTITRUST? COMPETITIVE FORCES
IN COMPUTING IN THE 1970S AND 1980S

Whether these steps and the pallor that shadowed IBM throughout the long
antitrust case really exerted such profound influence upon the course of events
in the computer industry remains a matter of conjecture. In the annals of the
software industry, the unbundling of software from hardware has often been
heralded as giving birth to the industry.[75] Several pioneering software firms of
the day certainly considered it a watershed event at the time, and their founders
recall it as such.[76] But the record is not quite so straightforward. In dropping
the bundled programs and software services, IBM lowered the price of associated hardware a mere 3 percent. Available statistics regarding the subsequent
development of the industry suggest that figure captures the immediate economic impact fairly accurately.[77]

Perhaps the bigger problem in assessing the influence of antitrust involves
the counterfactual argument, advanced forcefully by IBM counsel during the
subsequent trial, which holds that IBM would have unbundled even in the
absence of action by government. Competition from firms such as CDC and
Digital Equipment Corporation, which offered little programming or services,
would have compelled IBM to charge for them. Similar logic suggested that

IBM could not have long succeeded by tying what should be separate components into larger integrated systems through proprietary operating systems or
channels.[78]

In light of this defense, which has acquired considerable credence among
economists and other students of antitrust, the question of whether antitrust
altered events largely comes down to one of pace rather than direction. Would
IBM have permitted the inevitable forces of change to work so freely in the
absence of pressure from government? And if not, could it have mounted an
effective resistance, one which at least would have bought it some time to
adjust its strategy and cover its mistakes? Or might IBM, fearing reprisals from
government, actually have moved more radically than economic forces themselves would have dictated? Might IBM have erred, for instance, by not tying
components of its systems as tightly as warranted by grounds of efficiency?

We will likely never arrive at definitive answers to these questions. Publicly,
Cary maintained that IBM would conduct its business without concern for
antitrust. Whether managers actually carried out that directive is less clear.
During the years of the suit, IBM abandoned plans for a comprehensive product development effort comparable to System/360 and instead embraced a strategy that called for targeted focus on various market segments.[79] Computer
systems proliferated, as plants and divisions became known for particular products. Within each line, IBM pursued a vertically integrated approach, with
basic components and programming generated largely from internal sources. In
some cases integration brought a retreat from modularity, as elements such as
memory and control channels were incorporated into comprehensive proprietary designs. In others, such as disk storage units, printers, and terminals, IBM
looked to utilize common sources across the firm.[80]

Virtually all IBM products—peripherals and systems—faced clearly identi-
fiable rivals in their markets. Only rarely, however, did IBM face the same competitor in multiple market segments. The largest and most expensive computers,
aimed at major institutions with large routine data processing operations, were
targeted primarily by Japanese electronics manufacturers such as Hitachi and
Fujitsu.[81] Mid-size computers came from Boston's Route 128 and later from
Sun Microsystems, an archetypal firm of what was becoming known as Silicon
Valley.[82] Rivals in disk storage spun off from IBM's own facility in San Jose, at
the heart of the Valley.[83] IBM jockeyed with Xerox, its fellow East Coast giant,
for the printer and copier markets.[84] As IBM began to resemble a conglomer-
ate, it began to think less in terms of winning accounts from competitors and
more about producing machines that could match or surpass competitive offerings. In a situation quite reminiscent of the late 1950s, tensions mounted
between the centralized salesforce and the product development groups, and

268   top management grew frustrated attempting to forge compromises and overcome bureaucratic resistance to change.[85]

Out of this situation came two fateful decisions during the early 1980s. The
first involved IBM's move into the realm of personal computing, which startup
companies such as Apple had pioneered during the late 1970s. A frustrated
Cary, anxious to demonstrate that IBM could still react nimbly, placed responsibility for IBM's PC in an independent business unit (IBU).[86] Freed from the
usual corporate constraints, this small group of liberated IBMers chose to
obtain both the central processing component and the basic operating system
for the PC from outside sources. The former came from Intel, the latter from
Microsoft. IBM had, in effect, handed the advantages provided by its brand
reputation and established market presence to the two upstart suppliers. To
make matters worse, engineers also inadvertently designed the PC as what
would later become known as an open system, with nonproprietary input-output channels. Competitors could readily manufacture clones assembled from
alternative circuit boards, memories, disk drives, monitors, keyboards, and printers. As each of these components rapidly became a target of feverish competition, prices and profit margins plummeted. What had taken years to occur in
mainframe computing had happened virtually overnight with the PC.[87]

Though these developments initially reshaped computing in ways long
desired by antitrust prosecutors and other critics, concerns about antitrust do
not appear to have figured prominently in IBM's strategic choices regarding
the PC. IBM had not hesitated to build proprietary input-output channels into
computer systems targeted at other market segments during the 1970s, and the
PC designers had not intended for the channels in their machine to be nonproprietary. When their mistake came to light, they immediately scrambled
(unsuccessfully) to undo the gaffe with revised designs for subsequent models.[88] The choice to rely on outside suppliers, which in retrospect appears to
have been an even bigger mistake, seems to have been made with open eyes.
Participants recall that in final meetings of corporate staff and top management
prior to announcement of the PC, concerns were raised about excessive
dependence on outside suppliers. IBM stayed the course, not because it feared
antitrust action if it acted otherwise, but because management was anxious to
culminate the experimental venture and because a general feeling prevailed
that "you could not make big mistakes in small markets."[89] This assessment,
however ironic it might come to appear in light of the subsequent evolution of
computing, was entirely consistent with the approach IBM had taken to the
computer market since System/360.

The second fateful decision of the early 1980s occurred when IBM abandoned
leasing entirely and began offering all of its products on strictly a sales basis.[90] In

a world dominated by rapid product turnover in virtually every market segment, management presumed that IBM would be better off letting customers take ownership and assume the risk of possible obsolescence. The move culminated the steady drift toward a business model based on commodity sales rather than ongoing account management. As IBM soon discovered, however, leasing still provided a useful buffer against swings in demand. Coffers swelled for a time, as customers paid up front for machines they previously had leased while established leases continued to run their course. But when old leases expired and demand for new machines fell below expectations, IBM lost billions.[91]

In addition to bringing IBM to the brink of financial disaster, the two decisions contributed to a further strategic blunder. In developing the PC group as an independent business unit, IBM had not perceived that small computers might effectively turn the computer industry upside down, or inside out. As the transforming power of decentralized computing grew ever more apparent, IBM then faced the difficult choice of whether to let the new IBU continue to take the lead in executing IBM's move into the new realm, or whether to absorb the PC group within the larger structure of the parent company. Flushed with confidence from the surge in revenue in mainframe computing, and reasoning that the vast resources available within IBM could better perform tasks such as developing a new operating system, linking the PCs more effectively in a network, and tying them to centralized databases, management chose the latter course. Unfortunately for IBM, the marriage proved awkward and ineffective. Its efforts consistently trailed those of Microsoft, Cisco, and Oracle, firms that had sprung to life with the emergence of distributed computing and intuitively grasped its essential character.[92]

REORIENTING IBM: OPEN SYSTEMS AND THE GERSTNER ERA

As IBM at last confronted the sort of disruptive crisis that had periodically plagued other large firms, it turned outside for new leadership, for the first time since the elder Watson had arrived in 1914. The new CEO, Louis Gerstner, brought a drastically different perspective to staid, bureaucratic Big Blue. Gerstner had cut his teeth working for the management consulting company McKinsey before achieving prominence as head of RJR Nabisco, a firm forged through a highly chronicled hostile takeover. Part of the new breed of flamboyant and highly compensated chief executives, Gerstner had breathed new life into its brands while drastically trimming costs.[93]

At IBM, Gerstner set off on much the same course. He slashed the total workforce to just over half its peak of the mid-1980s.[94] Product divisions and research incurred especially deep cuts, while spending on corporate advertising actually

270   increased, as Gerstner looked to shore up IBM's image among the large business consumers he believed constituted his firm's core market sector. The campaign's slogan, "solutions for a small planet," conveyed what Gerstner considered
IBM's greatest asset: its reputation for providing comprehensive, reliable services to business managers instituting change across large organizations. The new
emphasis restored sales and customer service to a preeminence they had not
enjoyed since the days of the elder Watson.

In one important respect, moreover, Gerstner actually outdid Watson: he
allowed IBM salespeople and business consultants to use non-IBM products
in assembling systems for specific customers. This move effectively mobilized
IBM's salespeople and business consultants as key agents of innovative activity. Rather than trying to coordinate various product divisions in an effort to
arrive at a single comprehensive systems design, Gerstner looked for numerous groups within the sales and consulting forces to experiment with their own
distinctive assemblies of products that might come from any number of sources.
If this meant lots of PCs and few large computers, or the reverse, so be it. Product divisions must no longer engage in internecine struggles to gain a foothold
within comprehensive IBM programs and thus secure for themselves a large
guaranteed internal market. Rather, they must scramble to generate products
that the marketers and consultants, acting as free agents, deemed useful and
competitive. The move dramatically redrew the boundaries of the firm, both
internally and externally, and restructured the incentives to innovation.

While releasing the sales and consulting forces to contract freely with alternative suppliers, Gerstner also encouraged personnel involved in research and
development activities to seek markets and partners outside IBM. He created
special incentives for ambitious laboratory researchers and other technicians
to bypass the established product divisions and work directly with partners in
IBM marketing to commercialize their ideas. Gerstner also instructed IBM
researchers to license their patents and ideas to outside parties. Taking the step
Tom Watson had resisted in the early 1960s, Gerstner authorized the components facilities to sell their products on the open market and to enter cooperative agreements with outside companies, including former rivals such as Sun
Microsystems. Remarkably, IBM became one of the most vocal proponents of
so-called open systems, patched together from hardware and software supplied
by numerous sources. No single approach, argued Gerstner, should dominate
computing.

In adopting this stance, IBM was of course confronting Microsoft, whose
nearly ubiquitous Windows operating system had secured it a position in distributed computing much like that IBM had enjoyed in the mainframe era. In
a move rich with irony, IBM carried its battle with Microsoft into the antitrust

arena.[95] When the Justice Department accused Microsoft of leveraging its operating system unfairly by forcing manufacturers of PCs to place its Web browser
and other applications programs in prominent spots on the opening monitor display, IBM provided essential testimony in support of the claim. The evidence
IBM marshaled helped Justice secure a resounding victory. An initial ruling by
Judge Thomas Benfield Jackson would have compelled Microsoft to split itself
in two, with one part responsible for operating systems and another for applications programs. The threat harkened back to the 1956 consent decree, which
called for IBM to spin off its service bureau as a distinct subsidiary. Though Judge
Jackson's ruling was later severely modified upon appeal, keeping Microsoft
intact, the software giant did alter many of its more objectionable business practices. At century's end, it remained in the crosshairs of antitrust surveillance.

While lending assistance to the crusade against Microsoft, Gerstner quietly
renegotiated IBM's own agreements with the Justice Department. He secured
an important concession when government agreed to abandon the dictates of
the 1956 consent decree calling for IBM to establish separate pricing for various products and services.[96] Henceforth, IBM salespeople and business consultants would be free to assemble various mixes of hardware, software, and
services and offer them at a single, comprehensive price unique to each customer. IBM had, in effect, come full circle. Account management, with fully
bundled packages and no clear targets at which competitors could take aim,
once again characterized its core business practices. Owing largely to a resurgence of activity in services, revenue and employment reached levels many
observers had presumed IBM could no longer attain, though its profit margins
lagged well behind those of its own heyday and those of Microsoft, the firm
that had assumed its place in the antitrust arena.

CONCLUSION

At the close of the twentieth century, antitrust activity was often portrayed by
its many critics as unwarranted and unwise meddling with beneficent market
mechanisms. Perhaps no single case did more to bolster that view than the long
suit begun against IBM in 1969. When at last dismissed as without merit, after
absorbing staggering amounts of legal resources, the case became a ready symbol for those seeking to highlight the potential waste and futility that might
result from such wholesale intrusion into the operations of private enterprise.
The suit also generated a sophisticated body of economic analysis that has continued to influence antitrust policy and case law ever since.

Yet in tracing IBM's long odyssey with American antitrust law, what seems
most striking is not the stark opposition between antitrust intervention and

272 market mechanisms, but rather their frequent symbiosis. Government persistently pushed in precisely the direction that IBM and its industry were being drawn by the workings of the market and the ongoing process of technological change those market processes stimulated. Antitrust policy toward the industry exhibited a remarkable coherence across the twentieth century. In consistently pressing to break apart linkages whose necessity seemed tenuous, government looked to preserve or even stimulate market competition. Fearful that inherited legacies or unwarranted bundling might interfere with the play of market forces and slow their solvent effects, an impatient government intervened. In at least some instances, these actions mirrored those of managers themselves, who also worried about the possibility that bundles and legacies might insulate groups within their enterprises from the inevitable forces of change.

Whether those actions by government significantly altered IBM or the computer industry remains an open question. The issue comes down largely to a matter of pace. Did antitrust remove impediments and accelerate the currents of change? Or did it merely introduce disruptive eddies into the flow? Certainly many participants in the story considered antitrust extraordinarily important at the time. At several junctures, IBM clearly altered its strategic course, or at least significantly speeded up the pace of its moves, in direct response to antitrust intervention. Competitors often felt heartened by those interventions and gained tangible benefits from them. The overall performance of the American computer industry, which has outpaced the rest of the world as well as virtually every other domestic industry, hardly suggests that antitrust has inflicted serious damage.

When viewed from a comparative international perspective, antitrust in fact jumps out as perhaps the most distinctive feature of the American landscape. Virtually all industrial nations of the postwar era identified computing as a field of enormous national significance, both militarily and economically. Governments the world over poured resources into the industry and pursued policies intended to foster healthy and vibrant domestic champions such as IBM. None spent more lavishly than the United States itself, with its postwar affluence and overriding concern with containing communism. Yet alone among nations, the United States simultaneously took steps to curb the power of its leading firm. From its granting of the Social Security contract onward, the American government consistently challenged IBM under antitrust even as it turned to the firm for critical technologies. The pattern held through the Depression, the Cold War, and the economic challenges of mounting global competition. While fighting national struggles, Americans consistently embraced an economic policy intended to break apart concentrated power and preserve market forces.

1. Richard S. Tedlow, *The Watson Dynasty: The Fiery Reign and Troubled Legacy of IBM's Founding Father and Son* (New York: HarperBusiness, 2003), 260. In 1970, IBM ranked fifth in the Fortune 500, behind General Motors, Esso, Ford, and General Electric. Its market capitalization of $41.5 billion on January 2, 1970, was twice that of GM, three times that of Esso, and six times that of GE. IBM's valuation on that date exceeded that of AT&T by more than $14 billion. Kevin Maney, *The Maverick and His Machine: Thomas Watson, Sr. and the Making of IBM* (New York: John Wiley & Sons, 2003).

2. Steven W. Usselman, "IBM and Its Imitators: Organizational Capabilities and the Emergence of the International Computer Industry," *Business and Economic History* 22 (Winter 1993): 1–35, reprinted in *Industrial Research and Innovation in Business*, ed. David E. H. Edgerton (London: Edward Elgar, 1996), 452–86; and Steven W. Usselman, "Fostering a Capacity for Compromise: Business, Government, and the Stages of Innovation in American Computing," *Annals of the History of Computing* 18 (Summer 1996): 30–39.

3. Paul Carroll, *Big Blues: The Unmaking of IBM* (New York: Crown, 1993).

4. *United States v. International Business Machines Corporation* 1956 U.S. District LEXIS 3992; 1956 Trade Cas. (CCH) P68, 245. The suit was filed by the Department of Justice on January 21, 1952, and settled on January 25, 1956.

5. Franklin M. Fisher, James W. McKie, and Richard B. Mancke, *IBM and the U.S. Data Processing Industry: An Economic History* (New York: Praeger, 1983); and Franklin M. Fisher, John J. McGowan, and Joen E. Greenwood, *Folded, Spindled, and Mutilated: Economic Analysis and U.S. v. IBM* (Cambridge, Mass.: MIT Press, 1983).

6. Thomas J. Watson Jr. and Peter Petre, *Father, Son & Co.: My Life at IBM and Beyond* (New York: Bantam Books, 1990), 230–33.

7. On these business practices, see Tedlow, *The Watson Dynasty*; Maney, *The Maverick and His Machine*; Robert Sobel, *IBM: Colossus in Transition* (New York: Times Books, 1981); and Emerson W. Pugh, *Building IBM: Shaping an Industry and Its Technology* (Cambridge, Mass.: MIT Press, 1995).

8. Elizabeth Wasserman and Patrick Thibodeau, "Microsoft, IBM Face Off," *Info World* 21 (June 14, 1999): 30. The case has been the subject of extraordinary journalistic coverage. Three compilations are Joel Brinkley and Steve Lohr, *U.S. v. Microsoft: The Inside Story of the Landmark Case* (New York: McGraw-Hill, 2001); Richard B. McKenzie, *Trust on Trial: How the Microsoft Case Is Reframing the Rules of Competition* (Cambridge, Mass.: Perseus, 2001); and Ken Auletta, *World War 3.0: Microsoft and Its Enemies* (New York: Random House, 2001). For more academic analyses of issues germane to the case, see Jerry Ellig, ed., *Dynamic Competition and Public Policy: Technology, Innovation, and Antitrust Issues* (Cambridge and New York: Cambridge University Press, 2001).

9. Maney, *The Maverick and His Machine*; and James W. Cortada, *Before the Computer: IBM, NCR, Burroughs, and Remington Rand and the Industry They Created, 1865–1956* (Princeton: Princeton University Press, 1993).

10. *International Business Machines Corp. v. United States* 298 U.S. 131 (1936).

11. *International Business Machines Corp. v. United States* 298 U.S. 131 (1936).

12. Donald Dewey, *Monopoly in Economics and Law* (Chicago: Rand McNally, 1959), esp. 171–78; William Letwin, *Law and Economic Policy in America: The Evolution of the Sherman Antitrust Act* (New York: Random House, 1965), esp. 115–16 and 151–52, and Hans B. Thorelli, *The Federal Antitrust Policy: Organization of an American Tradition* (Baltimore: Johns Hopkins University Press, 1955), esp. 62 and 325. Ironically, in blocking firms from acquiring competitors and their patents, antitrust law often served as an inducement to the formation of internal corporate research and development programs that enabled firms in some cases to assert stronger control over technologies. See David A. Hounshell, "Industrial Research and Manufacturing Technology," in *Encyclopedia of the United States in the Twentieth Century*, ed. Stanley Kutler (New York: Scribner's, 1996), 831–57; and Leonard S. Reich, *The Making of American Industrial Research: Science and Business at GE and Bell, 1876–1926* (New York: Cambridge University Press, 1985).

13. For a case citing the Congressional debate, see *International Business Machines Corp. v. United States* 298 U.S. 131 (1936).

14. *Motion Picture Patents Co. v. Universal Film Manufacturing Co.* 243 U.S. 502 (1917) and *United Shoe Machinery Co. v. United States* 258 U.S. 451 (1922). For a recent discussion of these and related cases, see Keith N. Hylton, *Antitrust Law: Economic Theory and Common Law Evolution* (New York: Cambridge University Press, 2003), 285. As Hylton points out, the Court's concern with limiting the leverage of patents during this period can also be seen in its insistence that patent holders could not enter price maintenance agreements with secondary consumers of products incorporating their patents. A patentee could control the price charged by a licensee who manufactured the patented product, but not that charged by someone who incorporated the licensee's product into another product or process. See Hylton, *Antitrust Law*, 254–56 and 261. The relevant cases are *Bauer & Cie v. O'Donnell* 229 U.S. 1 (1913) and *Straus v. Victor Talking Machine Co.* 243 U.S. 490 (1917).

15. The justices emphasized that their opinion was not predicated upon the high profit margins IBM earned on its card sales.

16. *Zenith Radio Corp. v. Radio Corp. of America, et al.* 106 F. Supp. 561, June 13, 1952; Reich, *The Making of Industrial Research*; and August W. Giebelhaus, *Business and Government in the Oil Industry: A Case Study of Sun Oil, 1876–1945* (Greenwich, Conn.: JAI Press, 1980), 198–270.

17. In addition to taking extensive testimony, the committee commissioned a report on the topic: Walton Hamilton, *Patents and Free Enterprise*, TNEC monograph no. 31 (Washington, D.C.: U.S. Government Printing Office, 1941). This monograph and the committee hearings pertaining to this topic are examined in Larry Owens, "Patents, the 'Frontiers' of American Invention, and the Monopoly Committee of 1939: Anatomy of a Discourse," *Technology and Culture* 32 (Oct. 1991): 1076–93. See also George E. Folk, *Patents and Industrial Progress: A Summary, Analysis, and Evaluation of the Record on Patents of the Temporary National Economic Committee* (New York: Harper and Brothers, 1942), for a rebuttal to Hamilton funded by the National Association of Manufacturers. On TNEC and its place in the larger history of antitrust, see Ellis W. Hawley, *The New Deal and the Problem of Monopoly: A Study in Economic Ambivalence.* (Princeton: Princeton University Press, 1966), esp. 368–70 and 404–71.

18. Donald Dewey, *The Antitrust Experiment in America* (New York: Columbia University Press, 1990).

19. On the idea of mandatory licensing, see esp. Folk, *Patents and Industrial Progress.*

20. Joseph A. Schumpeter, *Capitalism, Socialism and Democracy*, 3rd ed. (New York: Harper & Row, 1950), esp. 78–106.

21. Alfred E. Kahn, "Fundamental Deficiencies of the American Patent Law," *American Economic Review* 30 (Sep. 1940): 475–91.

22. On Bush, see G. Pascal Zachary, *Endless Frontier: Vannevar Bush and the Engineering of the American Century* (Cambridge, Mass.: MIT Press, 1999).

23. Owens, "Patents," 1084 and 1092.

24. Zachary, *Endless Frontier*; and Daniel Lee Kleinman, *Politics on the Endless Frontier: Postwar Research Policy in the United States* (Durham, N.C.: Duke University Press, 1995).

25. *International Salt Co. v. United States* 332 U.S. 392 (1947); Hylton, *Antitrust Law*, 286–92.

26. *United States v. Griffith* 334 U.S. 100 (1948) was especially important in this regard.

27. *United States v. United Shoe Machinery Corp.* 110 F. Supp. 295 (D. Mass 1953). The court's reasoning has itself come under intense scrutiny by scholars of antitrust, but my concern here is merely with establishing the mind-set that prevailed at the time, not on assessing the merits of the economic reasoning involved.

28. See Chapter 4 in this volume, by Kenneth Lipartito.

29. Michael Riordan and Lillian Hoddeson, *Crystal Fire: The Birth of the Information Age* (New York: W. W. Norton, 1997).

30. On the significance of World War II for American business, including IBM, see Thomas K. McCraw, *American Business, 1920–2000: How It Worked* (Wheeling, Ill.: Harlan Davidson, 2000), 73–102.

31. *IBM Annual Report* (1946).

32. JoAnne Yates, *Structuring the Information Age: Life Insurance and Technology in the Twentieth Century* (Baltimore: Johns Hopkins University Press, 2005); and Arthur L. Norberg, *Computers and Commerce: A Study of Technology and Management at Eckert-Mauchly Computer Company, Engineering Research Associates, and Remington Rand, 1946–1957* (Cambridge, Mass.: MIT Press, 2005).

33. Charles J. Bashe, Lyle R. Johnson, John H. Palmer, and Emerson W. Pugh, *IBM's Early Computers* (Cambridge, Mass.: MIT Press, 1986), 135–64; and Pugh, *Building IBM*, 167–74 and 185–86.

34. David A. Hounshell, "Assets, Organizations, Strategies, and Traditions: Organizational Capabilities and Constraints in the Remaking of Ford Motor Company, 1946–1962," in *Learning by Doing in Markets, Firms, and Countries*, eds. Naomi R. Lamoreaux, Daniel M. G. Raff, and Peter Temin (Chicago: University of Chicago Press, 1999), 185–208; and David Halberstam, *The Reckoning* (New York: Morrow, 1986).

35. On IBM's successful efforts in this market, see Bashe et al., *IBM's Early Computers*, 68–72, 84–86, and Atsushi Akera, "IBM's Early Adaptation to Cold War Markets: Cuthbert Hurd and His Applied Science Field Men," *Business History Review* 76 (Winter 2002): 767–802.

36. Norberg, *Computers and Commerce.*

37. Maney, *The Maverick and His Machine*; Tedlow, *The Watson Dynasty*; and Watson Jr. and Petre, *Father, Son & Co.*

38. For more detailed discussion, based on numerous documents in IBM's corporate archives, see Steven W. Usselman, "Learning the Hard Way: IBM and the Sources of Innovation in Early Computing," in *Financing Innovation in the United States, 1870 to the Present*, eds. Naomi R. Lamoreaux and Kenneth L. Sokoloff (Cambridge, Mass.: MIT Press, 2007), 317–63.

39. Yates, *Structuring the Information Age*, 113–50, and Norberg, *Computers and Commerce*, 73–116 and 167–208. On IBM's response, see Bashe et al., *IBM's Early Computers*, 102–3 and 114–16; Usselman, "Learning the Hard Way"; and "Summary of Products Committee Meeting Held on October 30th and 31st, 1947" T. J. Watson Sr. Papers, IBM Corporate Archives, Somers, New York.

40. Thomas J. Watson Jr., interview by Steven W. Usselman and Richard Wight, March 27, 1986, Armonk, New York.

41. For two recent perspectives on this bill and its effects, see David M. Hart, "Antitrust and Technological Innovation in the U.S.: Ideas, Institutions, Decisions, and Impacts, 1890–2000," *Research Policy* 30 (July 2001): 923–37, and Louis Galambos, "The Monopoly Enigma, the Reagan Administration's Antitrust Experiment, and the Global Economy," in *Constructing Corporate America: History, Politics, and Culture*, eds. Kenneth Lipartito and David Sicilia (New York: Oxford University Press, 2004), 149–67.

42. Norberg, *Computers and Commerce*, 205–208.

43. Bashe et al., *IBM's Early Computers*, 372–95.

44. B. N. Slade, interview by Steven Usselman, Poughkeepsie, New York, November 4, 1985.

45. "The IBM-Texas Instruments, Inc. Relationship," Chapter 10 of Components Division Procurement Role in System/360, Legal Papers, IBM Archives. This material includes copies of the IBM-TI agreement, December 23, 1957; a memo from P. N. Whittaker to J. W. Birkenstock et al., January 23, 1958 (PT10275), explaining its origins and the philosophy behind it; and the letter officially amending the agreement on May 5, 1960 (PT10276). On the technical dimensions of transistor work at IBM and TI, see Bashe et al., *IBM's Early Computers*, 390–95 and 399–406.

46. Emerson W. Pugh, *Memories That Shaped an Industry: Decisions Leading to IBM System/360* (Cambridge, Mass.: MIT Press, 1984).

47. Usselman, "IBM and Its Imitators"; and Bashe et al., *IBM's Early Computers*.

48. Maney, *The Maverick and His Machine*, 385–92; Tedlow, *The Watson Dynasty*, 193; Watson Jr. and Petre, *Father, Son & Co.*, 215–20; Watson Jr. interview by Usselman and Wight.

49. *United States v. International Business Machines Corporation* 1956 U.S. District LEXIS 3992; 1956 Trade Cas. (CCH) P68, 245.

50. On the significance of the service bureau, see Steven W. Usselman, "Public Policies, Private Platforms: Antitrust and American Computing," in *Information Technology Policy: An International History*, ed. Richard Coopey (Oxford, U.K.: Oxford University Press, 2004), 97–120.

51. IBM Organization Chart, May 20, 1959, with accompanying remarks by Thomas J. Watson Jr. and T. Vincent Learson, Organization Papers, IBM Corporate Secretary's Office, Armonk, New York.

52. Bashe et al., *IBM's Early Computers*, 544–50.

53. IBM officially announced formation of a separate Components Division on March 31, 1961. Numerous reports and letters in the Corporate Management Committee Papers, IBM Corporate Secretary's Office, Armonk, New York pertain to the formation of this division. See esp. R. H. Bullen to A. K. Watson, December 8, 1964. See also W. B. McWhirter to T. V. Learson, June 29, 1960, with R. H. Bullen to file, n.d., attached; Learson to A. L. Williams, July 1, 1960; Bullen to file, July 5, 1960; Learson to Bullen, August 3, 1960; and Bullen to T. J. Watson Jr., March 16, 1961.

54. IBM Organization Chart, May 20, 1959, with accompanying remarks by Thomas J. Watson Jr. and T. Vincent Learson, Organization Papers, IBM Corporate Secretary's Office, Armonk, New York.

55. Among those expressing this opinion was Mervin Kelly, head of research at Bell Labs, whom Watson had hired as a consultant. Watson Jr. interview by Usselman and Wight, 1985; Bullen to file, July 5, 1960; and B. L. Havens to E. R. Piore, June 1, 1961, CMC Files.

56. IBM Organization Chart, May 20, 1959, with accompanying remarks by Thomas J. Watson Jr. and T. Vincent Learson, Organization Papers, IBM Corporate Secretary's Office, Armonk, New York.

57. See note 53.

58. For persistent concern on the part of Watson about measuring the performance of the division against outside suppliers, either by selling its products on the open market or by obtaining them from a second source, see T. J. Watson to T. V. Learson, June 14, 1963; A. H. Eschenfelder to Watson, March 3, 1964; Watson to Paul Knapland, February 19 and December 3, 1965; and Knaplund to Watson, February 24, 1965; Watson Jr. Papers.

59. These developments can be traced in the files of the Corporate Management Committee, the Executive Conference Papers, and the Watson Jr. Papers. A watershed moment came in early January, 1962, when the group executive responsible for the System/360 initiative, Vin Learson, reported on the program to Watson. T. V. Learson to T. J. Watson Jr. and A. L. Williams, January 15, 1962, Watson Jr. Papers. On escalation of System/360 into worldwide effort across all machines, see T. J. Watson Jr. to T. V. Learson and A. K. Watson, January 15, 1963, and A. L. Williams to Dr. E. R. Piore, January 22, 1963, Williams Papers, IBM Corporate Archives, Div. 10/602, Box 70856.

60. For a recent treatment, see Alfred D. Chandler Jr., *Inventing the Electronic Century: The Epic Story of the Consumer Electronics and Computer Industries* (New York: Free Press, 2001).

61. T. A. Wise, "IBM's $5,000,000,000 Gamble," *Fortune* 74 (Sep. 1966): 118–23, 224, 226, and 228. The idea may have come to the reporter from the common joke at the time that IBM Co. stood for "I bet my company." See also T. A. Wise, "The Rocky Road to the Marketplace," *Fortune* 75 (Oct. 1966): 138–43, 199, 201, 205–206, and 211–12.

62. *IBM Annual Reports* (1965–1969).

63. Usselman, "IBM and Its Imitators."

64. At an executive conference in May of 1966, after hearing reports from Cary and Opel, Tom Watson identified the "major lesson" of System/360 as "At our size, we can't go 100 percent with anything new again." Executive Conference Papers, IBM Corporate Secretary's Office, Armonk, New York. "When we have learned from one experience like the 360," he wrote to top-level staff a few months later, "it is necessary that we confirm

278   our learning by clearly stating we will never do it again and, accordingly, it should be our policy in the future never to announce a new technology which will require us to devote more than 25% of our production to that technology and equipment dependent upon it during the first year of major production of that technology." Watson to R. H. Bullen, F. T. Cary, W. C. Hume, G. E. Jones, T. V. Learson, A. K. Watson, and A. L. Williams, September 9, 1966, Watson Papers.

65. T. V. Learson to T. J. Watson Jr. and A. L. Williams, January 15, 1962, Watson Jr. Papers.

66. On emulators, see Fisher et al., *IBM and the U.S. Data Processing Industry*, 89 and 118, and Gerald W. Brock, *The U.S. Computer Industry: A Study in Market Power* (Cambridge, Mass.: Ballinger, 1975), 15. Their importance in altering IBM strategy associated with System/360 is readily apparent in IBM corporate files pertaining to the launch of the new line.

67. Fisher et al., *IBM and U.S. Data Processing Industry*, 93–94; Pugh, *Building IBM*, 296–97; and Watson Jr. and Petre, *Father, Son & Co.*, 382–85.

68. Fisher et al., *IBM and the U.S. Data Processing Industry*, 159–67; and Fisher et al., *Folded, Spindled, and Mutilated*, 284–88.

69. Fisher et al., *IBM and the U.S. Data Processing Industry*, 286–303 and 327–39; and Fisher et al., *Folded, Spindled, and Mutilated*, 331–36.

70. Filed January 16, 1969, the case was settled on January 8, 1982. Fisher et al., *Folded, Spindled, and Mutilated*, 1 and 353–69; and Watson Jr. and Petre, *Father, Son & Co.*, 376–89.

71. Watson Jr. and Petre, *Father, Son & Co.*, 387–89 and *IBM Annual Report* (1972), 3 and 26–27.

72. Watson Jr. and Petre, *Father, Son & Co.*, 381.

73. *IBM Annual Report* (1968), 5–6, and *IBM Annual Report* (1969), 5–6.

74. Watt S. Humphrey, "Reflections on a Software Life," in *In the Beginning: Personal Reflections of Software Pioneers*, ed. Robert L. Glass (Los Alamitos, Calif.: IEEE Computer Society, 1999), 29–53; and Watson Jr. and Petre, *Father, Son & Co.*, 381. For documents pertaining to the unbundling of software, see Box 3, IBM Legal Papers, Accession 1980, Hagley Museum and Library, Wilmington, Delaware, esp. PX 2717.

75. See R. N. Langlois and D. C. Mowery, "The Federal Government Role in the Development of the U.S. Software Industry," in *The International Computer Software Industry: A Comparative Study of Industry Evolution and Structure*, ed. D. C. Mowery (New York: Oxford University Press, 1996), 53–85; and Kira R. Fabrizio and David C. Mowery, "The Federal Role in Financing Major Innovations: Information Technology During the Postwar Period," in Lamoreaux and Sokoloff, *Financing Innovation in the United States*, 283–316.

76. Glass, *In the Beginning*; and Walter F. Bauer, "Software Markets in the 70's," in *Expanding Use of Computers in the 70's: Markets, Needs, and Technology*, ed. Fred Gruenberger (New York: Prentice-Hall, 1971), 53–57.

77. Usselman, "Public Policies, Private Platforms."

78. Fisher et al., *IBM and the U. S. Data Processing Industry*; and Fisher et al., *Folded, Spindled, and Mutilated*; Brock, *The U.S. Computer Industry*; and Franklin M. Fisher, "Innovation and Monopoly Leveraging," in Ellig, *Dynamic Competition and Public Policy*, 138–59.

79. One initiative, known as Future Systems (FS), is discussed in Pugh, *Building* **279**
IBM, 307–17. The head of this initiative, Bo Evans, later attempted to resurrect the strategy by leveraging the Reduced Instruction Set Computer (RISC) architecture into a
new basic platform for all IBM computers. See Carroll, *Big Blues*, 201. RISC technology, though developed within IBM, ultimately reached the computer market through
upstart Sun Microsystems, a Silicon Valley firm. See AnnaLee Saxenian, *Regional
Advantage: Culture and Competition in Silicon Valley and Route 128* (Cambridge, Mass.:
Harvard University Press, 1994). For additional perspective on Evans and his strategic
thinking, see Frederick P. Brooks Jr., *The Mythical Man-Month: Essays on Software
Twentieth Anniversary Edition* (Reading, Mass.: Addison-Wesley, 1995).

80. Pugh, *Building IBM*, 301–22; and *IBM Annual Reports*.

81. Martin Fransman, *The Market and Beyond: Cooperation and Competition in
Information Technology Development in the Japanese System* (Cambridge, U.K.: Cambridge University Press, 1990).

82. Saxenian, *Regional Advantage*.

83. David G. McKendrick, Richard F. Doner, and Stephan Haggard, *From Silicon
Valley to Singapore: Location and Competitive Advantage in the Hard Disk Drive Industry* (Palo Alto, Calif.: Stanford Business Books, 2000).

84. Michael Hiltzik, *Dealers of Lightning: Xerox PARC & the Dawn of the Computer Age* (New York: HarperCollins, 1999).

85. Carroll, *Big Blues*; and Doug Garr, *IBM Redux: Lou Gerstner and the Business
Turnaround of the Decade* (New York: HarperBusiness, 1999).

86. Carroll, *Big Blues*, 18–44, surveys IBM decisions regarding the PC.

87. Richard N. Langlois, "External Economies and Economic Progress: The Case
of the Microcomputer Industry," *Business History Review* 66 (Spring 1992): 1–50.

88. Carroll, *Big Blues*, 120–59.

89. Humphrey, "Reflections"; and Carroll, *Big Blues*, 22–23.

90. Carroll, *Big Blues*, 59–60.

91. After earning record profits in 1990, IBM declared losses of $2.9 billion in 1991,
$5 billion in 1992 (including $5.4 billion in the fourth quarter), and $8.1 billion in 1993.
See Garr, *IBM Redux*, 13.

92. See Carroll, *Big Blues*, for an excellent and engaging account of the conflicting
approaches to software development at IBM and Microsoft. For broader perspectives on
software development, see Detlev J. Hoch, Cyriac R. Roeding, Gert Purkert, and Sandro K. Lindner, *Secrets of Software Success: Management Insights from 100 Software Firms
Around the World* (Boston: Harvard Business School Press, 2000); and Martin Campbell-
Kelly, *A History of the Software Industry: From Airline Reservations to Sonic the Hedgehog* (Cambridge, Mass.: MIT Press, 2003). On Oracle and Cisco, see David A. Kaplan,
The Silicon Boys: And Their Valley of Dreams (New York: William Morrow, 1999).

93. My treatment of IBM during the Gerstner era is based largely upon Garr, *IBM
Redux*, supplemented with contemporary news coverage.

94. Before the turnaround that began in 1994, IBM employment stood at just over
200,000. Garr, *IBM Redux*, 282.

95. See note 8 for sources on IBM and the Microsoft case.

96. Ira Sager and Diane Brady, "Big Blue's Blunt Bohemian," *Business Week*, June
14, 1999, 107–8.

9

CREDIT AND THE MATURE MARKET
FOR AUTOMOBILES[1]

Sally H. Clarke

Installment credit represented a major innovation in the U.S. automobile market. Prior to the 1910s, few lenders wrote such loans. One suspects that they held back because motor cars were rudimentary and unreliable, and might not outlive the loan's maturity dates. But by the 1920s, vehicles had become sturdier and durable, and many finance companies offered installment loans. Although Henry Ford proved slow to utilize credit financing, General Motors (GM) created the General Motors Acceptance Corporation (GMAC) in 1919. Car buyers made down payments of one-third the purchase price of a new car—often using a trade-in—and paid off the remaining two-thirds of the new car's price with monthly installments.[2]

Thanks in part to credit financing, the U.S. automobile market became a mass market, with slightly more than half of all households owning an automobile by the late 1920s. The percentage of households possessing cars remained flat during the Great Depression and World War II. Indeed, in 1948, 46 percent of families did not own a vehicle. But then in the next two decades the market expanded rapidly. By 1965, most families (79 percent) owned at least one car, and nearly a quarter of households owned two or more vehicles.[3] Automobile ownership had become a fact of life for the vast majority of households.

This chapter examines the relationship between credit financing and the expansion of the automobile market during the twenty years after World War II.[4] I argue that there was another innovation with respect to credit financing during the 1950s. In this case, lenders made available loans with smaller down payments and much longer terms for repayment. Several of the chapters in *The Challenge of Remaining Innovative* argue that innovations may at times be discontinuous, and in the case of automobile credit, the abrupt change reflected policies of the Board of Governors of the Federal Reserve System (the Fed).

Rather than focus on the internal boundaries of firms, I study the external environment in which automakers produced and sold vehicles. Here the state played an active role in shaping the conditions under which credit was critical to the expansion of the automobile market. The state changed its policies thanks to pressure from automobile dealers. They mediated between manufacturers and consumers, and they ultimately had responsibility for the sale of cars. We do not normally think of ordinary citizens lobbying the Fed, but during the 1950s automobile dealers launched a "pressure campaign" to compel Fed officials to change credit policies.[5]

The liberalization of credit during the 1950s did not apply to all Americans, however. A second change in credit financing dated to the mid-1970s, and it also represented a sharp break with the past. On this occasion, the break was prompted by various consumers, especially women, who wanted greater access to credit markets. They pressured Congress to pass new legislation banning credit discrimination and opening credit financing to all segments of consumers.[6]

To set the stage for the changes in credit financing, I first examine Americans' spending patterns during the postwar years. Certainly that was a period of unusual prosperity in the United States, especially as compared to other industrialized nations.[7] One might reason that Americans did not need credit financing to purchase cars because they enjoyed such high incomes. I examine but also question this supposition. I also examine the financial burden imposed by the automobile and how it was associated with major changes in families' budgets.

MARKETING AUTOMOBILES: A QUANTITATIVE PORTRAIT, 1945–1965

The Big Three staked their profitability on the expansion of the automobile market, which meant selling cars to consumers further down the income ladder. The 1950 census found that car ownership was correlated with income. Two-thirds of the 37 percent of families with incomes below $3,000 (representing the poorest families) did not own automobiles; by contrast, eight in ten of the 22 percent of families in the richest income bracket with incomes of $5,000 or more owned vehicles.[8] The results make sense because the distribution of income was so skewed. In 1954, for example, the richest 5 percent of families held 20 percent of all income, and the richest fifth of families claimed 45 percent of all income. The bottom 40 percent of the population held just 16 percent of all income that year.[9]

To be sure, the distribution of income during the 1950s and early 1960s was less unequal than in other historical eras. During the 1920s, for instance, the richest 5 percent of the population held 33 percent of income and the richest

10 percent claimed 44 percent. By the mid-1950s, the richest 10 percent's share of income had shrunk to 32 percent. But by the late 1990s, their share of income once again topped 40 percent.[10] By relative comparison to the 1920s or the late 1990s, the 1950s offered households an unusual era in which incomes were less skewed, but comparing groups within different income brackets for the mid-decade, the overall pattern was still highly skewed, with the richest Americans claiming far larger incomes than citizens further down the income ladder. By implication, most households did not buy new cars but instead purchased used cars. For 1954, the Fed reported that more than 20 percent of consumers bought automobiles, but car dealers sold twice as many used as new vehicles.[11]

In addition, during the 1950s more families opted to buy a second car, as the percentage of families with two or more vehicles rose from 7 to 15 percent.[12] Historian Tom McCarthy made the valuable point that the Ford Motor Company began featuring women in advertisements during the 1950s so as to promote the purchase of a second car. One advertising theme focused on women being left at home when their husbands went to work. Certainly, during the 1950s, more households purchased a second car, and during the 1960s this trend accelerated, as observed by historian Gary Cross, who noted that families with at least two automobiles rose to 29 percent as of 1970.[13] Still, two-car ownership was closely linked to a family's income. In 1955, the Fed wrote that "two-car" families were mostly found among the "upper income groups." The 1965 Survey of Consumer Finances reaffirmed this point as it reported that in the richest fifth of the income distribution some 51 percent of households had at least two vehicles as compared to just 13 percent of families in the second to the bottom quintile and 19 percent of families in the middle quintile.[14]

What accounted for the expansion of the auto market? One possible answer is the postwar prosperity. Certainly Americans enjoyed unprecedented incomes during the 1950s. Overall, family incomes increased 129 percent from an average level of $2,335 in 1929 to $5,356 in 1954. (Incomes increased 41 percent once adjusted for inflation.)[15] Yet prices for automobiles rose *pari passu*. Chevrolet offers a useful example as the entry-level automobile with the largest market share. In 1929, Chevrolet's price tag ranged from $525 to $675; in 1948, its price had doubled to $1,280, as one GM executive testified to a congressional subcommittee; and in 1954 the price increased 80 percent to $2,301.[16] Whereas family incomes had increased 2.2 times between 1929 and 1954, the price of a Chevrolet had increased between three and four times. For the 1950s, car prices kept pace with gains in family incomes. The Fed reported nearly a 50 percent increase in new car prices, on average, and a 58 percent increase in used car prices for the decade, whereas family incomes (unadjusted for inflation) rose roughly 53 percent.[17] Put another way, automobile historian James Flink reported that "[t]he average wholesale price of a car increased from $1,270 in 1950 to $1,822 by 1960, twice the rise of all wholesale costs during the decade."[18]

During the 1960s, prosperity played a greater role in helping families purchase automobiles than it had during the 1950s. From 1959 to 1969, new car prices rose on average 17 percent, and used car prices gained 15 percent. Americans' incomes increased by a much larger amount. Average incomes of households rose 79 percent (unadjusted for inflation) from 1959 to 1969, and median incomes increased 76 percent.[19]

New or used, automobiles imposed a hefty cost on household budgets. Flink observed that "transportation costs rose from 9.5 percent to 11.5 percent of personal disposable income" between 1950 and 1960. Autos did not constitute the entire share of transportation costs. In 1960, they accounted for 13.7 percent of consumption expenditures and "other travel and transportation" accounted for 1.5 percent, bringing the total cost of transportation to 15.3 percent of consumption expenditures. Because consumption expenditures were a portion of disposable income, autos' share of disposable income was smaller than their share of consumption expenditures.[20] The figures also were reported for all households, not just those households that owned vehicles. For instance, adjusting figures for the 76 percent of families who owned automobiles in 1960, the cost of autos was 18 percent of family expenditures on average.[21] Economist Clair Brown reported that for different groups of Americans transportation claimed at least 18 percent of their budgets by 1973, which she attributed to both a larger percentage of families with cars and the larger number of "two-car" families.[22] Autos' financial burden may be expressed in different ways, but the important point is that cars absorbed a significant part of family outlays.[23]

Automobiles' expense also varied with families' incomes. For 1960, families who owned vehicles in the very richest income bracket committed the smallest percentage of their expenditures to autos, 12 percent. In the next income bracket, autos consumed 16 percent of budgets in 1960. Most families who owned cars in the middling income brackets spent roughly 18 percent of their budgets on autos in 1960. At the bottom of the income distribution, autos accounted for 19 and 21 percent of family budgets for those reporting cars in 1960.[24] By 1972–73, when the vast majority of families owned cars, their share of consumer expenditures averaged 18.5 percent for all income groups. Looking at low-income Americans, the percentage share was actually lower than the average, but for middle-income groups the share was roughly 18 percent, and wealthier Americans devoted roughly 20 percent of their consumer expenditures to autos.[25]

284   Although cars imposed a large burden on households' annual budgets, changes in the terms of credit made those burdens more manageable. As was the case during the 1920s and 1930s, many families relied on installment credit to finance automobiles during the postwar years. In 1950, 57 percent of used cars and 46 percent of new car sales involved installment financing. In subsequent years, credit remained important to the sale of new and used cars. From 1954 through 1970, the percentage of new car sales involving loans never dipped below 60 percent. For used cars, between 47 and 65 percent of sales required financing between 1950 and 1970.[26] Overall, the amount of installment debt rose dramatically. In 1929, automobile paper amounted to $1.4 billion (or $15.9 billion adjusted for inflation in 2005 dollars); it plunged in the early years of the Great Depression, but rebounded to $1.5 billion in 1939 (or $21.0 billion in real dollars). Installment paper dwindled during the war years, but with the peacetime economy, auto paper rose to almost $5 billion in 1949 ($37.6 billion in real dollars), and it reached $16 billion by 1959 ($109.7 billion in real dollars). Total installment credit followed a similar pattern, jumping from $39.8 billion (real dollars) in 1929 to $262.8 billion (real dollars) in 1959.[27] Part of this increase reflected a greater proportion of indebted households. Whereas 24 percent of households held installment debts in 1935–36, 47 percent were indebted by 1956.[28]

      In the years when more families assumed debts, the terms for installment loans became more generous. First, during the mid-1950s, down payments shrank. Whereas in 1953 seven in ten contracts carried a down payment of more than a third of a new car's price, in 1955 half of auto contracts were written with a down payment of less than a third of the car's price. For used cars, 54 percent of down payments were under a third of the car's price by 1957.[29] Second, lenders lengthened the duration of contracts. During the 1920s, the vast majority of installment loans had been written for twelve months or less. In 1934, the time frame for new and used cars still followed this pattern. By 1937, credit terms had begun to ease: just 22 percent of new car loans and 51 percent of used car loans were written for twelve months or less. Then, after the wartime interruption, lenders continued to stretch out the duration of loans. In 1953, 83 percent of new car loans ran for more than nineteen months and 86 percent of used car loans were over twelve months. In 1957, 44 percent of new car loans were repaid over more than a thirty-month period. In 1960, two-thirds of used car loans were written for more than twenty-four months.[30] Compared to the 1920s, maturities had doubled or tripled, while the size of down payments relative to the total loan had diminished, thereby reducing a car's month-to-month burden on budgets. Overall repayments on installment loans averaged 20 to 22 percent of debtors' personal income during the 1950s, much like the figure in

1935–36 of 23 percent.[31] Put another way, economists Geoffrey Moore and Philip Klein reflected on auto loans in the postwar era: "repayment periods have increased roughly in proportion to the larger debt obligations, leaving payments per month about the same."[32]

The new credit terms helped sell new and used cars. Noting the 45 percent increase in automobile credit in 1954 over 1953 (even though new auto sales in 1954 dipped below their 1953 level), the *Federal Reserve Bulletin* explained, "credit sales have been stimulated by lower down payments and longer maturities, particularly on new cars."[33] One in four consumer "spending units" (the Fed's technical term that closely approximated families) had bought a new or used car, and a majority needed to obtain a car loan.[34] The use of credit increased as families' incomes or their liquid assets (such as savings accounts) diminished. For 1954, 54 percent of spending units with more than $5,000 in annual income took out an auto loan to buy a new car, as compared to 70 percent with less than $5,000 who relied on installments for their new car purchases. Differences according to liquid assets were more striking. The Fed reported, "In 1954, nearly 90 per cent of new car purchasers having liquid assets of less than $500 financed their cars by credit and only 30 per cent of those with liquid assets of $2,000 or more."[35]

I do not argue that credit was the only factor at work in the sale of autos.[36] Aside from credit, family incomes had made the purchase of autos possible. In looking at incomes, I make no claims that families were fixed within a specific bracket of the income distribution: Americans moved up and down the income ladder during their life cycles.[37] Instead, I find that credit remained important for easing the financial burden of automobiles for those families that found themselves on the lower rungs of the income ladder. Some families were on the lower rungs simply because they were young, but many were on the lower rungs as a result of class distinctions, other social hierarchies, illness, injuries, or bouts of unemployment.[38]

What was important was that the liberalized credit helped reduce the auto's financial burden for any family who had access to credit, and as it turned out, families with fewer financial assets tended to take out loans. Consider GM's price ladder. To buy a Cadillac in 1954, shoppers paid the total price up front, in cash in 89 percent of all cases. For Chevrolet, 45 percent of buyers paid the entire price in cash. Buick, Oldsmobile, and Pontiac fell in between these two extremes. Ford showed a similar pattern. Nearly 75 percent of Lincoln buyers paid the total price in cash, whereas only 46 percent of Ford buyers did so.[39] For the purchase of new cars, Moore and Klein report in a table of "the median ratio of monthly payment to monthly income" that payments in 1953 came to 30 percent or more for consumers with an annual income of less than $3,000,

21 percent for those with incomes of $3,000 to $4,200, 16 percent for those with incomes of $4,200 to $6,000, and 12 percent for those with incomes of $6,000 to $12,000. Consumers with incomes in excess of $12,000 devoted 6 percent of their incomes to installment payments.[40]

In the two decades after World War II, the Big Three automakers marketed brands according to a price ladder that targeted the richest 30 to 40 percent of the population. As Detroit hoped, new car buyers were tempted to trade in their existing vehicles before they had worn out. The Fed reported, for instance, that in 1954 one in four new car buyers traded in his or her vehicle in "regular intervals."[41] The average wait time was 3.2 years.[42] Their shopping patterns supplied the used cars that in turn cascaded to families further down the income ladder. Again, credit for used cars helped dealers sell vehicles, especially to families with smaller incomes or fewer liquid assets.

Although automakers adhered to the basic elements of the marketing strategy articulated by Alfred Sloan for the 1920s—styling, installment credit, and the price pyramid—the quantitative portrait emphasized two changes.[43] First, in contrast to the years from 1910 to 1930 when falling costs and prices had triggered the market's expansion, during the 1950s auto prices rose in line with Americans' incomes.[44] Second, the terms for installment loans eased during the postwar years. It is beyond the scope of this chapter to examine why automakers raised car prices during the 1950s whereas they had reduced prices during the 1910s and 1920s. What was important was the change in credit terms. Here dealers played a critical mediating role between automakers and officials at the Federal Reserve Board.

THE STATE: THE FED'S LIBERALIZED CREDIT TERMS

During World War II and the early postwar years, the Fed focused on consumer credit as part of its effort to control inflation. Officials established a set of regulations, known as Regulation W, to curb installment purchases. Then, after World War II, the Fed initiated two additional periods of credit controls; yet in contrast to the war years, when dealers supported Regulation W, after the war the rules inspired protest. Automobile dealers led a "pressure campaign," charging that the Fed's credit policies discriminated against low-income consumers.

Under Executive Order 8843, dated August 9, 1941, the Fed established restrictions on the maximum length of maturities as well as the minimum down payment for various goods. In the case of automobiles, buyers were required to make a down payment of one-third of the purchase price and pay off the loan within eighteen months (later modified to fifteen months).[45] Controls went into effect in September 1941 and were removed in November 1947. But with the "inflationary threat" looming, President Truman proposed new controls, and Regulation W went back into operation from September 1948 through June 1949. Its third and last period of operation came with the Korean War. Under the Defense Production Act, the Fed again regulated consumer credit markets from September 1950 through May 1952, and for the period from October 16, 1950, through May 7, 1952, the restrictions were stiff: the Fed continued to require that the down payment equal a third of the purchase price (in cash or trade-in) and it restricted maturities on car loans to fifteen or eighteen months.[46]

Dealers protested the second and third phases of the regulation. Linking credit regulation to lost sales, they sent hundreds of letters to the Fed complaining that the regulation should be relaxed or, better still, dismantled. The dealers took particular aim at the distributional consequences of the restrictions on the size of down payments and the length of maturities. In December 1948, a dealer from Salem, Oregon, wrote, "Our lowest priced new car is $2594.00 and Regulation W in its present form eliminates at least 50 percent of our prospective buyers and therefore makes it impossible for us to make sufficient sales to remain in business." Another dealer wrote that in his "farming district," someone "does not buy a car for pleasure—but to go to work and earn a livelihood." He proposed extending the maturity on auto loans from 18 "to at least 24 months."[47] Again in 1951 and 1952, dealers protested Regulation W. A Chrysler dealer complained, "Regulation W created class buying" since it "discriminates against the majority of people." They need auto transport but "are unable to meet the drastic and limited monthly terms, thus, such persons cannot improve their transportation." Regulation W, he added, "hit the automobile business very hard," and the dealer complained that he and others might well be "squeezed" out of the market. A California dealer repeated this basic point: "Regulation W discriminates against the working man whom it is supposed to protect, because it permits only people of means or with high incomes to purchase new cars, due to the high monthly payments required." By extension, the dealer fretted, "Regulation W works a tremendous hardship on automobile dealers and their employees because of their inability to sell cars under its restrictive terms."[48]

In 1952, a Fed official in Houston observed that dealers for the most part understood the goal of limiting inflation. He advised retailers "that there was no injustice done with Regulation W when automobiles were in short supply and you could sell all you could get without any extended terms. On both these points most everyone agreed. However, they [the dealers] bring out the fact that automobiles are no longer in short supply and the high prices that cars are now being sold for makes it difficult to sell them on the present market with

288 this limiting control." The official added, "As it now stands, I can't help but agree with them." In other words, the shift from a seller's to a buyer's market had put dealers' own prosperity at risk. Writing in February 1952, the president of Willys-Overland Motors, Inc. observed that "[t]he take-home price of the lowest priced automobile today is in the neighborhood of $2,000, approximately double what it was at the end of the war, and nearly three times what it was before the war." Under the new pricing conditions, he explained, many families could not cover the entire down payment, and even if they could their monthly payments of "approximately $70 a month" would be too much for their budgets.[49] Neither writer asked why auto prices had risen relative to Americans' incomes. In other words, rather than fault the credit restrictions, dealers or managers could have asked why manufacturers had not established a different price ladder of brands such that the price of new vehicles better matched the distribution of Americans' incomes, obviating the need for larger loans with longer maturities.

The Fed lifted Regulation W in the middle of 1952. The next year the economy dipped into a recession, but it revived in late 1954. With the economic rebound in 1955, auto sales surged and installment credit rose nearly 25 percent. The increase so captured Americans' attention that President Eisenhower requested Fed officials to investigate credit markets. Wanting to determine whether installment credit had a destabilizing influence on the monetary policy of the economy, the Fed conducted a four-part study, paying special attention to automobile paper. In its sections on credit regulation and again in part III, "Views on Regulation," the Fed recounted the various objections to credit controls, including dealers' complaints about the discriminatory consequences of minimum down payments and maximum maturities. Fed officials for the most part agreed with these complaints: they concluded that the limits for credit controls were felt more by families at the low end of the income distribution (or distribution of liquid assets) rather than farther up the ladder.[50]

THE FAMILY: WOMEN'S WORK AND WOMEN'S CREDIT

During the Cold War, policymakers in the United States sought to demonstrate the superiority of the U.S. economy to the Soviet system by emphasizing American families' increased consumption. Indeed, according to the Cold War ideology, the United States was or was soon to be a "classless society" thanks to the benefits of consumption.[51] When Vice President Richard Nixon undertook the famous "Kitchen Debate," he not only argued that consumption demonstrated the superiority of capitalism, but claimed as well that women in the United States as the family consumer-shoppers were better off than their Soviet counterparts.

Historian Elaine Tyler May recalled the debate: "Nixon and Khrushchev revealed some basic assumptions of their two systems. . . . Khrushchev countered Nixon's boast of comfortable American housewives with pride in productive Soviet female workers: in his country they did not have that 'capitalist attitude toward women.' Nixon clearly did not understand that the Communist system had no use for full-time housewives, for he replied: 'I think that this attitude toward women is universal. What we want is to make easier the life of our housewives.'"[52] U.S. housewives enjoyed a higher standard of living than Soviet wives, yet contrary to Nixon's dictum, during these years, married women with young children joined the paid workforce. Their decisions coincided with changes in family budgets, but also came in the context of segmented labor and credit markets.[53]

One of the most noticeable changes for families during the postwar years was the move to the suburbs. Suburbs had existed prior to the war, and states had funded roads. But after 1945 the federal government encouraged this move and a concomitant surge in auto sales in at least two important regards. First, the Federal Housing Administration along with the Veterans Administration provided low-interest loans for home mortgages and thus encouraged the wide-spread pattern of families moving to the suburbs and counting on their automobiles for transportation. Second, the vast network of highways, facilitated by the 1956 Highway Act, made automobile travel more feasible for households. The 1965 Survey of Consumer Finances indicated that families living in the suburbs were more likely to own one or two cars than were families that inhabited cities. The contrast was especially distinctive between the very largest cities and their surrounding suburbs.[54]

A third factor also assisted the move to the suburbs: families reallocated their budgets, spending much more for transportation, especially automobiles. This analysis emerges from the work of Clair Brown in tracking family budgets over the course of the twentieth century.[55] She examined families in different groups, separated roughly by income but also by occupation. In the example of white laborer households, which were composed mostly of unskilled workers, food's share fell from 34.5 percent in 1935 to 18.9 percent by 1973. The same trend applied to the other three groups.[56] Although food was the largest and most important item in the budget that shrank in size, the share of the budget devoted to fuel and lighting also diminished.[57] In addition, family incomes rose through this period. Brown writes, "real disposable income (per person) grew only 15%" during the 1950s, but rose "an amazing 55%" in the next thirteen years.[58] Even with these gains in income and lower operating expenses, autos still posed a significant cost. White laborers reported nearly a threefold increase in real expenditures on transportation between 1935 and 1950 and another 67

290   percent increase by 1960.[59] It is important to keep in mind that the figures for expenditures on automobiles as a percentage of family expenditures were just that, average figures.[60] As rates of auto ownership increased, a larger proportion of families devoted funds to the purchase of new and used vehicles as well as their upkeep, and so average costs increased. Looking once again at working-class families ("laborers"), by 1973, 89 percent of these families were car owners, and they devoted 18 percent of their budgets to transportation.[61]

      Families varied in how they earned the money to purchase their cars. Some consumers counted on the higher pay built into the cost of living adjustments (COLAs) of union contracts. The so-called "Treaty of Detroit" set the standard for other oligopolistic industries in which union workers received similar pay increases and benefits. Yet outside oligopolistic markets and even in the automobile industry outside of the Big Three automakers, workers did not receive pay packages on a par with the United Automobile Workers at GM or Ford. COLAs were rare in most industries, and so the earnings in other parts of the manufacturing sector, such as clothing and retail industries, fell to less than 65 percent of the autoworkers' average weekly earnings by 1960.[62] Although during World War II some labor leaders had promoted an "egalitarian wage," this objective diminished with the Korean War and the post-1965 inflation.[63] Although their settlement with the UAW allowed the Big Three to raise auto prices along with wages for their workers, employees in other segments of the economy were not as well remunerated and thus not as well prepared to finance the higher-priced automobiles.[64] A second option for these workers was to take advantage of the longer-term loans, but families could also elect a third option: both husbands and wives, mothers and fathers could work for pay.

      Indeed, in the postwar years, growing numbers of wives joined the workforce, even those with children under school age.[65] Between 1950 and 1960, wives with children younger than six increased their participation in the workforce from 12 to 19 percent, and wives with children six years of age or older increased their rates from 28 to 39 percent.[66] Brown writes, "In 1973, one-third of mothers with young children (under six years) and one-half of mothers with older children (6–18 years) were employed." Overall, the percentage of married women working for paid jobs roughly doubled from 22 percent in 1948 to 42 percent by 1973. Thus women's added incomes helped to purchase big-ticket items, and offered an alternative to credit financing.[67] This added income could be especially important as many families opted to buy not just one but two cars in the 1960s. The Survey of Consumer Finances for 1965 showed a steady uptick in the percentage of families with two or more cars as wives earned more income. Whereas just 22 percent of families owned two cars if the wife had an income below $500, 42 percent of families owned two cars in which the wife earned more than $5,000.[68]

Women who joined the workforce took jobs typically segregated along gender lines. Historian Nelson Lichtenstein put the issue in broad terms: "The weakness of the postwar welfare state and the extreme fragmentation inherent in the American system of industrial relations did much to redivide the American working class into a unionized segment . . . and a still larger stratum, predominantly young, minority, and female, that was left out in the cold."[69] The segmentation applied as well to credit markets. Although dealers pressed their case for easing credit terms so as not to discriminate on the basis of income, neither Fed officials nor retailers questioned credit discrimination associated with race, age, and sex. In these cases, discrimination was not directly linked to the Fed's efforts to stabilize the economy's performance through restrictions on the terms of credit, but simply as found in lenders' decisions to grant some borrowers credit but not others.

During the 1970s, studies documented numerous forms of credit discrimination.[70] In 1972, the National Commission on Consumer Finance identified five problems women faced: "1. Single women have more trouble obtaining credit than single men. . . . 2. Creditors generally require a woman upon marriage to reapply for credit, usually in her husband's name. Similar reapplication is not asked of men when they marry. . . . 3. Creditors are often unwilling to extend credit to a married woman in her own name. . . . 4. Creditors are often unwilling to count the wife's income when a married couple applies for credit. . . . 5. Women who are divorced or widowed have trouble re-establishing credit. Women who are separated have a particularly difficult time, since the accounts may still be in the husband's name."[71] Because loan applicants were often evaluated based on a credit system with points assigned for certain answers, subtle distinctions worked against women. For instance, legal scholar Maureen Ellen Lally reported that extra points were given for having a telephone in the applicant's name but in most households the telephone was in "the husband's name."[72] The "baby letter" was also used by creditors, as Lally noted: "Other lenders require a baby letter, a statement of the wife and her pediatrician that she is on a program of birth control with no plans of pregnancy." And Gail Reizenstein recalled women being asked to "sign an affidavit 'swearing not to endanger their ability to repay their debts by having children.'" One couple reported that the bank did not demand the usual baby letter but went so far as to request that the wife sign "an affidavit proving she had a hysterectomy."[73] Single women also faced problems, as Lally complained: "The poor risk status of the single woman is founded upon a belief that she is inherently unstable and incapable of handling her own affairs."[74] Regarding divorced women receiving alimony, Lally noted that they faced the problem that creditors viewed alimony as "an unstable source of income" and often would not consider it.[75]

Although data are scarce comparing female and male borrowers, the studies that I have found lend support that women were actually better credit risks.[76] In one study, economist David Durand observed, "Some credit men have expressed surprise that women should appear to be the better risks, and they have suggested that these results may be due to the indirect effect of other factors." John Chapman reported on a study comparing male and female borrowers, which showed that single and married women composed less than 24 percent of the sample of total borrowers, suggesting the relatively limited access of women to credit. But even this figure may be too generous. Excluding the category for "others" (for which the relative proportion of women cannot be determined) then women constituted less than 17 percent of "good loans" and less than 8 percent of "bad loans."[77] Economists Geoffrey H. Moore and Philip A. Klein in their consumer study found that although men and women received similar terms for automobile loans in the 1950s, the number of loans written for men far outpaced those for women by more than a ten-to-one ratio, which may have reflected both the effects of credit discrimination and women being a smaller portion of potential buyers. Historian Louis Hyman in his study of consumer credit reported that women were hard pressed to get auto paper.[78]

By the early 1970s, the National Commission on Consumer Finance heard a variety of complaints about credit that gave personal accounts to the bare-bones numbers. One married couple reported in 1972 that because the husband was a college student, the wife wanted to apply for the loan herself, but the bank refused and demanded that she find someone to "co-sign" the loan.[79] Another woman reported, "I am married but have a job and need transportation. I tried a credit union and two small loan companies to finance a car. All said it would be my husband's credit, not mine, that they would go on. My husband has been ill for several years and naturally has not worked steady. On the other hand, I work seven days a week at two jobs (one full time, one part time). And I think it very unfair they will not take that as a fact."[80] As reported in 1973, a divorced woman found that her efforts to obtain loans to buy a car came to naught because "the bank would count neither her child support nor her at-home telephone jobs as income." Thus "without a car, she couldn't take her children to an acceptable, inexpensive day care center and take a job outside her home."[81] The National Commission on Consumer Finance also cited a study conducted by the St. Paul Department of Human Rights in which two testers visited lenders.

> A man and a woman with virtually identical qualifications applied for a $600 loan to finance a used car without the signature of the other spouse. Each applicant was the wage earner, and the spouse was in school. Eleven of the banks visited by the woman "either strictly required the husband's signature or stated it was their prefer-

ence although they would accept an application and possibly make an exception to the general policy." When the same banks, plus two additional banks that would make no commitment to the female applicant, were visited by the male interviewer, six said that they would prefer both signatures but would make an exception for him; one insisted on both signatures; and six "told the male interviewer that he, as a married man, could obtain the loan without his wife's signature."[82]

Women also reported problems obtaining insurance along with car loans. As discussed at the hearings, "insurance companies put divorced and separated women into the 'high risk' pool for auto insurance."[83]

Married and single women both faced problems, albeit of somewhat different natures. Married women complained about the loss of their credit identity. As recounted in the 1972 report of the National Commission on Consumer Finance, Jorie Luteloff Friedman testified that when she married her husband she notified stores in order to receive "new credit cards with my new name and address. . . . Otherwise I maintained the same status—the same job, the same salary, and presumably, the same credit rating. The response of the stores was swift. One store closed my account immediately. All of them sent me application forms to open a new account—forms that asked for my husband's name, my husband's bank, my husband's employer. There was no longer any interest in me, my job, my bank, or my ability to pay my own bills."[84] Single women were often treated as "unstable" credit risks. This view came out in a 1974 case study of mortgages in Hartford, Connecticut. Lenders often assumed that "the female is inherently unstable and incapable of conducting her own affairs. She allegedly needs the protection of a male, usually a husband or father. In the lending industry the myth translated into a reluctance to grant a woman a mortgage loan outright and often, a requirement of an assumption or a male cosigner."[85]

In 1974 Congress acknowledged the long-standing climate of discrimination when its members passed the Equal Credit Opportunity Act (ECOA). Problems of credit discrimination were not confined to women. Hyman pointed to the discrimination many African Americans faced who lived in poor city areas.[86] A commonly overlooked problem in lending discrimination concerned retired Americans. Reizenstein noted that "discrimination against the elderly was the abuse most frequently documented in the hearings held to amend the 1974 Act."[87] In 1976, Congress added several categories to the ECOA such that it prohibited credit discrimination "on the basis of sex or marital status," as well as "race, color, religion, national origin, age, receipt of income from public assistance programs, and good faith exercise of rights under the Consumer Protection Act of 1968."[88] Speaking before these hearings, Barbara Jordan, representative from the state of Texas, recalled that Will Rogers had found a "solution for traffic jams: keep every car off the road until it was paid for." That cure, she noted,

294   "illustrates the extent to which our economy is dependent upon credit. The credit principle is inherent in the modern American way of life. More than half of the families in this Nation make use of consumer credit. The use of credit helps in emergency situations, management of money and provides a higher standard of living."[89]

Women had entered the segmented job market and had been excluded from consumer credit markets at a time when families reallocated their expenditures and purchased automobiles. Ironically, the very significance of automobiles and the need for credit to purchase them lent support for women and many other groups of consumers demanding passage of the ECOA. With the bill's enactment, a new chapter opened for the Fed: through a set of policies called Regulation B, its officials sought to eliminate numerous unfair lending practices. Among the Fed's actions, it prohibited the use of baby letters and points in credit scoring systems for having a telephone in one's name.[90] It also banned a common practice among creditors whereby they required women upon marriage to close out their credit accounts and open new accounts in their husbands' names. Thus, unless married women bought cars with cash, the credit for the purchase depended on their husbands' credit record.[91]

Although a full analysis is beyond the scope of this chapter, this review of women and credit suggests questions for the subsequent transformation of the auto market. The ECOA came about as the automobile market was undergoing significant changes. During the 1960s, automakers began to segment the market and introduced a variety of new models, especially compacts. James Rubenstein reported that "[t]he number of distinct models" rose "from 243 in 1950 and 244 in 1960 to 275 in 1970." By 1970, market watchers placed small cars at "one-third of the market" and by 1979 this share rose to 50 percent.[92] To what extent women were car buyers, especially of the second family car during the 1960s, I cannot say. But certainly Americans became more dependant on cars; as historian Gary Cross wrote, "There were 3.74 Americans for every car in 1950; that figure dropped to 2.9 in 1960 and 1.86 by 1980."[93] As I discussed earlier, for two-car households, wives' extra incomes contributed to car purchases, but families that bought a second car were nevertheless concentrated among the richer households regardless of whether wives worked.

Looking at market data for the years immediately after the passage of the ECOA, auto researchers reported that women claimed 24 percent of used car sales in 1982, as compared to "only 14 percent in 1979."[94] New car sales showed a similar pattern, although the numbers were reported in different ways. Some studies indicated that by the mid-1980s women accounted for 35 or 40 percent of new car sales, up from 25 percent in 1978.[95] Women were also said to have "influenced" eight in ten "new car purchases."[96] These figures should be taken cautiously, though, as it is not clear whether they represent sales to women as buyers of a second family car or to women as independent buyers. Another study lowered the overall magnitude of women's importance for car sales but not the general trend toward more influence by female buyers. This study in *Automotive News* found that as of the late 1980s "women are solely responsible for new car purchase and use in 28 percent of the cases, up from 19 percent in 1984."[97]

Although the general press did not track female car buyers much prior to 1970, articles about women buyers began to appear with much greater frequency in the 1980s.[98] Among the stories on women were those that concerned credit. J. D. Power & Associates (a market research organization) reported that two-thirds of women surveyed complained about difficulties in buying cars, including "[p]roblems in getting credit and patronizing attitudes of salesmen."[99] In 1985, Chevrolet announced a plan under which GMAC would advertise in women's magazines a program for women to apply for auto loans without visiting their car dealer. If approved, GM's manager was quoted as saying, the loan would be "good at any Chevy dealer." GM developed the plan because executives concluded that "financing is especially important to women."[100] In 1988, Chevy again used marketing to reach women buyers with pamphlets that among other topics addressed "credit and warranties."[101]

This analysis is too brief to determine how important access to credit with the ECOA had been for opening the auto market for women as well as other consumers.[102] Was it possible for women to have been major buyers during the 1960s, when, as Cross notes, nearly a third of families became "two-car" households?[103] One might also speculate as to whether auto dealers had proven to be exceptions to the general pattern of credit discrimination. Alternatively, family purchases of a second car may have relied on the husband's credit standing, although this access would not have applied to single women. But the timing of the ECOA in the middle of the 1970s certainly marked the upsurge of women as buyers, as noted in the press. The 1970s is usually singled out for the energy crisis and Americans' efforts to buy small cars to save on fuel costs. But perhaps the narrative was more complex. The decade not only marked high fuel costs and the arrival of imports, especially those from Japan, but also continued the trends both of women joining the paid workforce and of more divorces. Perhaps these trends along with the ECOA contributed to the increased demand by consumers with more modest incomes for economy cars. In 1980 one article reported that various "small cars" claimed "half of the US auto market for the first time in 1979." Among its notes, the article pointed to "working women" who favored "a small car instead of an intermediate or full-size car."[104] Ten years later Toyota reported

296   that "56 percent of their total sales are to women."[105] More research will be required to sort out these different trends, but this analysis suggests that one factor to be included in future studies is the role of credit financing.

CONCLUSION

Between the end of World War II and the mid-1960s, the automobile became entrenched as a daily necessity for the vast majority of Americans. The market's expansion came in part as families reaped the benefits of postwar prosperity, especially the gains in income they earned during the 1960s. But this account alone is insufficient to explain the market's expansion because car prices rose roughly as fast as Americans' incomes during the 1950s. Moreover, autos accounted for a large share of family budgets, especially for those living at the lower end of the income ladder. Instead, I have argued that credit proved crucial to the auto market's expansion. In contrast to the 1920s, during the 1950s lenders reduced the size of down payments and doubled and then tripled maturities for auto loans. Whereas vehicles imposed a substantially higher financial burden, families made smaller down payments and stretched out the length of the maturities on their loans. Not all families could count on postwar bonuses, good union jobs, or liberalized credit terms. The 1950s and 1960s saw the rapid movement of married women, including mothers with young children, into the workforce. Their incomes helped purchase expensive consumables such as automobiles.

Several chapters in this volume call attention to the changing institutional framework in which markets develop. In the particular history of the U.S. automobile market, there were two changes in the institutional framework for credit. The first came during the 1950s as car dealers lobbied the Fed to dismantle the restrictive credit controls of Regulation W. When the Fed complied, the liberalized credit terms assisted the sale of cars to families further down the income ladder. A second institutional change came almost twenty years later. Under pressure from the women's movement, Congress passed legislation banning credit discrimination on account of sex or marital status and then other forms of discrimination such as that based on age and race or color. The Fed once again played a key role in implementing fair lending policies, known as Regulation B. These policies, in turn, compelled lenders to make credit available to women as well as other Americans for the purchase of automobiles and other expensive durables.

The Challenge of Remaining Innovative has argued collectively that the use and management of information remains central to the process of innovation. This chapter affirms that basic point but also cautions that economic actors filter information through cultural lenses.[106] During the 1950s, car dealers protested Regulation W, claiming it discriminated on the basis of class or income. But their letters did not focus on other ways that creditors limited access to loans on the basis of sex, age, or race. That legal scholars objected during the 1970s to the view that single women were "unstable" or that married women should have their credit ratings subsumed under their husbands' reflected a period in our history when cultural attitudes shifted and Congress responded with the ECOA and the Fed initiated Regulation B.

NOTES

1. This chapter is based on excerpts from chapter 7 of my book plus some additional information on credit discrimination as related mostly to women. See Sally H. Clarke, *Trust and Power: Consumers, the Modern Corporation, and the Making of the United States Automobile Market* (New York: Cambridge University Press, 2007). The subject of automobile markets covers a range of issues, and even narrowing the subject to autos and women or postwar credit still leaves open many possible angles of analysis. This chapter offers a limited perspective. I do not discuss other topics associated with credit discrimination, such as discrimination in home mortgages or discrimination faced by African American farmers. On the subject of the postwar auto expansion, I do not pursue the role of marketing in this expansion but recognize that it played a role along with credit. Some readers interested in women and gender will find that this subject leads to questions that go beyond credit financing. A good introduction to the subject of women and car driving is found in Margaret Walsh, "Gendering Mobility: Women, Work and Automobility in the United States," *History* 93, no. 311 (July 2008): 376–95. Walsh is undertaking an extensive study of women and automobiles. On postwar developments in the auto market, readers should also consult recent studies such as James M. Rubenstein, *Making and Selling Cars: Innovation and Change in the U.S. Automotive Industry* (Baltimore: Johns Hopkins University Press, 2001); Avner Offer, *The Challenge of Affluence: Self-Control and Well-Being in the United States and Britain Since 1950* (Oxford: Oxford University Press, 2006), 193–229; Tom McCarthy, *Auto Mania: Cars, Consumers, and the Environment* (New Haven: Yale University Press, 2007); and Gary Cross, *An All-Consuming Century: Why Commercialization Won in Modern America* (New York: Columbia University Press, 2000). A good introduction to credit discrimination and mortgages is found in Helen F. Ladd, "Evidence of Discrimination in Mortgage Lending," *Journal of Economic Perspectives* 12 (Spring 1998): 41–62. A valuable introduction to many issues concerning the postwar history of the United States is found in James T. Patterson, *Grand Expectations: The United States, 1945–1974* (New York: Oxford University Press, 1996). I would like to note to readers that this chapter includes some corrections and some new material as compared to my book. I mistakenly omitted sources for the changes in car prices during the 1950s, and that information is now included in note 17. In addition, I extended my analysis of changes in car prices and family incomes into the 1960s. This exercise yielded the conclusion that postwar prosperity was more important during the 1960s than during the 1950s in

298   helping families purchase new and used automobiles. I would also like to point out that in my book I focused simply on the expansion of the auto market, but historian Tom McCarthy brings out the important point of sales of a second vehicle to families in the postwar period. I address this issue briefly. See McCarthy, *Auto Mania*, 148–51. See also Walsh, "Gendering Mobility." In this revised version of my chapter, I have tried to call attention to the many relevant data series in the new version of *Historical Statistics*. See Susan B. Carter et al., *Historical Statistics of the United States: Millennial Edition Online* (New York: Cambridge University Press, 2006), available at http://hsus.cambridge .org.ezproxy.lib.utexas.edu/HSUSWeb/HSUSEntryServlet (accessed Aug. 23, 2008).

    2. I review these issues in *Trust and Power*, 116–17, 121, and 255–56. It is important to note that Martha Olney has argued that the origins of installment credit came in manufacturer-dealer relations and not in selling cars to consumers. Olney found that installment credit was used to help dealers finance inventories. In addition, Olney traced the dramatic growth of installment credit for autos and other consumer durables during the 1920s and estimated the percentage of car sales that used financing. See Martha L. Olney, *Buy Now, Pay Later: Advertising, Credit, and Consumer Durables in the 1920s* (Chapel Hill: University of North Carolina Press, 1991), 86–134, esp. 92–97. On Henry Ford, see David A. Hounshell, *From the American System to Mass Production, 1800–1932: The Development of Manufacturing Technology in the United States* (Baltimore: Johns Hopkins University Press, 1984), 277 and 293; and Olney, *Buy Now, Pay Later*, 127–28. On the relationship between credit and car sales, see Alfred P. Sloan Jr., *My Years with General Motors*, ed. John McDonald with Catherine Stevens (Garden City, N.Y.: Doubleday, 1964), 162–63 and 302–12. Rubenstein also identified the link between a car's durability and the use of installment credit in *Making and Selling Cars*, 276. The marketing battle between GM and Ford is treated in Richard S. Tedlow, *New and Improved: The Story of Mass Marketing in America* (New York: Basic Books, 1990), 112–81. The terms of loans in the 1920s are reported in Geoffrey H. Moore and Philip A. Klein, *The Quality of Consumer Instalment Credit*, National Bureau of Economic Research (New York: Columbia University Press, 1967), 9 and 12. Olney also reported similar terms for installment loans in *Buy Now, Pay Later*, 109 and 113.

    3. For 1948, the first postwar year available, see U.S. Bureau of the Census, *Historical Statistics of the United States: From Colonial Times to the Present* (Washington, D.C.: U.S. Government Printing Office, 1976), series Q175, 717. For comparable figures on the automobile market's growth during the postwar years, see Lizabeth Cohen, *A Consumers' Republic: The Politics of Mass Consumption in Postwar America* (New York: Knopf, 2003), 123. For earlier data on the percentage of households with cars, see Stanley Lebergott, *Pursuing Happiness: American Consumers in the Twentieth Century* (Princeton, N.J.: Princeton University Press, 1993), 130. Walsh traced the increased number of cars registered in the United States in "Gendering Mobility," 377–78. In his essay on transportation, Louis Cain similarly noted the rapid increase in auto ownership after World War II. See Louis P. Cain, "Transportation," in Carter et al., *Historical Statistics of the United States: Millennial Edition Online*, http://hsus.cambridge.org/HSUSWeb/ search/searchessaypath.do?id=Df.Ess.01 (accessed July 28, 2008). The online version of *Historical Statistics* also reports data for car ownership and credit financing for the

period from 1947 to 1970. See Carter et al., *Historical Statistics of the United States: Millennial Edition Online*, series Df330–Df332, http://hsus.cambridge.org/HSUSWeb/search/searchTable.do?id=Df330-338 (accessed Aug. 3, 2008).

4. Numerous scholars of consumption, including Jackson Lears, Regina Lee Blaszczyk, Roland Marchand, and Susan Strasser, have noted the importance of installment credit for the sale of automobiles or other products but have not examined the nature of credit or its impact on the shape of consumer markets. There are important exceptions. One is Lendol Calder's valuable institutional and cultural study of credit during the years prior to World War II. Economic historian Martha Olney has provided the most detailed review of installment credit for the years prior to World War II. In her work, Olney charted the growth of credit, identified its importance for smoothing production flows from manufacturers to car dealers, and examined the terms of installment loans, especially their high rates. Lizabeth Cohen has also identified the importance of credit in her study of postwar consumption. Cohen called special attention to the subsidies returning GIs (who were often white and male) received as compared to other groups of Americans, notably women. She also singled out problems of credit discrimination faced by women and other groups of Americans. Louis Hyman has also written about credit discrimination, including the problems faced by women but also those for African Americans living in poor, urban neighborhoods. See Louis Hyman, "Ending Discrimination, Legitimating Debt: The Political Economy of Race, Gender, and Credit Access, 1968–1974," unpublished paper, Business History Conference, April 12, 2008. Hyman's analysis is more fully developed in *Debtor Nation: Changing Credit Practices in 20th Century America* (Princeton: Princeton University Press, forthcoming), chapter 7. A valuable article comparing consumer credit in the United States and West Germany in the immediate postwar years is Jan Logemann, "Different Paths to Mass Consumption: Consumer Credit in the United States and West Germany During the 1950s and '60s," *Journal of Social History* 41 (Spring 2008): 525–58. A recent study of the dramatic increase in national debt and the burden it places on Americans is found in Andrew L. Yarrow, *Forgive Us Our Debts: The Intergenerational Dangers of Fiscal Irresponsibility* (New Haven: Yale University Press, 2008). See also T. J. Jackson Lears, *Fables of Abundance: A Cultural History of Advertising in America* (New York: Basic Books, 1994); Regina Lee Blaszczyk, *Imagining Consumers: Design and Innovation from Wedgwood to Corning* (Baltimore: Johns Hopkins University Press, 2000); Susan Strasser, *Satisfaction Guaranteed: The Making of the American Mass Market* (New York: Pantheon Books, 1989); Lendol Calder, *Financing the American Dream: A Cultural History of Consumer Credit* (Princeton: Princeton University Press, 1999); Olney, *Buy Now, Pay Later*, 86–134; and Cohen, *A Consumers' Republic*, esp. 137–48, 156–60, 166–73, 281–83, and 369–70.

5. The Fed used the phrase "pressure campaign" to catalog its letters. See note 47.

6. On credit discrimination and the coming of the Equal Credit Opportunity Act, see also Hyman, "Ending Discrimination, Legitimating Debt"; and Hyman, *Debtor Nation*, chapter 7. Although I do not examine credit discrimination that African American families experienced, Hyman treated this subject in *Debtor Nation*, chapters 5 and 7.

7. Angus Maddison, *Dynamic Forces in Capitalist Development* (New York: Oxford University Press, 1991), Table 1.1, 6–7.

8. U.S. Department of Labor, Bureau of Labor Statistics, *Study of Consumer Expenditures, Incomes and Savings: Statistical Tables, Urban U.S.–1950*, volume 18 (Philadelphia: University of Pennsylvania, Wharton School of Finance and Commerce, 1957), Tables 1-1 and 1-10, 2 and 11. Fed officials similarly wrote, "Automobile ownership is closely related to income. Early this year, nine-tenths of the spending units with incomes of $5,000 or more owned cars as compared with only two-fifths of those with incomes of less than $3,000. Ownership of two or more cars was largely confined to the upper income groups." See "1955 Survey of Consumer Finances: Purchases of Durable Goods in 1954," *Federal Reserve Bulletin* 41 (May 1955): 467.

9. U.S. Bureau of the Census, *Historical Statistics of the United States*, series G319–G336, 301–302.

10. On the distribution of income across historical eras, see Carter et al., *Historical Statistics of the United States: Millennial Edition Online*, series Be28–29, http://hsus.cambridge.org/HSUSWeb/search/searchTable.do?id=Be27-29 (accessed July 26, 2008). Many issues related to the distribution of income are covered by Peter H. Lindert in his essay "The Distribution of Income and Wealth," Chapter Be - Economic Inequality and Poverty, in Carter et al., *Historical Statistics of the United States: Millennial Edition Online*, http://hsus.cambridge.org/HSUSWeb/search/searchessaypath.do?id=Be.ESS.01 (accessed July 26, 2008). One of the early treatments of the history of the distribution of income was Jeffrey G. Williamson and Peter H. Lindert, *American Inequality: A Macroeconomic History* (New York: Academic Press, 1980). A less technical account is offered by Frank Levy in *The New Dollars and Dreams: American Incomes and Economic Change* (New York: Russell Sage Foundation, 1998). See also Claudia Goldin and Robert A. Margo, "The Great Compression: The Wage Structure in the United States at Mid-Century," *Quarterly Journal of Economics* 107 (Feb. 1992): 1–34. Peter Lindert summarized the increased inequality since 1970 and assessed this inequality in relation to race and gender in "Three Centuries of Inequality in Britain and America," *Handbook of Income Distribution*, vol. 1 (Amsterdam, N.Y.: Elsevier, 2000): 167–216, esp. 201–204.

11. "1955 Survey of Consumer Finances: Purchases of Durable Goods in 1954," 466–67, 474. Gary Cross reported a similar conclusion for data from 1957 in an article in *U.S. News & World Report*. In 1957, most new cars were purchased by Americans in the top income brackets, even the cheapest-priced brand, the Chevrolet; conversely, most Americans purchased used cars. See Gary Cross, *An All-Consuming Century: Why Commercialism Won in Modern America* (New York: Columbia University Press, 2000), 89; and "New Cars: Who Buys Them, How They're Paid For," *U.S. News & World Report* (Mar. 14, 1958): 84–86.

12. Data for households with two or more cars are reported in Clarke, *Trust and Power*, 239. See also Carter et al., *Historical Statistics of the United States: Millennial Edition Online*, series Df332, http://hsus.cambridge.org/HSUSWeb/search/searchTable.do?id=Df330-338 (accessed Aug. 3, 2008).

13. See McCarthy, *Auto Mania*, 148–51. On the percentage of two-car families, see Cross, *An All-Consuming Century*, 182–83. Walsh also discussed the role of women and the trend in car ownership in "Gendering Mobility," 377–78.

14. See "1955 Survey of Consumer Finances: Purchases of Durable Goods in 1954," 467; and George Katona, Eva Mueller, Jay Schmiedeskamp, and John A. Sonquist, *1965 Survey of Consumer Finances*, Monograph No. 42 (Ann Arbor, Mich.: University of Michigan, 1966), Table 4-10, 79.

15. Figures are calculated from U.S. Bureau of the Census, *Historical Statistics*, series G308–G309, 301.

16. For Chevrolet prices, see "Chevrolet Prices: 1920–31 Incl.," File 83-1.4-9, and "Profits, Prices and Products," Statement and Discussion Before Subcommittee on Profits of the Joint Committee on the Economic Report by M. E. Coyle, Executive Vice President of GM, Washington, D.C., 12/20/48, File 83-1.9-51, Daniel C. Wilkerson Collection, Richard P. Scharchburg Archives, Kettering University, Flint, Michigan. See also "General Motors Pricing Policy *Better Value at the Same Price or Equal Value at Reduced Price* Chevrolet History 1925–1939," and "Chevrolet," both in File 87-4.18-56, Box 20, GMC/Proving Grounds Collection, Richard P. Scharchburg Archives, Kettering University, Flint, Michigan. For Chevrolet's price in 1954, see Clarke, *Trust and Power*, 245. Daniel Raff and Manuel Trajtenberg calculated an unweighted index of automobile prices in current and constant dollars, and illustrate the indices with two diagrams covering the years from 1906 to 1940. Daniel M. G. Raff and Manuel Trajtenberg, "Quality-Adjusted Prices for the American Automobile Industry: 1906–1940," in *The Economics of New Goods*, eds. Timothy F. Bresnahan and Robert J. Gordon (Chicago: University of Chicago Press, 1997), 78–79.

17. Changes in car prices were calculated from 1950 to 1959 using median prices for new and used cars. In 1950, the median price for new cars was $2,110, and it rose to $3,120 in 1959. For used cars, the median price rose from $550 to $870. For new car prices, the trend showed a steady upward tilt, whereas for used car prices, the median price rose sharply from $550 in 1950 to $900 in 1953 but then fell to $600 in 1955 before climbing again to $870 in 1959. Overall, the Fed noted in 1959, "List prices of automobiles have advanced at a higher rate over the postwar period as a whole than have prices of most other consumer goods." See "Consumer Durable Goods Recovery," *Federal Reserve Bulletin* 45 (Jan. 1959): 3. For data on car prices, see "1953 Survey of Consumer Finances. Part II. Purchases of Durable Goods in 1952 and Buying Patterns for 1953," *Federal Reserve Bulletin* 39 (July 1953): 698; "1955 Survey of Consumer Finances: Technical Appendix," *Federal Reserve Bulletin* 41 (May 1955): 474; "1958 Survey of Consumer Finances: Purchases of Durable Goods," *Federal Reserve Bulletin* 44 (July 1958): 771; "1959 Survey of Consumer Finances: The Financial Position of Consumers," *Federal Reserve Bulletin* 45 (July 1959): 718; and Survey Research Center, Institute for Social Research, *1960 Survey of Consumer Finances* (Ann Arbor, Mich.: University of Michigan, 1961), 34. Figures for income are calculated from U.S. Bureau of the Census, *Historical Statistics*, series G308–G309, 301. For the 1950s, the median income for households rose from $2,990 in 1950 to $4,759 in 1959 for a 59 percent gain or to $4,970 in 1960 for a 66 percent increase. See Carter et al., *Historical Statistics of the United States: Millennial Edition Online*, series Be10, http://hsus.cambridge.org/HSUSWeb/search/searchTable.do?id=Be1-18 (accessed July 26, 2008).

18. James J. Flink, *The Automobile Age* (Cambridge, Mass.: MIT Press, 1988), 287.

19. In contrast to the 1950s, for the 1960s car prices were reported as average prices (not median prices). In 1959, the average price of a new car was $3,150, as compared to $3,690 in 1969. The average price of a used car was $1,020 in 1959 versus $1,170 in 1969. (Used car prices were unusually high in 1959. In 1958, the average used car price was $850, resulting in a 38 percent increase by 1969. Yet incomes still outpaced this increase in prices.) Data for car prices are reported in Survey Research Center, Institute for Social Research, 1960 *Survey of Consumer Finances*, 34; and George Katona, Lewis Mandell, and Jay Schmiedeskamp, 1970 *Survey of Consumer Finances* (Ann Arbor, Mich.: University of Michigan, 1971), 54–55. The average and the median incomes of households are reported in Carter et al., *Historical Statistics of the United States: Millennial Edition Online*, series Be9 and Be10, http://hsus.cambridge.org/HSUSWeb/search/searchTable.do?id=Be1-18 (accessed July 26, 2008).

20. Flink, *The Automobile Age*, 287; and U.S. Bureau of Labor Statistics, U.S. Department of Commerce, "Consumer Expenditures and Income: Total United States, Urban and Rural, 1960–61," BLS Report No. 237-93 (Feb. 1965): 11.

21. In the 1960 survey, 76 percent of those surveyed reported owning cars. The data were reported as "average income, expenditures and savings," and automobiles accounted for 13.7 percent of total "expenditures for current consumption." Because these figures are average data for all those surveyed, I divided the figure by .76 to arrive at the share of costs for automobiles among those households that did own a car. This exercise leads to the conclusion that autos accounted for 18 percent of household budgets. My figures for 1960 are calculated from data reported in U.S. Bureau of Labor Statistics, U.S. Department of Labor, "Consumer Expenditures and Income," 11. Looking at what she called "major commodity groups" (but not items such as food and lighting), Martha Olney reported consumer expenditures for various durable goods, including new and used cars, in *Buy Now, Pay Later*, Table A.6, 213–21. Her analysis indicates that auto expenditures' share of total consumer expenditures was roughly 40 percent in the postwar era. The data are also reported in Carter et al., *Historical Statistics of the United States: Millennial Edition Online*, series Cd411–Cd412, http://hsus.cambridge.org/HSUSWeb/table/showtablepdf.do?id=Cd411-423 (accessed July 28, 2008). See also Lee A. Craig, "Consumer Expenditures," Chapter Cd - Consumer Expenditures, in Carter et al., *Historical Statistics of the United States: Millennial Edition Online*, http://hsus.cambridge.org/HSUSWeb/search/searchessaypath.do?id=Cd.ESS.01, (accessed July 28, 2008).

22. Clair Brown, "Consumption Norms, Work Roles, and Economic Growth, 1918–1980," in *Gender in the Workplace*, eds. Clair Brown and Joseph A. Perchman (Washington, D.C.: Brookings Institution, 1987), Table 4, 24–25, Table 6, 30, and 33.

23. In tracing the expanded consumer interest in "recreation and travel," Gary Cross also noted the relative fall in expenditures for autos. He offered a lower figure for the cost of ownership, noting "the share of household budgets for cars dropped from 6.2 percent to 4.6 percent between 1955 and 1960." Cross based his comment on an article in *U.S. News & World Report*, but he did not include the cost of "[g]asoline, tires, repairs." Looking at the cost of automobiles in terms of their purchase, operation, and maintenance, the data reported in *U.S. News & World Report* indicated that consumers spent 12.6 percent of their dollars on cars in 1955 and 10.9 percent in 1960. Total trans-

portation costs came to 13.9 percent in 1955, 12 percent in 1960. The journal's article did not provide a full citation for its source of data, and so it is not possible to follow up on the nature of the data in this case. Cross, *An All-Consuming Century*, 179; and "Big Change in Buying Habits—What It Means to Business," *U.S. News & World Report* (Feb. 6, 1961): 73–75.

24. To repeat, figures are adjusted in terms of the expenditure for automobiles relative to those families who owned automobiles. U.S. Bureau of Labor Statistics, U.S. Department of Labor, *Study of Consumer Expenditures, Incomes and Savings*, 3, 11; and U.S. Bureau of Labor Statistics, U.S. Department of Labor, "Consumer Expenditures and Income," 11. If the figures are not adjusted in terms of the percentage of families who owned cars, then wealthier groups show a higher percentage of expenditures for transportation. See Brown, "Consumption Norms, Work Roles, and Economic Growth, 1918–1980," Table 4, 24–25.

25. Calculations are made from the table on consumer expenditures in Carter et al., *Historical Statistics of the United States: Millennial Edition Online*, series Cd600 and Cd609, http://hsus.cambridge.org/HSUSWeb/table/showtablepdf.do?id=Cd597-616 (accessed July 28, 2008). Brown indicates that "low income" Americans spent 11.5 percent of their budgets on transportation in 1973, but wealthier groups spent 17 or 18 percent on transportation. See Clair Brown, *American Standards of Living, 1918–1988* (Oxford, U.K.: Blackwell, 1994), Table 6.1, 270.

26. Clarke, *Trust and Power*, Table 7.4, 253. See also Carter et al., *Historical Statistics of the United States: Millennial Edition Online*, series Df336 and Df338, http://hsus.cambridge.org/HSUSWeb/table/showtablepdf.do?id=Df330-338 (accessed Aug. 3, 2008).

27. Clarke, *Trust and Power*, Table 7.5, 254. On trends in consumer credit prior to World War II, see Olney, *Buy Now, Pay Later*, 87–90. Hyman analyzed debt and postwar Americans' ideas about and patterns of consumption in *Debtor Nation*, chapter 5.

28. Moore and Klein, *The Quality of Consumer Instalment Credit*, Table 7, 21. Cohen traced the growth of indebtedness in *A Consumers' Republic*, 123–24.

29. Moore and Klein, *The Quality of Consumer Instalment Credit*, 12.

30. Data are reported for sales finance companies. For banks, Moore and Klein reported that in 1956 38 percent of loans for new vehicles were made for more than 31 months. By 1960, the figure was 63 percent. Moore and Klein, *The Quality of Consumer Instalment Credit*, 9 and 15. Cross also noted the general trend toward longer maturities for auto loans in *An All-Consuming Century*, 92.

31. Moore and Klein, *The Quality of Consumer Instalment Credit*, 24–25. See also U.S. Board of Governors of the Federal Reserve System, *Consumer Instalment Credit* (Washington, D.C.: U.S. Government Printing Office, 1957), part 1, vol. 1, "Growth and Import," 135–39.

32. Moore and Klein, *The Quality of Consumer Instalment Credit*, 15. Olney traced changes in the terms of installment credit for autos and other durable goods in *Buy Now, Pay Later*, esp. 113.

33. "Growth of Consumer Instalment Credit," *Federal Reserve Bulletin* 41 (Dec. 1955): 1312–13. See also U.S. Board of Governors of the Federal Reserve System, *Consumer Instalment Credit*, part 1, vol. 1, "Growth and Import," 130–39. Used car sales rose

304 in 1954. On auto sales, see "1955 Survey of Consumer Finances: Purchases of Durable Goods in 1954," 466.

34. "1955 Survey of Consumer Finances: Purchases of Durable Goods in 1954," 466–67. The vast majority of these spending units were families, but the spending unit could have been a single individual or an extended family. See also "1955 Survey of Consumer Finances: Technical Appendix," 471; and Irving Schweiger, "Size Distribution of Family Income, Savings, and Liquid Asset Holdings in 1945," June 24, 1947, Folder 565.5, Box 2532, Records of the Board of Governors of the Federal Reserve System, Entry 1, Record Group 82, National Archives, College Park, Maryland. (hereafter cited as RG 82, NA). The Fed focused on spending units, as it reported in 1947, because "they represent consumer units of economic decisions, actions, and plans better than families (which in some cases contain more than one spending unit) or individuals." "Survey of Consumer Finances, Part I. Expenditures for Durable Goods and Investments," *Federal Reserve Bulletin* 33 (June 1947): 647.

35. The exact figures for the relationship between liquid assets and new car sales, as reported in Table 2, showed that loans were required by 87 percent of buyers with liquid assets under $500, 64 percent with liquid assets between $500 and $2,000, and 30 percent with liquid assets over $2,000. A similar pattern applied to used car sales. The Fed reported, "70 per cent of the used car purchasers with $500 or less of liquid assets used credit as compared with 20 per cent of the purchasers with liquid assets of $2,000 or more." The actual figures in Table 2 were slightly different. They indicated that for those consumers with liquid assets of $2,000 or more, 22 percent used credit financing, whereas 51 percent of those with liquid assets of $500 to $1,999 used credit financing and 68 percent of those with fewer than $500 in liquid assets needed an auto loan. The pattern for incomes was different, as the percentage of credit financing was similar for all income groups in buying used cars. Liquid assets proved a clearer predictor of credit financing, whereas the nature of income in buying used cars may have had less predictive value given that fewer households with incomes below $3,000 owned cars. "1955 Survey of Consumer Finances," 466–67. Moore and Klein also linked credit to variations in liquid assets. See Moore and Klein, *The Quality of Consumer Instalment Credit*, 29. Among recent historical studies, Hyman also discussed liquid assets and debt in *Debtor Nation*, chapter 5.

36. Richard Easterlin examined the question of Americans' materialism in his recent collection of essays. See Richard A. Easterlin, *The Reluctant Economist: Perspectives on Economics, Economic History, and Demography* (New York: Cambridge University Press, 2004), 32–53.

37. One study based on the Survey of Consumer Finances reported significant variations in income in a three-year period. See Ralph B. Bristol Jr., "Factors Associated with Income Variability," *American Economic Review* 48 (May 1958): 279–90. See also George Katona, *The Powerful Consumer: Psychological Studies of the American Economy* (New York: McGraw-Hill, 1960), 183–84.

38. The new edition of *Historical Statistics* includes data looking at wage gaps among different groups of Americans as well as differences in unemployment rates. In 1967 (the earliest year for which data is reported), African American men earned $19,503 (in 1994 dollars) as compared to $30,195 for white men; African American women

earned $13,040, whereas white women earned $17,470. Overall, all women earned on average $16,943 in contrast to the average for all men of $29,322. See Carter et al., *Historical Statistics of the United States: Millennial Edition Online*, series Ba4512–Ba4514, and Ba4514–Ba4518, http://hsus.cambridge.org/HSUSWeb/table/showtablepdf.do?id =Ba4512-4520 (accessed July 28, 2008). See also Robert A. Margo, "Wages and Wage Inequality," Chapter Ba-Labor, in Carter et al., *Historical Statistics of the United States: Millennial Edition Online*, http://hsus.cambridge.org/HSUSWeb/search/searchessaypath .do?id=Ba.ESS.04 (accessed July 2008). On unemployment, different groups of Americans suffered different rates of unemployment. In the year 1960, for instance, the unemployment rate for white men was just 4.5 percent. African American men had an unemployment rate twice as high, or 8.9 percent. White women had a low unemployment rate of 4.9 percent, but African American women faced an unemployment rate of 8.8 percent. See Carter et al., *Historical Statistics of the United States: Millennial Edition Online*, series Ad783, Ad784, Ad788, and Ad789, http://hsus.cambridge.org/HSUS Web/table/showtablepdf.do?id=Ad782-791 (accessed July 28, 2008).

39. Market Research Division, *U.S. News & World Report* and Benson & Benson, Inc., *The People Buying New Automobiles Today: A Marketing Report on 18 Studies Among the Families Who Bought New 1954 Models of Each of the Principal Makes of Automobiles . . . Conducted by Market Research Division, U.S. News and World Report [and] Benson & Benson, Inc., Princeton, New Jersey* ([Washington, D.C.]: U.S. News Publishing Corporation, 1955), 32, Baker Old Class Collection, Baker Library Historical Collections, Harvard Business School, Boston, Massachusetts.

40. Moore and Klein, *The Quality of Consumer Instalment Credit*, 92.

41. "1955 Survey of Consumer Finances," 466.

42. Market Research Division,*U.S. News & World Report* and Benson & Benson, Inc., *The People Buying New Automobiles Today: A Marketing Report on 18 Studies Among the Families Who Bought New 1954 Models of Each of the Principal Makes of Automobiles . . . Conducted by Market Research Division, U.S. News and World Report [and] Benson & Benson, Inc., Princeton, New Jersey* ([Washington, D.C.]: U.S. News Publishing Corporation, 1955), 29, Baker Old Class Collection, Baker Library Historical Collections, Harvard Business School, Boston, Massachusetts. Americans also reported owning their new cars for roughly three to four years during the mid- to late 1960s. See Katona et al., *1970 Survey of Consumer Finances*, 58.

43. The price pyramid is discussed in Rubenstein, *Making and Selling Cars*, 204–206. For styling, readers should consult C. Edson Armi, *The Art of American Car Design: The Profession and Personalities* (University Park: Pennsylvania State University Press, 1990); David Gartman, *Auto Opium: A Social History of American Automobile Design* (London: Routledge, 1994); and Richard Martin, "Fashion and the Car in the 1950s," *Journal of American Culture* 20, no. 3 (1997): 51–66. On marketing in general, a valuable study is Thomas Frank, *Conquest of Cool: Business Culture, Counterculture, and the Rise of Hip Consumerism* (Chicago: University of Chicago Press, 1997). One view of women and design is offered by Shelley Nickles, "'Preserving Women': Refrigerator Design as Social Process in the 1930s," *Technology and Culture* 43 (Oct. 2002): 693–727. See also Shelley Nickles, "More Is Better: Mass Consumption, Gender, and Class Identity in Postwar America," *American Quarterly* 54 (Dec. 2002): 581–622.

44. The Bureau of Labor Statistics stated, "Probably the most significant factor in the growth of car ownership was a drop of 40 percent in factory-delivered prices of cars" during the 1920s. U.S. Department of Labor, *How American Buying Habits Change* (Washington, D.C.: U.S. Government Printing Office, 1959), 186. On the relationship between prices and technology, see Raff and Trajtenberg, "Quality-Adjusted Prices for the American Automobile Industry: 1906–1940," 71–101.

45. U.S. Board of Governors of the Federal Reserve System, *Consumer Instalment Credit*, part 1, vol. 1, "Growth and Import," 289–96.

46. U.S. Board of Governors of the Federal Reserve System, *Consumer Instalment Credit*, part 1, vol. 1, "Growth and Import," 292–302, esp. Table 54, 292–94, and Table 55, 300. The National Bureau of Economic Research investigated the question as to whether changes in credit terms would affect consumers' buying habits. See Wallace P. Mors, *Consumer Credit Finance Charges: Rate Information and Quotation* (New York: Columbia University Press, 1965), 45–52.

47. E. U. Teague to Thomas B. McCabe, chairman, Board of Governors, December 16, 1948; and George Brown, Freehold Willys, Inc., to Board of Governors, December 16, 1948, both in Folder 502.111, "Automobiles—Pressure Campaign to Permit Longer Maturities," Box 2325, Entry 1, RG 82, NA.

48. H. W. Robinson, Harry Sommers, Inc., to Walter F. George, no date [stamped May 2 1952]; and Dick S. Heffern, Secretary, Motor Car Dealers Association of Orange County, to Board of Governors of the Federal Reserve System, April 18, 1952, both in Folder 502.111 "Automobiles (Jul 1951–1954) Consumer Durable Goods—Credit Control," Box 2316, Entry 1, RG 82, NA. Logemann in his review of consumer credit during the 1950s drew a similar conclusion about Regulation W as being "discriminatory." See Logemann, "Different Paths to Mass Consumption," 535.

49. Ross Stewart to W. H. Holloway, Federal Reserve Bank of Dallas, April 8, 1952, and Ward M. Canaday, president, Willys-Overland Motors, Inc., to William M. Martin Jr., chairman, Board of Governors, February 11, 1952, both in Folder 501.111 "Automobiles (Jul 1951–1954) Consumer Durable Goods—Credit Control," Box 2316, Entry 1, RG 82, NA.

50. U.S. Board of Governors of the Federal Reserve System, *Consumer Instalment Credit*, part 1, vol. 1, "Growth and Import," vols. 1–2, 315–24; and U.S. Board of Governors of the Federal Reserve System, *Consumer Instalment Credit*, part 3, "Views on Regulation," esp. 112–34 and 168–83. See also Logemann, "Different Paths to Mass Consumption," 535.

51. "Classless society" was a term Nixon used. Cohen, *A Consumers' Republic*, 124–27. See also the discussion in Elaine Tyler May, *Homeward Bound: American Families in the Cold War Era*, rev. ed. (New York: Basic Books, 1999), 10–18.

52. May, *Homeward Bound*, 12. See also Cohen, who wrote that the postwar consumption "promised the socially progressive end of economic inequality without requiring politically progressive means of redistributing existing wealth." Cohen, *A Consumers' Republic*, 125–27, quote 127.

53. Brown, "Consumption Norms, Work Roles, and Economic Growth, 1918–1980," 47–48. See also Brown, *American Standards of Living*, 273–74.

54. Mark Rose discussed the Highway Act of 1956 in *Interstate: Express Highway Politics, 1941–1956* (Lawrence, Kans.: Regents Press of Kansas, 1979). See also Mark H.

Rose, "Reframing American Highway Politics, 1956–1995," *Journal of Planning History* 2 (Aug. 2003): 212–36. For an account of VA loans, see Kenneth T. Jackson, *Crabgrass Frontier: The Suburbanization of the United States* (New York: Oxford University Press, 1985), 195–206. On the diffusion of electrical networks, see Mark H. Rose, *Cities of Light and Heat: Domesticating Gas and Electricity in Urban America* (University Park: Pennsylvania State University Press, 1995); and Richard F. Hirsh, *Power Loss: The Origins of Deregulation and Restructuring in the American Electric Utility System* (Cambridge, Mass.: MIT Press, 1999). See also Lizabeth Cohen, *Making a New Deal: Industrial Workers in Chicago 1919–1939* (New York: Cambridge University Press, 1990), 273–77. For an assessment of who had access to these resources as well as their distributional consequences, see Cohen, *A Consumers' Republic*. For an account of class distinctions in the suburbs, see Becky M. Nicolaides, *My Blue Heaven: Life and Politics in the Working-Class Suburbs of Los Angeles, 1920–1965* (Chicago: University of Chicago Press, 2002). Data about the funding for highways are reported in Carter et al., *Historical Statistics of the United States: Millennial Edition Online*, series Df225–Df242, http://hsus.cambridge.org/HSUSWeb/table/showtablepdf.do?id=Df225-242 (accessed July 28, 2008). Walsh linked the growth of suburbs with women's work and use of cars in "Gendering Mobility," 390–92. Data on families with cars living in cities as compared to suburbs are found in Katona et al., *1965 Survey of Consumer Finances*, Table 4-10, 80. The table indicated that for the twelve largest cities, fewer than 15 percent of families owned at least two cars in 1965, whereas 39 percent of families in the surrounding suburbs owned at least two cars. In other cities, the percentage of families with at least two vehicles was larger, averaging 22 percent in 1965.

55. This pattern is based on data in Table 4 and is part of her general analysis. Brown details the decline in spending for food and the rise in spending for autos plus what she calls "recreation" and "personal insurance." See Brown, "Consumption Norms, Work Roles, and Economic Growth, 1918–1980," 24–25, 48. See also Brown, *American Standards of Living*.

56. Brown, "Consumption Norms, Work Roles, and Economic Growth, 1918–1980," Table 4, 24–25. See also Brown, *American Standards of Living*, 9, 106, 188, and 270.

57. Among white laborer families, the cost of fuel and lighting came to 8.5 percent in 1935 but slipped to 4.4 percent in 1973. Brown, *American Standards of Living*, 106, 188, and 270; and Brown, "Consumption Norms, Work Roles, and Economic Growth, 1918–1980," Table 4, 24–25. On the subject of heat and lighting, see also Rose, *Cities of Heat and Light*.

58. Brown, *American Standards of Living*, 268.

59. Real auto expenditures are calculated from data in Brown, "Consumption Norms, Work Roles, and Economic Growth, 1918–1980," Tables 4 and 5, 24–25, 28–29, and for real disposable income, see 23. In general, on auto expenditures, see Brown, "Consumption Norms, Work Roles, and Economic Growth, 1918–1980," 42.

60. In its 2001 survey, the Bureau of Labor Statistics made the point, "Because not all consumer units purchased all items during the survey period, the mean expenditure for an item is usually considered lower than the expenditure by those consumer units that purchased it." Officials gave the example of the purchase of new cars. "For instance, reference table 1 shows average expenditures for new cars and trucks of only $1,628 because relatively few consumer units actually purchased a new vehicle. For example,

308 if about 7 percent of the households reported purchasing a new car or truck in 1999, the average expenditure on new cars and trucks for those households would be $23,257." U.S. Department of Labor, Bureau of Labor Statistics, "Consumer Expenditure Survey, 1998–99," *Bulletin* 955 (Nov. 2001): 40.

61. Brown, "Consumption Norms, Work Roles, and Economic Growth, 1918–1980," Tables 4 and 6, 24–25, and 30.

62. Nelson Lichtenstein, "From Corporatism to Collective Bargaining: Organized Labor and the Eclipse of Social Democracy in the Postwar Era," in *The Rise and Fall of the New Deal Order, 1930–1980*, eds. Steve Fraser and Gary Gerstle (Princeton: Princeton University Press, 1989), 122–52, esp. 141–45, and Table 5.1, 145.

63. Lichtenstein, "From Corporatism to Collective Bargaining," 144.

64. Lichtenstein, "From Corporatism to Collective Bargaining," 132–33, 141, and 145. For other treatments of labor issues in the postwar era, consult Thomas Sugrue, *The Origins of the Urban Crisis: Race and Inequality in Postwar Detroit* (Princeton: Princeton University Press, 1996); Nelson Lichtenstein, *The Most Dangerous Man in Detroit: Walter Reuther and the Fate of American Labor* (New York: Basic Books, 1995); Christopher L. Tomlins, *The State and the Union: Labor Relations, Law, and the Organized Labor Movement in America, 1880–1960* (New York: Cambridge University Press, 1985); and Meg Jacobs, *Pocketbook Politics: Economic Citizenship in Twentieth-Century America* (Princeton: Princeton University Press, 2005).

65. May, *Homeward Bound*, 16. Dora L. Costa has reviewed the movement of women into the paid workforce in the United States and compared this development to those found in other countries. See Dora L. Costa, "From Mill Town to Board Room: The Rise of Women's Paid Labor," *Journal of Economic Perspectives* 14 (Fall 2000): 101–22. See also Brown, "Consumption Norms, Work Roles, and Economic Growth, 1918–1980."

66. Aside from Brown's data series, statistical data on the participation of women with children younger than six may be found in Carter et al., *Historical Statistics of the United States: Millennial Edition Online*, series Ba581, http://hsus.cambridge.org/HSUS Web/toc/tableToc.do?id=Ba579-582 (accessed Aug. 2, 2008); and Brown, "Consumption Norms, Work Roles, and Economic Growth, 1918–1980," Table 7, 31.

67. Brown, "Consumption Norms, Work Roles, and Economic Growth, 1918–1980," Table 7, 31; Brown, *American Standards of Living*, 273–74. Walsh discussed the increasing numbers of women who worked in "Gendering Mobility," 388–89. Richard Easterlin offered a different account, but with a similar timing of the movement of women into the paid workforce, in *The Reluctant Economist*, 205–18. As Cohen reported, autoworkers chose lifestyles after the war that reflected their social concerns and not simply their income levels. See Cohen, *A Consumers' Republic*, 161–62. In general, see Pierre Bourdieu, *Distinction: A Social Critique of the Judgement of Taste*, trans. Richard Nice (Cambridge, Mass.: Harvard University Press, 1984). My account of changes within the household facilitating the increased demand for consumer goods is similar to Jan de Vries's account of households during the early modern period. See Jan de Vries, "The Industrial Revolution and the Industrious Revolution," *Journal of Economic History* 54 (June 1994): 249–70. Data on the participation of married women in the job force are found in Carter et al., *Historical Statistics of the United States: Millennial Edition*,

series Ba579, http://hsus.cambridge.org/HSUSWeb/toc/tableToc.do?id=Ba579-582 (accessed Aug. 2, 2008).

68. Brown, "Consumption Norms, Work Roles, and Economic Growth, 1918–1980," 33. Walsh discussed the trend of owning two cars and the role of working women in "Gendering Mobility," 384–87. The data on "two-car" families and wives' incomes are reported in Katona et al., *1965 Survey of Consumer Finances*, Table 4-11, 81. It is worth noting that Table 4-11 reported that women with incomes over $5,000 constituted just three percent of those surveyed, and households with "[n]o wife, or wife has no earned income" represented 69 percent of those surveyed and just 18 percent of these families had two cars. (One is left with the sense that the survey was skewed in certain ways, such as under-representing the percentage of families with wives in the paid workforce. Still, as the table stands, it indicated that for the different categories of income, wives with more earnings were linked to the tendency of their families buying a second car.) See Katona et al., *1965 Survey of Consumer Finances*, 81.

69. Lichtenstein, "From Corporatism to Collective Bargaining," 144. Sugrue discussed labor market segmentation in *The Origins of the Urban Crisis*. On the subject of job discrimination, see, for example, Claudia Goldin, *Understanding the Gender Gap* (New York: Oxford University Press, 1990); and Pamela Walker Laird, *Pull: Networking and Success Since Benjamin Franklin* (Cambridge, Mass.: Harvard University Press, 2006), 92–136.

70. Louis Hyman discussed the National Commission on Consumer Finance and other issues related to the discrimination women faced in "Ending Discrimination, Legitimating Debt"; and Hyman, *Debtor Nation*, Chap. 7.

71. National Commission on Consumer Finance, *Consumer Credit in the United States: Report of the National Commission on Consumer Finance* (Washington, D.C.: U.S. Government Printing Office, December 1972), 152–53. The five problems were reported in Gail R. Reizenstein, "A Fresh Look at the Equal Credit Opportunity Act," *Akron Law Review* 14 (Fall 1980): 216. Hyman reviewed many of these problems in "Ending Discrimination, Legitimating Debt"; and Hyman, *Debtor Nation*, Chap. 7.

72. Maureen Ellen Lally, "Comments: Women and Credit," *Duquesne Law Review* 12 (Summer 1974): 864–66.

73. On the baby letter, see Lally, "Comments: Women and Credit," 869. Reizenstein, "A Fresh Look at the Equal Credit Opportunity Act," 217 and 217 note 18, and 227. Reizenstein quoted John W. Cairns on the affidavit. See John W. Cairns, "Credit Equality Comes to Women: An Analysis of the Equal Credit Opportunity Act," *San Diego Law Review* 13 (July 1976): 965. See also Cohen, *A Consumers' Republic*, 147. Georgia Dullea, "Women Demanding Equal Treatment in Mortgage Loans," *New York Times* (October 29, 1972), Proquest Historical Newspapers *The New York Times* (1851–2005): R1.

74. Lally, "Comments: Women and Credit," 875. See also Reizenstein, "A Fresh Look at the Equal Credit Opportunity Act," 226.

75. Lally, "Comments: Women and Credit," 867. See also Reizenstein, "A Fresh Look at the Equal Credit Opportunity Act," 225. Hyman similarly identified cultural factors as accounting for discriminatory practices. See Hyman, "Ending Discrimination, Legitimating Debt."

76. Cohen, *A Consumers' Republic*, 147–48. Hyman also makes this point in "Ending Discrimination, Legitimating Debt."

77. David Durand, *Risk Elements in Consumer Instalment Financing*, (New York: National Bureau of Economic Research, 1941), 74; and John M. Chapman and Associates, *Commercial Banks and Consumer Instalment Credit* (New York: National Bureau of Economic Research, 1940), Table B-3, 278.

78. Moore and Klein, *The Quality of Consumer Instalment Credit*, 48, 52; and Hyman, "Ending Discrimination, Legitimating Debt." According to the Survey of Consumer Finances, women represented roughly 10 percent of new car buyers in 1954, which is consistent with them being less than 10 percent of borrowers. It is something of a "chicken or egg" problem as to whether few women were car buyers because they could not get credit financing or whether few women showed up as borrowers because they had not entered the auto market. By the early 1970s, as I discuss, many women complained that it was the lack of access to credit. See Market Research Division, *U.S. News & World Report* and Benson & Benson, Inc., *The People Buying New Automobiles Today: A Marketing Report on 18 Studies Among the Families Who Bought New 1954 Models of Each of the Principal Makes of Automobiles . . . Conducted by Market Research Division, U.S. News and World Report [and] Benson & Benson, Inc., Princeton, New Jersey* ([Washington, D.C.]: U.S. News Publishing Corporation, 1955), 23, Baker Old Class Collection, Baker Library Historical Collections, Harvard Business School, Boston, Massachusetts.

79. Elizabeth M. Fowler, "Some Women Find Discrimination When Trying to Establish Credit," *New York Times* (May 15, 1972), Proquest Historical Newspapers *The New York Times* (1851–2005): 53.

80. National Commission on Consumer Finance, *Consumer Credit in the United States*, 153.

81. Aileen Jacobson, "Money, Women and Divorce," *Washington Post* (Dec. 9, 1973), Proquest Historical Newspapers *The Washington Post* (1877–1991), PO18, 37, 38.

82. National Commission on Consumer Finance, *Consumer Credit in the United States*, 153. The study was identified on page 287, note 4 simply as St. Paul Department of Human Rights, "Instalment Loan Survey of St. Paul Banks: Is There Sex Discrimination" (no date).

83. Jacobson, "Money, Women and Divorce," 38. Hyman discussed numerous cases of credit discrimination in *Debtor Nation*, chapter 7. One point he brought out was that many of the women who complained were solidly middle class.

84. National Commission on Consumer Finance, *Consumer Credit in the United States*, 152. See also Hyman, *Debtor Nation*, Chap. 7, esp. 388–89.

85. U.S. Commission on Civil Rights, "Mortgage Money: Who Gets It? A Case Study in Mortgage Lending Discrimination in Hartford, Connecticut," Clearinghouse Publication 48 (June 1974) reprinted in U.S. Congress, House of Representatives, *Credit Discrimination: Hearings Before the Subcommittee on Consumer Affairs, House Committee on Banking and Currency, Relating to Legislation Prohibiting Sex Discrimination in the Granting of Credit*, 93rd Cong., 2nd sess., part 1, June 20 and 21, 1974 (Washington, D.C.: U.S. Government Printing Office, 1974), 267.

86. Hyman, "Ending Discrimination, Legitimating Debt"; and Hyman, *Debtor Nation*, Chap. 7.

87. Reizenstein, "A Fresh Look at the Equal Credit Opportunity Act," 219–29, quote 228.

88. "Regulation B: Amendment," *Federal Reserve Bulletin* 63 (Jan. 1977): 89–95, quote 89. In general on credit discrimination, see the National Consumer Law Center, *Credit Discrimination*, 3rd ed. (Boston: National Consumer Law Center, 2002). See also "Equal Credit Opportunity," *Federal Reserve Bulletin* 63 (Feb. 1977): 101–107.

89. Statement of Representative Barbara Jordan, *Credit Discrimination: Hearings Before the Subcommittee on Consumer Affairs, House Committee on Banking and Currency, Relating to Legislation Prohibiting Sex Discrimination in the Granting of Credit,* 472; and Donna Dunkelberger Geck, "Equal Credit: You Can Get There from Here—The Equal Credit Opportunity Act," *North Dakota Law Review* 52 (Winter 1975): 385.

90. Reizenstein, "A Fresh Look at the Equal Credit Opportunity Act," 217 and 217 note 18, 227–28. I have found the best overall discussion of Regulation B in National Consumer Law Center, *Credit Discrimination*. Hyman also emphasized the importance of the state in changing credit markets. See Hyman, "Ending Discrimination, Legitimating Debt"; and Hyman, *Debtor Nation*, Chap. 7.

91. Reizenstein, "A Fresh Look at the Equal Credit Opportunity Act," 227. On the example of married women's credit cards, see Lizabeth Cohen, "From Town Center to Shopping Center: The Reconfiguration of Community Marketplaces in Postwar America," *American Historical Review* 101 (Oct. 1996): 1073–75. See also Cohen, *A Consumers' Republic*, 147–48, 281–83, and 369–70. For an account of credit discrimination that African Americans faced, see Martha L. Olney, "When Your Word Is Not Enough: Race, Collateral, and Household Credit," *Journal of Economic History* 58 (June 1998): 408–31; and for recent years, Ian Ayres, "Fair Driving: Gender and Race Discrimination in Retail Car Negotiations," *Harvard Law Review* 104 (Feb. 1991): 817–72.

92. Rubenstein, *Making and Selling Cars*, 222, Charles E. Doyle, "Small Cars Crab Off Half the US Automobile Market," *Christian Science Monitor* (Apr. 28, 1980): B2. On market segmentation, see also Gartman, *Auto Opium*, 182–211; Cross, *An All-Consuming Century*; and Cohen, *A Consumers' Republic*, 292–344. On the small car, see especially McCarthy, *Auto Mania*, 207–30. McCarthy makes the point that the small car was embraced not simply as a response to the energy crisis.

93. Cross, *An All-Consuming Century*, 182–83. Walsh also makes this point in "Gendering Mobility," 377–78.

94. Bruce Horovitz, "Used-Car Sales Take Turn for Worse, Too," *Industry Week* (Nov. 29, 1982): 72.

95. For 35 percent of sales, see Charles R. Day Jr., "'Bow Tie' Now Pink Bow? Chevy Yearns to Be a Ladies' Marquee," *Industry Week* (Sep. 30, 1985): 21. For 40 percent of sales, as reported by the Ford Motor Company, see Ellen Roseman, "The Consumer Game Car Maker Driving to Win Over," *The Globe and Mail* (Canada) (Apr. 21, 1983): n.p. One article reported that in the summer of 1965 women accounted for 54 out of 234 new car sales at a Chicago dealership. If this study was representative, then it indicates women accounted for roughly 23 percent of sales. See Gerald D. Bell, "Self-Confidence and Persuasion in Car Buying," *Journal of Marketing Research* 4 (Feb. 1967): 46–52.

96. Roseman, "The Consumer Game Car Maker Driving to Win Over," n.p. See also Rubenstein, *Making and Selling Cars*, 293–98.

97. "Wooing Women Buyers: Women Are Involved in 80 Percent of New Vehicle Purchases," *Automotive News* (Aug. 22, 1988): E13.

312   98. For example, a general search of "women and car" in Lexis-Nexis Academic turned up 10 news items for the years before 1970 and 133 articles for the period from 1970 to 1975, whereas the number of articles with these key words jumped to 999 for the years 1975 to 1980. Most of these articles were not directly on the subject of women and cars, but the number of relevant articles increased from none before 1970 to a few by the late 1970s and more articles for the 1980s. Articles about women and marketing include Paul A. Eisenstein, "Women Are Big Force in Car Sales," *Christian Science Monitor* (Aug. 16, 1989): 9; Anneta Miller with David L. Gonzalez and Peter McAlevey, "What Cars Do Women Want?" *Newsweek* (Mar. 24, 1986): 54; Edward Clifford, "Sporty Cars Attracting More Women Buyers," *The Globe and Mail* (Canada) (Jan. 30, 1984): n.p.; Raymond Serafin, "Still Leading Herd: Ford Mustang Gallops into 25th Year," *Advertising Age* (Apr. 10, 1989): 30; Larry Kramer, "New Best Seller: A Pickup Truck; Women, Young People Make Ford Pickup Hot Vehicle; Boom Linked to Individuality of Women, Young People," *Washington Post* (Feb. 16, 1978): A1; Roger Neal, "Women's Prerogative: To Switch to Sport Utility Vehicles; Advertising Gears Up to Reach Women with Macho, Independent Image of New 4WD Category," *Adweek* (Apr. 7, 1986): n.p.; and Raymond Serafin, "Dealers Hold Key to Sales Success in Segment," *Advertising Age* (Sep. 15, 1986): S-16.

99. Serafin, "Dealers Hold Key to Sales Success in Segment," S-16.

100. Day, "'Bow Tie' Now Pink Bow?" 21.

101. Laura Clark with John Russell, "Chevy, VW Aim Direct-Mail Campaign at Women," *Automotive News* (Apr. 4, 1988): 63.

102. One possible avenue of research would be to examine data sets from the Survey of Consumer Finances to determine whether credit was key to car sales to women. This research strategy would depend on the availability and detail of the data series.

103. The one study that I have found of women buyers from 1965 suggested that they accounted for roughly 23 percent of all new car buyers, but this study was based on the experience of one dealer and so I hesitate to generalize from this case study. The study also did not explore the purchase of a second household car. See Bell, "Self-Confidence and Persuasion in Car Buying," 46–52. On the two-car household, see also Walsh, "Gendering Mobility," 384–87.

104. Doyle, "Small Cars Grab Off Half the US Automobile Market," B2. On the divorce trend, see, for instance, Jacobson, "Money, Women and Divorce"; and Walsh, "Gendering Mobility," 391–92. On women as buyers of small cars as well as independent households, see also Rubenstein, *Making and Selling Cars*, 296–98. On the energy crisis and small cars, see McCarthy, *Auto Mania*, 207–30.

105. Eisenstein, "Women Are Big Force in Car Sales," 9. Rubenstein also singled out Toyota for selling cars to women. See Rubenstein, *Making and Selling Cars*, 297–98.

106. Hyman also focused on culture in his analysis of credit discrimination. See Hyman, "Ending Discrimination, Legitimating Debt."

REFERENCE MATTER

Selected Bibliography

Note: For archival and most other primary sources, see the notes to the individual chapters.

Abernathy, William J., and Robert H. Hayes. "Managing Our Way to Economic Decline." *Harvard Business Review* 58 (July-Aug. 1980): 67–77.

Acs, Zoltan J., and David Audretsch. "Innovation in Large and Small Firms: An Empirical Analysis." *American Economic Review* 78 (Sep. 1988): 679–90.

Adams, Stephen B. "Regionalism in Stanford's Contribution to the Rise of Silicon Valley." *Enterprise and Society* 4 (Sep. 2003): 521–43.

———. "Stanford and Silicon Valley: Lessons on Becoming a High-Tech Region." *California Management Review* 48 (Fall 2005): 29–51.

Adams, Stephen B., and Orville R. Butler. *Manufacturing the Future: A History of Western Electric.* New York: Cambridge University Press, 1999.

Ahlstrand, Bruce, Joseph Lampel, and Henry Mintzberg. *Strategy Safari: A Guided Tour Through the Wilds of Strategic Management.* New York: The Free Press, 1998.

Akera, Atsushi. "IBM's Early Adaptation to Cold War Markets: Cuthbert Hurd and His Applied Science Field Men." *Business History Review* 76 (Winter 2002): 767–802.

Allen, Robert. "Collective Invention." *Journal of Economic Behavior and Organization* 4 (Jan. 1983): 1–24.

Almeida, Paul, and Bruce Kogut. "Localization of Knowledge and the Mobility of Engineers in Regional Networks." *Management Science* 45 (July 1999): 905–17.

Ansoff, H. Igor. *Corporate Strategy: An Analytic Approach to Business Policy for Growth and Expansion.* New York: McGraw-Hill, 1965.

———. "Critique of Henry Mintzberg's 'The Design School: Reconsidering the Basic Premises of Strategic Management.'" *Strategic Management Journal* 12 (Sep. 1991): 449–61.

Armi, C. Edson. *The Art of American Car Design: The Profession and Personalities.* University Park: Pennsylvania State University Press, 1990.

Arora, Ashish, Andrea Fosfuri, and Alfonso Gambardella. *Markets for Technology: The Economics of Innovation and Corporate Strategy.* Cambridge, Mass.: MIT Press, 2001.

316

Arrow, Kenneth J. "Economic Welfare and the Allocation of Resources for Invention." In *The Rate and Direction of Economic Activity*, Universities–National Bureau Committee for Economic Research, 609–25. Princeton: Princeton University Press, 1962.

Assimakopoulos, Dimitris, Sean Everton, and Kiyoteru Tsutsui. "The Semiconductor Community in the Silicon Valley: A Network Analysis of the SEMI Genealogy Chart (1947–1986)." *International Journal of Technology Management* 25 (Nos. 1 and 2, 2003): 181–99.

Athos, Anthony, and Richard Pascale. *The Art of Japanese Management: Applications for American Executives.* New York: Simon & Schuster, 1981.

Auletta, Ken. *World War 3.0: Microsoft and Its Enemies.* New York: Random House, 2001.

Ayres, Ian. "Fair Driving: Gender and Race Discrimination in Retail Car Negotiations," *Harvard Law Review* 104 (Feb. 1991): 817–72.

Barnes, Louis B., Abby J. Hansen, and C. Roland Christensen. *Teaching and the Case Method: Text, Cases, and Readings.* Boston: Harvard Business School Press, 1994.

Barry, David, and Michael Elmes. "Strategy Retold: Toward a Narrative View of Strategic Discourse." *Academy of Management Review* 22 (Apr. 1997): 429–52.

Bashe, Charles J., Lyle R. Johnson, John H. Palmer, and Emerson W. Pugh. *IBM's Early Computers.* Cambridge, Mass.: MIT Press, 1986.

Bauer, Walter F. "Software Markets in the 70's." In *Expanding Use of Computers in the 70's: Markets, Needs, and Technology*, edited by Fred Gruenberger, 53–57. New York: Prentice-Hall, 1971.

Bell, Gerald D. "Self-Confidence and Persuasion in Car Buying." *Journal of Marketing Research* 4 (Feb. 1967): 46–52.

Bernstein, Jeremy. *Three Degrees Above Zero.* New York: Charles Scribner's Sons, 1984.

"Big Change in Buying Habits—What It Means to Business," *U.S. News & World Report* (Feb. 6, 1961): 73–75.

Blaszczyk, Regina Lee. *Imagining Consumers: Design and Innovation from Wedgwood to Corning.* Baltimore: Johns Hopkins University Press, 2000.

Bode, H. W. *Synergy: Technical Integration and Technological Innovation in the Bell System.* Murray Hill, N.J.: Bell Laboratories, 1971.

Bork, Robert H. *The Antitrust Paradox: A Policy at War with Itself.* New York: Basic Books, 1978.

Boston Consulting Group, Limited, The. *Strategy Alternatives for the British Motorcycle Industry: A Report Prepared for the Secretary of State for Industry.* London: Her Majesty's Stationery Office, 1975.

Bourdieu, Pierre. *Distinction: A Social Critique of the Judgment of Taste*, translated by Richard Nice. Cambridge, Mass.: Harvard University Press, 1984.

Brinkley, Joel, and Steve Lohr. *U.S. v. Microsoft: The Inside Story of the Landmark Case.* New York: McGraw-Hill, 2001.

Bristol, Ralph B. Jr. "Factors Associated with Income Variability." *American Economic Review* 48 (May 1958): 279–90.

Brock, Gerald W. *The U.S. Computer Industry: A Study in Market Power.* Cambridge, Mass.: Ballinger, 1975.

Brooks, Frederick P. Jr. *The Mythical Man-Month: Essays on Software, Twentieth Anniversary Edition.* Reading, Mass.: Addison-Wesley, 1995.

Brown, Clair. "Consumption Norms, Work Roles, and Economic Growth, 1918–1980." In *Gender in the Workplace*, edited by Clair Brown and Joseph A. Perchman, 13–58. Washington, D.C.: Brookings Institution, 1987.

Brown, Clair. *American Standards of Living, 1918–1988*. Oxford, U.K.: Blackwell, 1994.

Brown, John K. *The Baldwin Locomotive Works, 1831–1915*. Baltimore: Johns Hopkins University Press, 1995.

Brusoni, Stefano, Keith Pavitt, and Andrea Prencipe. "Knowledge Specialization, Organization Coupling, and the Boundaries of the Firm: Why Do Firms Know More Than They Make?" *Administrative Science Quarterly* 46 (Dec. 2001): 597–621.

Buderi, Robert. *Engines of Tomorrow: How the World's Best Companies Are Using Their Research Labs to Win the Future*. New York: Simon & Schuster, 2000.

Buzzell, Robert D. "Note on the Motorcycle Industry—1975: Teaching Note." Boston: Harvard Business School Press, 1985.

Cairns, John W. "Credit Equality Comes to Women: An Analysis of the Equal Credit Opportunity Act." *San Diego Law Review* 13 (July 1976): 960–77.

Calder, Lendol. *Financing the American Dream: A Cultural History of Consumer Credit*. Princeton: Princeton University Press, 1999.

Campbell-Kelly, Martin. *A History of the Software Industry: From Airline Reservations to Sonic the Hedgehog*. Cambridge, Mass.: MIT Press, 2003.

Carlson, W. Bernard. *Innovation as a Social Process: Elihu Thomson and the Rise of General Electric, 1870–1900*. New York: Cambridge University Press, 1991.

———. "The Coordination of Business Organization and Technological Innovation Within the Firm: A Case Study of the Thomson-Houston Electric Company." In *Coordination and Information*, edited by Naomi R. Lamoreaux and Daniel M. G. Raff, 55–94. Chicago: University of Chicago Press, 1995.

———. "Innovation and the Modern Corporation: From Heroic Invention to Industrial Science." In *Science in the Twentieth Century*, edited by John Krige and Dominique Pestre, 203–26. Australia: Harwood Academic Publishers, 1997.

Carroll, Paul. *Big Blues: The Unmaking of IBM*. New York: Crown, 1993.

Carter, Susan B., et al. *Historical Statistics of the United States: Earliest Times to the Present, Millennial Edition*. New York: Cambridge University Press, 2006, 5 vols.

———. *Historical Statistics of the United States: Millennial Edition Online* (New York: Cambridge University Press, 2006). http://hsus.cambridge.org/HSUSWeb/search/searchTable.do?id=Dc761-780 (accessed March 31, 2008).

Carty, John J. "The Relation of Pure Science to Industrial Research." *Science* 44 (Oct. 13, 1916): 511–18.

Castilla, Emilio J., Hokyu Hwang, Ellen Granovetter, and Mark Granovetter. "Social Networks in Silicon Valley." In *The Silicon Valley Edge: A Habitat for Innovation and Entrepreneurship*, edited by Chong-Moon Lee, William F. Miller, Marguerite Gong Hancock, and Henry S. Rowen, 218–47. Stanford, Calif.: Stanford University Press, 2000.

Ceruzzi, Paul E. *A History of Modern Computing*. Cambridge, Mass.: MIT Press, 1998.

Chandar, Nandini, and Paul J. Miranti. "Networks and Uncertainty: Forecasting, Budgeting and Production Planning at the Bell System." Paper presented at the Fifth Accounting History International Conference, Banff, Canada, August 10, 2007.

318   Chandler, Alfred D. Jr. *Strategy and Structure: Chapters in the History of the American Industrial Enterprise*. Cambridge, Mass.: MIT Press, 1962.

————. *The Visible Hand: The Managerial Revolution in American Business*. Cambridge, Mass: Harvard University Press, 1977.

————. *Inventing the Electronic Century: The Epic Story of the Consumer Electronics and Computer Industries*. New York: Free Press, 2001.

————. *Shaping the Industrial Century: The Remarkable Story of the Evolution of the Modern Chemical and Pharmaceutical Industries*. Cambridge, Mass.: Harvard University Press, 2005.

Channon, Derek F. *The Strategy and Structure of British Enterprise*. Boston: Harvard Business School, 1973.

Chapman, John M., and Associates. *Commercial Banks and Consumer Instalment Credit*. New York: National Bureau of Economic Research, 1940.

Chesbrough, Henry W. *Open Innovation: The New Imperative for Creating and Profiting from Technology*. Boston: Harvard Business School Press, 2003.

Chposky, James, and Ted Leonsis. *Blue Magic: The People, Power, and Politics Behind the IBM Personal Computer*. New York: Facts on File, 1988.

Christensen, Clayton M. *The Innovator's Dilemma: When New Technologies Cause Great Firms to Fail*. Boston: Harvard Business School Press, 1997.

————. "The Rules of Innovation." *Technology Review* 105 (June 2002): 33–38.

Christiansen, E. Tatum, and Richard T. Pascale. "Honda (A)," Case # 9-384-049. Boston: Harvard Business School Press, 1989.

————. "Honda (B)," Case # 9-384-050. Boston: Harvard Business School Press, 1989.

Clark, Laura, with John Russell. "Chevy, VW Aim Direct-Mail Campaign at Women." *Automotive News* (Apr. 4, 1988): 63.

Clarke, Sally H. *Regulation and the Revolution in United States Farm Productivity*. New York: Cambridge University Press, 1994.

————. "Negotiating Between the Firm and the Consumer: Bell Labs and the Development of the Modern Telephone." In *The Modern Worlds of Business and Industry*, edited by Karen Merrill, 161–82. Turnhout, Belg.: David Brown, 1998.

————. *Trust and Power: Consumers, the Modern Corporation, and the Making of the United States Automobile Market*. New York: Cambridge University Press, 2007.

Clifford, Edward. "Sporty Cars Attracting More Women Buyers." *The Globe and Mail* (Canada) (Jan. 30, 1984): n.p.

Cohen, Lizabeth. *Making a New Deal: Industrial Workers in Chicago, 1919–1939*. New York: Cambridge University Press, 1990.

————. "From Town Center to Shopping Center: The Reconfiguration of Community Marketplaces in Postwar America." *American Historical Review* 101 (Oct. 1996): 1050–81.

————. *A Consumers' Republic: The Politics of Mass Consumption in Postwar America*. New York: Knopf, 2003.

Cohen, Wesley M., and Daniel A. Levinthal. "Absorptive Capacity: A New Perspective on Learning and Innovation." *Administrative Science Quarterly* 35 (Mar. 1990): 128–52.

Collins, Jim, and Jerry I. Porras. *Built to Last: Successful Habits of Visionary Companies*. New York: HarperBusiness, 1994.

"Consumer Durable Goods Recovery." *Federal Reserve Bulletin* 45 (Jan. 1959): 1–6.   319

Cooper, Carolyn C. *Shaping Invention: Thomas Blanchard's Machinery and Patent Management in Nineteenth-Century America*. New York: Columbia University Press, 1991.

Cortada, James W. *Before the Computer: IBM, NCR, Burroughs, and Remington Rand and the Industry They Created, 1865–1956*. Princeton: Princeton University Press, 1993.

Costa, Dora L. "From Mill Town to Board Room: The Rise of Women's Paid Labor." *Journal of Economic Perspectives* 14 (Fall 2000): 101–22.

Croley, Steven P., and John D. Hanson. "Rescuing the Revolution: The Revived Case for Enterprise Liability." *Michigan Law Review* 91 (Feb. 1993): 683–797.

Cross, Gary. *An All-Consuming Century: Why Commercialism Won in Modern America*. New York: Columbia University Press, 2000.

Davis, Lance E. "The Investment Market, 1870–1914: The Evolution of a National Market." *Journal of Economic History* 25 (Sep. 1965): 355–99.

Day, Charles R. Jr., "'Boy Tie' Now Pink Bow? Chevy Yearns to Be a Ladies' Marquee." *Industry Week* (Sep. 30, 1985): 21.

de Vries, Jan. "The Industrial Revolution and the Industrious Revolution." *Journal of Economic History* 54 (June 1994): 249–70.

Deal, Terence E., and Allan A. Kennedy. *Corporate Cultures: The Rites and Rituals of Corporate Life*. Reading, Mass.: Addison-Wesley, 1982.

Demsetz, Harold. "Industry Structure, Market Rivalry, and Public Policy." *Journal of Law and Economics* 16 (Apr. 1973): 1–9.

Dennis, Michael. "Accounting for Research: New Histories of Corporate Laboratories and the Social History of American Science." *Social Studies in Science* 17 (Aug. 1987): 479–518.

Dewey, Donald. *Monopoly in Economics and Law*. Chicago: Rand McNally, 1959.

———. *The Antitrust Experiment in America*. New York: Columbia University Press, 1990.

Dosi, Giovanni, Renato Giannetti, and Pier Angelo Toninelli, eds. *Technology and Enterprise in Historical Perspective*. New York: Oxford University Press, 1992.

Doyle, Charles E. "Small Cars Grab Off Half the US Automobile Market." *Christian Science Monitor* (Apr. 28, 1980): B2.

Dullea, Georgia. "Women Demanding Equal Treatment in Mortgage Loans," *New York Times* (October 29, 1972), Proquest Historical Newspapers *The New York Times* (1851–2005): R1.

Durand, David. *Risk Elements in Consumer Instalment Financing*. New York: National Bureau of Economic Research, 1941.

Dyer, Davis, and Daniel P. Gross. *The Generations of Corning: The Life and Times of a Global Corporation*. New York: Oxford University Press, 2001.

Easterlin, Richard A. *The Reluctant Economist: Perspectives on Economics, Economic History, and Demography*. New York: Cambridge University Press, 2004.

Edgerton, David. "Industrial Research in the British Photographic Industry, 1879–1939." In *The Challenge of New Technology: Innovation in British Business Since 1850*, edited by Jonathan Liebenau, 106–34. Brookfield, Vt.: Gower, 1988.

320 Eisenstein, Paul A. "Women Are Big Force in Car Sales." *Christian Science Monitor* (Aug. 16, 1989): 9.

Eisner, Marc Allen. *Antitrust and the Triumph of Economics: Institutions, Enterprise, and Policy Change.* Chapel Hill, N.C.: University of North Carolina Press, 1991.

Ellig, Jerry, ed. *Dynamic Competition and Public Policy: Technology, Innovation, and Antitrust Issues.* New York: Cambridge University Press, 2001.

Epstein, R. J. *A History of Econometrics.* Amsterdam: North Holland, 1987.

"Equal Credit Opportunity." *Federal Reserve Bulletin* 63 (Feb. 1977): 101–107.

Etzkowitz, Henry. "The Making of an Entrepreneurial University: The Traffic Among MIT, Industry, and the Military, 1860–1960." In *Science, Technology, and the Military,* edited by Everett Mendelsohn, Merritt Roe Smith, and Peter Weingart, 515–40. London: Kluwer, 1988.

Fabrizio, Kira R., and David C. Mowery. "The Federal Role in Financing Major Innovations: Information Technology During the Postwar Period." In *Financing Innovation in the United States,* edited by Naomi R. Lamoreaux and Kenneth L. Sokoloff, 283–316. Cambridge, Mass.: MIT Press, 2007.

Fagen, M. D., ed. *A History of Engineering and Science in the Bell System: The Early Years (1875–1925).* New York: AT&T Bell Laboratories, 1975.

———. ed. *A History of Engineering and Science in the Bell System: Communication Sciences (1925–1980).* New York: AT&T Bell Laboratories, 1984.

Fear, Jeffrey. "Thinking Historically About Organizational Learning." In *Handbook of Organizational Learning and Knowledge,* edited by Meinholf Dierkes, Ariane Berthoin Antal, John Child, and Ikujiro Nonaka, 162–91. New York: Oxford University Press, 2001.

Ferrie, Joseph P. "Longitudinal Data for the Study of American Geographic, Occupational, and Financial Mobility, 1850–1990." Unpublished paper, Faculty of Economics, Northwestern University, 2004.

Fishback, Price, and Shawn Everett Kantor. *Prelude to the Welfare State: The Origins of Workers' Compensation.* Chicago: University of Chicago Press, 2000.

Fisher, Franklin M. "Innovation and Monopoly Leveraging." In *Dynamic Competition and Public Policy,* edited by Jerry Ellig, 138–59. New York: Cambridge University Press, 2001.

Fisher, Franklin, John J. McGowan, and Joen E. Greenwood. *Folded, Spindled, and Mutilated: Economic Analysis and U.S. v. IBM.* Cambridge, Mass: MIT Press, 1983.

Fisher, Franklin M., James W. McKie, and Richard B. Mancke. *IBM and the U.S. Data Processing Industry: An Economic History.* New York: Praeger, 1983.

Fisk, Catherine L. "Removing the 'Fuel of Interest' from the 'Fire of Genius': Law and the Employee-Inventor, 1830–1930." *University of Chicago Law Review* 65 (Fall 1998): 1127–98.

Fitzgerald, Deborah Kay. *The Business of Breeding: Hybrid Corn in Illinois, 1890–1940.* Ithaca, N.Y.: Cornell University Press, 1990.

Flink, James J. *America Adopts the Automobile, 1895–1910.* Cambridge, Mass.: MIT Press, 1970.

———. *The Automobile Age.* Cambridge, Mass.: MIT Press, 1988.

Folk, George E. *Patents and Industrial Progress: A Summary, Analysis, and Evaluation*   321
of the Record on Patents of the Temporary National Economic Committee. New York:
Harper and Brothers, 1942.

Fowler, Elizabeth M. "Some Women Find Discrimination When Trying to Establish
Credit." *New York Times* (May 15, 1972), Proquest Historical Newspapers *The New
York Times* (1851–2005): 53.

Frank, Thomas. *Conquest of Cool: Business Culture, Counterculture, and the Rise of
Hip Consumerism.* Chicago: University of Chicago Press, 1997.

Fransman, Martin. *The Market and Beyond: Cooperation and Competition in Information Technology Development in the Japanese System.* Cambridge, U.K.: Cambridge
University Press, 1990.

Freyer, Tony A. *Antitrust and Global Capitalism, 1930–2004.* New York: Cambridge
University Press, 2006.

Fry, Thornton C. *Probability and Its Engineering Uses.* New York: D. Van Nostrand, 1928.

Fujimoto, Takahiro. *The Evolution of a Manufacturing System at Toyota.* New York:
Oxford University Press, 1999.

Galambos, Louis. *Competition and Cooperation: The Emergence of a National Trade
Association.* Baltimore: Johns Hopkins University Press, 1966.

———. "The American Economy and the Reorganization of the Sources of Knowledge,"
In *The Organization of Knowledge in Modern America, 1860–1920,* edited by Alexandra Olsen and John Voss, 269–82. Baltimore: Johns Hopkins University Press, 1979.

———. *America at Middle Age: A New History of the United States in the Twentieth
Century.* New York: New Press, 1982.

———. "Technology, Political Economy, and Professionalization: Central Themes of
the Organizational Synthesis." *Business History Review* 57 (Winter 1983). 471–93.

———. "Theodore N. Vail and the Role of Innovation in the Modern Bell System."
Business History Review 66 (Spring 1992): 95–126.

———. "End of the Century Reflections: Weber and Schumpeter with Karl Marx Lurking in the Background." *Industrial and Corporate Change* 5 (Jan. 1996): 925–93.

———. "The Monopoly Enigma, the Reagan Administration's Antitrust Experiment,
and the Global Economy." In *Constructing Corporate America: History, Politics, and
Culture,* edited by Kenneth Lipartito and David Sicilia, 149–67. New York: Oxford
University Press, 2004.

Galambos, Louis, and Joseph Pratt. *The Rise of the Corporate Commonwealth: U.S.
Business and Public Policy in the Twentieth Century.* New York: Basic Books, 1987.

Galambos, Louis, with Jane Eliot Sewell. *Networks of Innovation: Vaccine Development
at Merck, Sharp & Dohme, and Multford, 1895–1995.* New York: Cambridge University Press, 1995.

Galambos, Louis, and Jeffrey L. Sturchio. "Pharmaceutical Firms and the Transition
to Biotechnology: A Study in Strategic Innovation." *Business History Review* 72
(Summer 1998): 250–78.

Gans, Joshua, and Scott Stern. "The Product Market and the Market for 'Ideas': Commercialization Strategies for Technology Entrepreneurs." *Research Policy* 32 (Feb.
2003): 333–50.

322 Garnet, Robert W. *The Telephone Enterprise: The Evolution of the Bell System's Horizontal Structure, 1876–1900*. Baltimore: Johns Hopkins University Press, 1985.

Garr, Doug. *IBM Redux: Lou Gerstner and the Business Turnaround of the Decade*. New York: HarperBusiness, 1999.

Gartman, David. *Auto Opium: A Social History of American Automobile Design*. London: Routledge, 1994.

Geck, Donna Dunkelberger. "Equal Credit: You Can Get There From Here—The Equal Credit Opportunity Act." *North Dakota Law Review* 52 (Winter 1975): 381–409.

Gehani, Narain. *Bell Labs: Life in the Crown Jewel*. Summit, N.J.: Silicon Press, 2003.

Genesove, David, and Wallace P. Mullin. "The Sugar Institute Learns to Organize Information Exchange." In *Learning by Doing in Markets, Firms, and Countries*, edited by N. R. Lamoreaux, D.M.G. Raff, and P. Temin, 103–36. Chicago: University of Chicago Press, 1999.

George, Claude S. Jr. *History of Management Thought*. Englewood Cliffs, N.J.: Prentice-Hall, 1968.

Ghemawat, Pankaj. "Competition and Business Strategy in Historical Perspective." *Business History Review* 76 (Spring 2002): 37–74.

Giebelhaus, August W. *Business and Government in the Oil Industry: A Case Study of Sun Oil, 1876–1945*. Greenwich, Conn.: JAI Press, 1980.

Gillmor, C. Stewart. *Fred Terman at Stanford: Building a Discipline, a University, and Silicon Valley*. Stanford, Calif.: Stanford University Press, 2004.

Gilson, Ronald J. "The Legal Infrastructure of High Technology Industrial Districts: Silicon Valley, Route 128, and Covenants Not to Compete." *New York University Law Review* 74 (June 1999): 575–629.

Goldin, Claudia. *Understanding the Gender Gap*. New York: Oxford University Press, 1990.

Goldin, Claudia, and Robert A. Margo. "The Great Compression: The Wage Structure in the United States at Mid-Century." *Quarterly Journal of Economics* 107 (Feb. 1992): 1–34.

Goold, Michael. "Design, Learning and Planning: A Further Observation on the Design School Debate." *California Management Review* 38 (Summer 1996): 94–95.

Graham, Margaret B. W. *RCA and the VideoDisc: The Business of Research*. New York: Cambridge University Press, 1986.

———. "Less Transfer Than Transformation: The Formation of Corning's Avon Laboratory." Paper presented at The Johns Hopkins Conference on Organizing for Innovation, Baltimore, October 26, 2002.

———. "From Satellite Laboratory to Partner: The Formation of Corning's Fontainebleau Laboratory." Paper presented at the Business History Conference, Lowell, Massachusetts, June 26–28, 2003.

———. "Financing Fiber: Corning's Invasion of the Telecommunications Market." In *Financing Innovation in the United States*, edited by Naomi R. Lamoreaux and Kenneth L. Sokoloff, 247–82. Cambridge, Mass.: MIT Press, 2007.

Graham, Margaret B. W., and Bettye H. Pruitt. *R&D for Industry: A Century of Technical Innovation at Alcoa*. New York: Cambridge University Press, 1990.

Graham, Margaret B. W., and Alec T. Shuldiner. *Corning and the Craft of Innovation.*   **323**
New York: Oxford University Press, 2001.

Griliches, Zvi. "Research Costs and Social Return: Hybrid Corn and Related Innovations." *Journal of Political Economy* 66 (Oct. 1958): 419–31.

"Growth of Consumer Instalment Credit." *Federal Reserve Bulletin* 41 (Dec. 1955): 1311–16.

Gruenberger, Fred, ed., *Expanding the Use of Computers in the 70's: Markets, Need,
and Technology.* New York: Prentice-Hall, 1971.

Guillén, Mauro F. *Models of Management: Work, Authority, and Organization in a
Comparative Perspective.* Chicago: University of Chicago Press, 1994.

Gunderson, Les C., and Donald B. Keck. "Optical Fibres: Where Light Outperforms
Electrons." *Technology Review* 86 (May-June 1983): 32–44.

Halberstam, David. *The Reckoning.* New York: Morrow, 1986.

Hamasaki, Shogo. "Patent Races with Information Spillovers: Covenants Not to Compete and the Silicon Valley Advantage." Unpublished paper, Faculty of Economics,
University of California, Los Angeles, 2008.

Hamilton, Walton. *Patents and Free Enterprise.* TNEC monograph no. 31. Washington,
D.C.: U.S. Government Printing Office, 1941.

Harrington, Jerry. "The Midwest Agricultural Chemical Association: A Regional Study
of an Industry on the Defensive." *Agricultural History* 70 (Spring 1996): 415–38.

Hart, David M. *Forged Consensus: Science, Technology, and Economic Policy in the
United States, 1921–1953.* Princeton: Princeton University Press, 1998.

———. "Antitrust and Technological Innovation in the U.S.: Ideas, Institutions, Decisions, and Impacts, 1890–2000." *Research Policy* 30 (July 2001): 923–37.

Hawley, Ellis W. *The New Deal and the Problem of Monopoly: A Study in Economic
Ambivalence.* Princeton: Princeton University Press, 1966.

———. "Three Facets of Hooverian Associationalism: Lumber, Aviation, and Movies,
1921–1930." In *Regulation in Perspective: Historical Essays,* edited by Thomas K.
McCraw, 95–123. Cambridge, Mass.: Harvard University Press, 1981.

Hecht, Jeffrey. *City of Light.* New York: Oxford University Press, 1999.

Helfat, Constance E., ed. *The SMS Blackwell Handbook of Organizational Capabilities: Emergence, Development, and Change.* Malden, Mass.: Blackwell, 2003.

Helper, Susan. "Strategy and Irreversibility in Supplier Relations: The Case of the U.S.
Auto Industry." *Business History Review* 65 (Winter 1991): 781–824.

Henderson, Rebecca, Luigi Orsenigo, and Gary P. Pisano. "The Pharmaceutical Industry and the Revolution in Molecular Biology: Interactions Among Scientific, Institutional, and Organizational Change." In *Sources of Industrial Leadership: Studies of
Seven Industries,* edited by David C. Mowery and Richard R. Nelson. New York:
Cambridge University Press, 1999, 267–311.

Hendry, David F., and Mary S. Morgan, eds. *The Foundations of Econometric Analysis.*
Cambridge, U.K.: Cambridge University Press, 1995.

Heppenheimer, T. A. "What Made Bell Labs Great." *Invention and Technology* 12 (Summer 1996): 46–57.

Hicks, Diana, Tony Breitzman, Dominic Olivastro, and Kimberly Hamilton. "The
Changing Composition of Innovative Activity in the U.S.—A Portrait Based on
Patent Analysis." *Research Policy* 30 (Apr. 2001): 681–703.

324   Hiltzik, Michael. *Dealers of Lightning: Xerox PARC & the Dawn of the Computer Age.* New York: HarperCollins, 1999.

Hirsh, Richard F. *Power Loss: The Origins of Deregulation and Restructuring in the American Electric Utility System.* Cambridge, Mass.: MIT Press, 1999.

Hoch, Detlev J., Cyriac R. Roeding, Gert Purkert, and Sandro K. Lindner *Secrets of Software Success: Management Insights from 100 Software Firms Around the World.* Boston: Harvard Business School Press, 2000.

Hoddeson, Lillian Hartmann. "The Entry of Quantum Theory of Solids into the Bell Telephone Laboratories, 1925–40: A Case-Study of the Industrial Application of Fundamental Science," *Minerva* 18 (Sep. 1980): 423–47.

———. "The Discovery of the Point-Contact Transistor." *Historical Studies in the Physical Sciences* 12, Part 1 (1981): 41–76.

Hooks, Gregory. *Forging the Military-Industrial Complex: World War II's Battle of the Potomac.* Urbana, Ill.: University of Illinois Press, 1991.

Horovitz, Bruce. "Used-Car Sales Take Turn for Worse, Too." *Industry Week* (Nov. 29, 1982): 72.

Hounshell, David A. *From the American System to Mass Production, 1800–1932: The Development of Manufacturing Technology in the United States.* Baltimore: Johns Hopkins University Press, 1984.

———. "Continuity and Change in the Management of Industrial Research: The Du Pont Company, 1902–1980." In *Technology and Enterprise in Historical Perspective,* edited by Giovanni Dosi, Renato Giannetti, and Pier Angelo Toninelli, 231–60. New York: Oxford University Press, 1992.

———. "Du Pont and the Management of Large–Scale Research and Development." In *Big Science: The Growth of Large-Scale Research,* edited by Peter Galison and Bruce Hevly, 236–61. Stanford, Calif.: Stanford University Press, 1992.

———. "The Evolution of Industrial Research in the United States." In *Engines of Innovation,* edited by Richard S. Rosenbloom and William J. Spencer, 13–85. Boston: Harvard Business School Press, 1996.

———. "Industrial Research and Manufacturing Technology." In *Encyclopedia of the United States in the Twentieth Century,* edited by Stanley Kutler, 831–57. New York: Charles Scribner's Sons, 1996.

———. "Assets, Organizations, Strategies, and Traditions: Organizational Capabilities and Constraints in the Remaking of Ford Motor Company, 1946–1962." In *Learning by Doing in Markets, Firms, and Countries,* edited by Naomi R. Lamoreaux, Daniel M. G. Raff, and Peter Temin, 185–208. Chicago: University of Chicago Press, 1999.

Hounshell, David A., and John Kenly Smith Jr. *Science and Corporate Strategy: Du Pont R&D, 1902–1980.* New York: Cambridge University Press, 1988.

Hughes, Thomas Parke. *Elmer Sperry: Inventor and Engineer.* Baltimore: Johns Hopkins University Press, 1971.

———. *American Genesis: A Century of Invention and Technological Enthusiasm.* New York: Viking, 1989.

Humphrey, Watt S. "Reflections on a Software Life." In *In the Beginning: Personal Reflections of Software Pioneers,* edited by Robert L. Glass, 29–53. Los Alamitos, Calif.: IEEE Computer Society, 1999.

Hyde, Alan. *Working in Silicon Valley: Economic and Legal Analysis of a High-Velocity*   325
Labor Market. Armonk, N.Y.: M. E. Sharpe, 2003.

Hylton, Keith N. *Antitrust Law: Economic Theory and Common Law Evolution.* New York: Cambridge University Press, 2003.

Hyman, Louis. "Ending Discrimination, Legitimating Debt: The Political Economy of Race, Gender, and Credit Access, 1968–1974." Unpublished paper, Business History Conference, April 12, 2008.

———. *Debtor Nation: Changing Credit Practices in 20th Century America.* Princeton: Princeton University Press, forthcoming.

Hynes, James. *Publish and Perish: Three Tales of Tenure and Terror.* New York: Picador, 1997.

Institute for Social Research, Survey Research Center. *1960 Survey of Consumer Finances.* Ann Arbor: University of Michigan Press, 1961.

Israel, Paul. *Machine Shop to Industrial Laboratory: Telegraphy and the Changing Context of American Invention, 1830–1920.* Baltimore: Johns Hopkins University Press, 1992.

———. *Edison: A Life of Invention.* New York: John Wiley & Sons, 1998.

Jackson, Kenneth T. *Crabgrass Frontier: The Suburbanization of the United States.* New York: Oxford University Press, 1985.

Jacobs, Meg. *Pocketbook Politics: Economic Citizenship in Twentieth-Century America.* Princeton: Princeton University Press, 2005.

Jacobson, Aileen. "Money, Women and Divorce." *Washington Post* (Dec. 9, 1973), Proquest Historical Newspapers *The Washington Post* (1877–1991), PO18, 37, 38.

Johnson, E. A., and E. A. Katcher. *Mines Against Japan.* Silver Spring, Md.: Naval Ordnance Laboratory, 1973.

Johnson, Stephen B. "Three Approaches to Big Technology: Operations Research, Systems Engineering, and Project Management." *Technology and Culture* 38 (Oct. 1997): 891–919.

Jorde, Thomas M., and David J. Teece, eds., *Antitrust, Innovation, and Competitiveness.* New York: Oxford University Press, 1992.

Kagan, Robert. *Adversarial Legalism: The American Way of Law.* Cambridge, Mass.: Harvard University Press, 2001.

Kahn, Alfred E. "Fundamental Deficiencies of the American Patent Law." *American Economic Review* 30 (Sep. 1940): 475–91.

Kaplan, David A. *The Silicon Boys: And Their Valley of Dreams.* New York: William Morrow, 1999.

Kargon, Robert H. "Temple to Science: Cooperative Research and the Birth of the California Institute of Technology." *Historical Studies in the Physical Sciences* 8 (1977): 3–31.

Kargon, Robert, and Elizabeth Hodes. "Karl Compton, Isaiah Bowman, and the Politics of Science in the Great Depression." *Isis* 76 (Sep. 1985): 300–18.

Katona, George. *The Powerful Consumer: Psychological Studies of the American Economy.* New York: McGraw-Hill, 1960.

Katona, George, Lewis Mandell, and Jay Schmiedeskamp. *1970 Survey of Consumer Finances.* Ann Arbor, Mich.: University of Michigan, 1971.

326   Katona, George, Eva Mueller, Jay Schmiedeskamp, and John A. Sonquist. 1965 *Survey of Consumer Finances*, Monograph No. 42. Ann Arbor, Mich.: University of Michigan, 1966.

Kay, John. "The Structure of Strategy." *Business Strategy Review* 4 (Summer 1993): 17–37.

Kenney, Martin, ed. *Understanding Silicon Valley: The Anatomy of an Entrepreneurial Region*. Stanford, Calif.: Stanford University Press, 2000.

Kenney, Martin, and Richard Florida. *The Breakthrough Illusion: Corporate America's Failure to Move from Innovation to Mass Production*. New York: Basic Books, 1990.

Kerr, Clark. *The Uses of the University*. Cambridge, Mass.: Harvard University Press, 1963.

Khan, B. Zorina. "Property Rights and Patent Litigation in Early Nineteenth-Century America." *Journal of Economic History* 55 (Mar. 1995): 58–97.

Khan, B. Zorina, and Kenneth L. Sokoloff. "'Schemes of Practical Utility': Entrepreneurship and Innovation Among 'Great Inventors' During Early American Industrialization, 1790–1865." *Journal of Economic History* 53 (June 1993): 289–307.

———. "Two Paths to Industrial Development and Technological Change." In *Technological Revolutions in Europe: Historical Perspectives*, edited by Maxine Berg and Kristine Bruland, 292–313. Cheltenham, U.K.: Edward Elgar, 1998.

———. "Institutions and Technological Innovation During Early Economic Growth: Evidence from the Great Inventors of the United States, 1790–1930." In *Institutions, Development, and Economic Growth*, edited by Theo S. Eicher and Cecilia Garcia–Peñalosa, 123–58. Cambridge, Mass.: MIT Press, 2006.

Kirby, M. W., and R. Capey. "The Air Defense of Great Britain, 1920–1940: An Operational Perspective." *Journal of the Operational Research Society* 48 (June 1997): 555–68.

———. "The Area Bombing of Germany in World War II." *Journal of Operational Research Society* 48 (July 1997): 661–77.

Kirp, David L. *Shakespeare, Einstein, and the Bottom Line: The Marketing of Higher Education*. Cambridge, Mass.: Harvard University Press, 2003.

Klein, Jennifer. *For All These Rights: Business, Labor, and the Shaping of America's Public-Private Welfare State*. Princeton: Princeton University Press, 2003.

Kleinman, Daniel Lee. *Politics on the Endless Frontier: Postwar Research Policy in the United States*. Durham, N.C.: Duke University Press, 1995.

Kloppenburg, Jack Ralph Jr. *First the Seed: The Political Economy of Plant Biotechnology, 1492–2000*. New York: Cambridge University Press, 1988.

Knowles, Scott Gabriel. *Inventing Safety: Fire, Technology, and Trust in Modern America*. Ph.D. dissertation, Johns Hopkins University, 2003.

Knowles, Scott G., and Stuart W. Leslie. "'Industrial Versailles': Eero Saarinen's Corporate Campuses for GM, IBM and AT&T." *Isis* 92 (Mar. 2001): 1–33.

Koehn, Nancy. *Brand New: How Entrepreneurs Earned Consumers' Trust from Wedgwood to Dell*. Boston: Harvard Business School Press, 2001.

Koerner, Steve. "The Japanese Motor Cycle Industry, 1945 to 1960." Paper presented at the Association of Business Historians Conference, Chapel Hill, North Carolina, September 2, 1999.

Koerner, Steve, and Jun Otahara. "The Honda Motor Co. and the American Motorcycle Market." Paper presented at the Business History Conference, Wilmington, Delaware, April 20, 2002.

Kolko, Gabriel. *The Triumph of Conservatism.* New York: Free Press, 1963. 327

Kortum, Samuel, and Josh Lerner. "Assessing the Contribution of Venture Capital to Innovation." *RAND Journal of Economics* 31 (Winter 2000): 674–92.

Kramer, Larry. "New Best Seller: A Pickup Truck; Women, Young People Make Ford Pickup Hot Vehicle; Boom Linked to Individuality of Women, Young People." *Washington Post* (Feb. 16, 1978): A1.

Ladd, Helen F. "Evidence of Discrimination in Mortgage Lending." *Journal of Economic Perspectives* 12 (Spring 1998): 41–62.

Laird, Pamela Walker. *Pull: Networking and Success Since Benjamin Franklin.* Cambridge, Mass.: Harvard University Press, 2006.

Lally, Maureen Ellen. "Comments: Women and Credit." *Duquesne Law Review* 12 (Summer 1974): 863–90.

Lamoreaux, Naomi R. "Regulatory Agencies." *Encyclopedia of American Political History.* New York: Charles Scribner's Sons, 1984, 1107–1116.

———. *The Great Merger Movement in American Business, 1895–1904.* New York: Cambridge University Press, 1985.

———. "Reframing the Past: Thoughts About Business Leadership and Decision Making Under Uncertainty." *Enterprise and Society* 2 (Dec. 2001): 632–59.

Lamoreaux, Naomi R., Margaret Levenstein, and Kenneth L. Sokoloff. "Financing Invention During the Second Industrial Revolution: Cleveland, Ohio, 1870–1920." In *Financing Innovation in the United States,* edited by Naomi R. Lamoreaux and Kenneth L. Sokoloff, 39–84. Cambridge, Mass.: MIT Press, 2007.

Lamoreaux, Naomi R., and Daniel M. G. Raff, eds. *Coordination and Information: Historical Perspectives on the Organization of Enterprise.* Chicago: University of Chicago Press, 1995.

Lamoreaux, Naomi R., Daniel M. G. Raff, and Peter Temin, eds. *Learning by Doing in Markets, Firms, and Countries.* Chicago: University of Chicago Press, 1999.

———. "Beyond Markets and Hierarchies: Toward a New Synthesis of American Business History." *American Historical Review* 108 (Apr. 2003): 404–33.

Lamoreaux, Naomi R., Daniel M. G. Raff, Peter Temin, Charles F. Sabel, and Jonathan Zeitlin. "Symposium: Framing Business History." *Enterprise & Society* 5 (Sep. 2004): 353–403.

Lamoreaux, Naomi R., and Kenneth L. Sokoloff. "Inventors, Firms, and the Market for Technology in the Late Nineteenth and Early Twentieth Centuries." In *Learning by Doing in Markets, Firms, and Countries,* edited by Naomi R. Lamoreaux, Daniel M. G. Raff, and Peter Temin, 19–60. Chicago: University of Chicago Press, 1999.

———. "Market Trade in Patents and the Rise of a Class of Specialized Inventors in the 19th-Century United States." *American Economic Review* 91 (May 2001): 39–44.

———. "Intermediaries in the U.S. Market for Technology, 1870–1920." In *Finance, Intermediaries, and Economic Development,* edited by Stanley L. Engerman, Philip T. Hoffman, Jean-Laurent Rosenthal, and Kenneth L. Sokoloff, 209–46. New York: Cambridge University Press, 2003.

———. "The Market for Technology and the Organization of Invention in U.S. History." In *Entrepreneurship, Innovation, and the Growth Mechanism of the Free-Market Economies,* edited by Eytan Sheshinski, Robert J. Strom, and William J. Baumol, 213–43. Princeton: Princeton University Press, 2007.

328   ———, eds. *Financing Innovation in the United States: 1870 to the Present.* Cambridge, Mass.: MIT Press, 2007.

Langlois, Richard N. "External Economies and Economic Progress: The Case of the Microcomputer Industry." *Business History Review* 66 (Spring 1992): 1–50.

———. "The Vanishing Hand: The Changing Dynamics of Industrial Capitalism." *Industrial and Corporate Change* 12 (Apr. 2003): 351–85.

Langlois, R. N., and D. C. Mowery. "The Federal Government Role in the Development of the U.S. Software Industry." In *The International Computer Software Industry: A Comparative Study of Industry Evolution and Structure,* edited by D. C. Mowery, 53–85. New York: Oxford University Press, 1996.

Lardner, Harold. "The Origin of Operational Research." *Operations Research* 32 (Mar.-Apr. 1984): 465–75.

Larson, Henrietta, and Kenneth Wiggins Porter. *History of Humble Oil & Refining Company.* New York: Harper & Brothers, 1959.

Lassman, Thomas C. "Industrial Research Transformed: Edward Condon at the West-inghouse Electric and Manufacturing Company, 1935–1942." *Technology and Culture* 44 (Apr. 2003): 306–39.

Lazonick, William. *Business Organization and the Myth of the Market Economy.* New York: Cambridge University Press, 1991.

Lears, T. J. Jackson. *Fables of Abundance: A Cultural History of Advertising in America.* New York: Basic Books, 1994.

Lebergott, Stanley. *Pursuing Happiness: American Consumers in the Twentieth Century.* Princeton, NJ: Princeton University Press, 1993.

Lécuyer, Christophe. "Academic Science and Technology in the Service of Industry: MIT Creates a 'Permeable' Engineering School," *American Economic Review* 88, no. 23 (May 1998): 28–33.

———. *Making Silicon Valley: Innovation and the Growth of High-Tech, 1930–1970.* Cambridge, Mass.: MIT Press, 2006.

Lenoir, Timothy, with Christophe Lécuyer. "Instrument Makers and Discipline Builders: The Case of Nuclear Magnetic Resonance." In *Instituting Science: The Cultural Production of Scientific Disciplines,* edited by Timothy Lenoir, 239–92. Stanford, Calif.: Stanford University Press, 1997.

Leslie, Mitchell. "The Vexing Legacy of Lewis Terman." *Stanford Magazine,* July-Aug. 2000. www.stanfordalumni.org/news/magazine/2000/julaug/articles/terman.html (accessed September 25, 2008).

Leslie, Stuart W. *Boss Kettering: Wizard of General Motors.* New York: Columbia University Press, 1983.

———. *The Cold War and American Science: The Military-Industrial-Academic Complex at MIT and Stanford.* New York: Columbia University Press, 1993.

———. "The Biggest 'Angel' of Them All." In *Understanding Silicon Valley,* edited by Martin Kenney, 48–67. Stanford, Calif.: Stanford University Press, 2000.

———. "Blue Collar Science: Bringing the Transistor to Life in the Lehigh Valley." *Historical Studies in the Physical Sciences and the Biological Sciences* 32, Part 2 (2001): 71–113.

Leslie, Stuart W., and Robert H. Kargon. "Selling Silicon Valley: Frederick Terman's Model for Regional Advantage." *Business History Review* 70 (Winter 1996): 435–72.

Letwin, William. *Law and Economic Policy in America: The Evolution of the Sherman Antitrust Act*. New York: Random House, 1965.

Levenstein, Margaret. "Mass Production Conquers the Pool: Firm Organization and the Nature of Competition in the Nineteenth Century." *Journal of Economic History* 55 (Sep. 1995): 575–611.

Levi, Edward H. *An Introduction to Legal Reasoning*. Chicago: University of Chicago Press, 1949.

Levy, Frank. *The New Dollars and Dreams: American Incomes and Economic Change*. New York: Russell Sage Foundation, 1998.

Lewis, Michael. *Liar's Poker: Rising Through the Wreckage on Wall Street*. New York: W. W. Norton, 1989.

Lichtenstein, Nelson. "From Corporatism to Collective Bargaining: Organized Labor and the Eclipse of Social Democracy in the Postwar Era." In *The Rise and Fall of the New Deal Order, 1930–1980*, edited by Steve Fraser and Gary Gerstle, 122–52. Princeton: Princeton University Press, 1989.

———. *The Most Dangerous Man in Detroit: Walter Reuther and the Fate of American Labor*. New York: Basic Books, 1995.

Lindert, Peter H. "Three Centuries of Inequality in Britain and America." *Handbook of Income Distribution*, vol. 1. Amsterdam: N.Y.: Elsevier, 2000, 167–216.

Lipartito, Kenneth. *The Bell System and Regional Business: The Telephone in the South, 1877–1920*. Baltimore: Johns Hopkins University Press, 1989.

———. "Constructing Telephone Networks in Britain and America." Paper presented at the Conference on Constructing Markets, Shaping Production: The Historical Construction of Product Markets in Europe and America, Idöborg, Sweden, July 5–7, 2002.

———. "Picturephone and the Information Age: The Social Meaning of Failure." *Technology and Culture* 44 (Jan. 2004): 50–81.

Litterer, Joseph A. "Systematic Management: The Search for Order and Integration." *Business History Review* 35 (Winter 1961): 461–76.

Logemann, Jan. "Different Paths to Mass Consumption: Consumer Credit in the United States and West Germany During the 1950s and '60s." *Journal of Social History* 41 (Spring 2008): 525–58.

Lowen, Rebecca. *Creating the Cold War University: The Transformation of Stanford* (Berkeley and Los Angeles: University of California Press, 1997).

Lowood, Henry. *From Steeples of Excellence to Silicon Valley: The Story of Varian Associates and Stanford Industrial Park*. Palo Alto, Calif.: Varian Associates, 1987.

Luri, Jonathan. "Lawyers, Judges, and Legal Change, 1852–1916: New York as a Case Study." *Working Papers from the Regional Economic History Research Center* 3 (1980): 31–56.

MacArthur, C. W. *Operations Analysis in the U.S. Army Eighth Air Force in World War II*. Providence, R.I.: American Mathematical Society, 1990.

Maddison, Angus. *Dynamic Forces in Capitalist Development*. New York: Oxford University Press, 1991.

Magaziner, Ira C., and Mark Patinkin. *The Silent War: Inside the Global Business Battles Shaping America's Future*. New York: Random House, 1989.

Mair, Andrew. "Reconciling Managerial Dichotomies at Honda Motors." In *Strategy: Process, Content, Context—An International Perspective*, edited by Bob De Wit and Ron Meyer, 893–911. London: International Thompson Business Press, 1998.

330 ———. "Learning from Honda." *Journal of Management Studies* 36 (Jan. 1999): 25–44.

———. "Learning from Japan? Interpretations of Honda Motors by Strategic Management Theorists." Nissan Occasional Paper Series, Number 2, Nissan Institute of Japanese Studies, University of Oxford, 1999.

Maney, Kevin. *The Maverick and His Machine: Thomas Watson, Sr. and the Making of IBM.* New York: John Wiley & Sons, 2003.

Mansfield, Edwin, and Samuel Wagner. "Organizational and Strategic Factors Associated with Probabilities of Success in Industrial R&D." *Journal of Business* 48 (Apr. 1975): 179–98.

Marchand, Roland. *Creating the Corporate Soul: The Rise of Public Relations and Corporate Imagery in American Big Business.* Berkeley, Calif.: University of California Press, 1998.

Market Research Division, *U.S. News & World Report* and Benson & Benson, Inc. *The People Buying New Automobiles Today: A Marketing Report on 18 Studies Among the Families Who Bought New 1954 Models of Each of the Principal Makes of Automobiles . . . Conducted by Market Research Division, U.S. News and World Report [and] Benson & Benson, Inc., Princeton, New Jersey.* [Washington, D.C.]: U.S. News Publishing Corporation, 1955.

Marsch, Ulrich. "Strategies for Success: Research Organization in German Chemical Companies and IG Farben Until 1936." *History and Technology* 12 (Issue 1, 1994): 23–77.

Martin, Albro. *Enterprise Denied: The Origins of the Decline of American Railroads, 1897–1917.* New York: Columbia University Press, 1971.

Martin, Richard. "Fashion and the Car in the 1950s." *Journal of American Culture* 20, no. 3 (1997): 51–66.

Mass, William, and Andrew Robertson. "From Textiles to Automobiles: Mechanical and Organizational Innovation in the Toyoda Enterprises, 1895–1933." *Business and Economic History* 25 (Winter 1996): 1–37.

May, Elaine Tyler. *Homeward Bound: American Families in the Cold War Era.* Rev. ed. New York: Basic Books, 1999.

McCarthy, Tom. *Auto Mania: Cars, Consumers, and the Environment.* New Haven: Yale University Press, 2007.

McCloskey, Donald N. *If You're So Smart: The Narrative of Economic Expertise.* Chicago: University of Chicago Press, 1990.

McCloskey, J. F. "British Operational Research in World War II." *Operations Research* 35 (May-June 1987): 453–70.

———. "US Operational Research in World War II." *Operations Research* 35 (Nov.-Dec. 1987): 910–25.

McCraw, Thomas K. *Prophets of Regulation: Charles Francis Adams, Louis D. Brandeis, James M. Landis, and Alfred E. Kahn.* Cambridge, Mass.: Harvard University Press, 1984.

———. *American Business, 1920–2000: How It Worked.* Wheeling, Ill.: Harlan Davidson, 2000.

———. "Schumpeter's Business Cycles as Business History." *Business History Review* 80 (Summer 2006): 231–61.

————. *Prophet of Innovation: Joseph Schumpeter and Creative Destruction*. Cambridge, Mass.: Harvard University Press, 2007.

McEvoy, Arthur F. *The Fisherman's Problem: Ecology and Law in the California Fisheries, 1850–1980*. New York: Cambridge University Press, 1986.

McKendrick, David, Richard F. Doner, and Stephan Haggard. *From Silicon Valley to Singapore: Location and Competitive Advantage in the Hard Disk Drive Industry*. Palo Alto, Calif.: Stanford Business Books, 2000.

McKenna, Christopher D. *The World's Newest Profession: Management Consulting in the Twentieth Century*. New York: Cambridge University Press, 2006.

McKenzie, Richard B. *Trust on Trial: How the Microsoft Case Is Reframing the Rules of Competition*. Cambridge, Mass.: Perseus, 2001.

McMillan, G. Steven, and Diana Hicks. "Science and Corporate Strategy: A Bibliometric Update of Hounshell and Smith." *Technology Analysis & Strategic Management* 13 (Dec. 2001): 497–505.

Merkel, Philip L. "Going National: The Life Insurance Industry's Campaign for Federal Regulation After the Civil War." *Business History Review* 65 (Autumn 1991): 528–53.

Meyer-Thurow, Georg. "The Industrialization of Invention: A Case Study from the German Chemical Industry." *Isis* 73 (Sep. 1982): 363–81.

Micklethwait, John, and Adrian Wooldridge. *The Witch Doctors: Making Sense of the Management Gurus*. New York: Times Books, 1996.

Miller, Anneta, with David L. Gonzalez and Peter McAlevey. "What Cars Do Women Want?" *Newsweek* (Mar. 24, 1986): 54.

Mintzberg, Henry. "The Manager's Job: Folklore and Fact." *Harvard Business Review* 53 (Jul.-Aug. 1975): 49–61.

————. "Crafting Strategy." *Harvard Business Review* 65 (July-Aug. 1987): 66–75.

————. "The Design School: Reconsidering the Basic Premises of Strategic Management." *Strategic Management Journal* 11 (Mar.-Apr. 1990): 171–95.

————. "Learning 1, Planning 0: Reply to Igor Ansoff." *Strategic Management Journal* 12 (Sep. 1991): 463–66.

————. *The Rise and Fall of Strategic Planning*. New York: The Free Press, 1994.

————. "Learning 1, Planning 0." *California Management Review* 38 (Summer 1996): 92–93.

————. "Reply to Michael Goold." *California Management Review* 38 (Summer 1996): 96–99.

Mintzberg, Henry, Richard Pascale, Michael Goold, and Richard Rumelt. "The 'Honda Effect' Revisited." *California Management Review* 38 (Summer 1996): 78–117.

Miranti, Paul J. "Networks and Uncertainty: Forecasting, Budgeting and Production Planning at the Bell System." Paper presented at the Fifth Accounting History International Conference, Banff, Canada, August 10, 2007.

Misa, Thomas J. *A Nation of Steel: The Making of Modern America, 1865–1925*. Baltimore: Johns Hopkins University Press, 1995.

Miser, H. J., ed. *Operations Analysis in the Eighth Air Force: Four Contemporary Accounts*. Linthicum, Md.: Institute for Operations Research and Management Science, 1997.

332

Moore, Geoffrey H., and Philip A. Klein. *The Quality of Consumer Instalment Credit.* New York: Columbia University Press, 1967.

Moore, J. I. *Writers on Strategy and Strategic Management.* London: Penguin, 2001.

Morgan, Mary S. *The History of Econometric Ideas.* Cambridge, U.K.: Cambridge University Press, 1990.

Morone, Joseph G. *Winning in High-Tech Markets.* Boston: Harvard Business School Press, 1993.

Morris, Susan. "Organizing for Industrial Research: The Resource Networks of Small Enterprises." Paper presented at the Society for the History of Technology Annual Meeting, Toronto, Canada, October, 2002.

Mors, Wallace P. *Consumer Credit Finance Charges: Rate Information and Quotation.* New York: Columbia University Press, 1965.

Morse, Philip M., and George E. Kimball. *Methods of Operations Research.* Rev. ed. New York: The Technology Press of the Massachusetts Institute of Technology and John Wiley & Sons, 1954.

Morton, J. A. *Organizing for Innovation.* New York: McGraw-Hill, 1971.

Mowery, David C. "The Relationship Between Intrafirm and Contractual Forms of Industrial Research in American Manufacturing, 1900–1940." *Explorations in Economic History* 20 (Oct. 1983): 351–74.

———. "Industrial Research and Firm Size, Survival, and Growth in American Manufacturing, 1921–1946: An Assessment." *Journal of Economic History* 43 (Dec. 1983): 953–80.

———. "The Boundaries of the U.S. Firm in R&D." In *Coordination and Information,* edited by Naomi R. Lamoreaux and Daniel M. G. Raff, 147–76. Chicago: University of Chicago Press, 1995.

Mowery, David C., and Nathan Rosenberg. *Technology and the Pursuit of Economic Growth.* New York: Cambridge University Press, 1989.

Nash, George H. *Herbert Hoover and Stanford University.* Stanford, Calif.: Hoover Institution Press, 1988.

National Commission on Consumer Finance. *Consumer Credit in the United States: Report of the National Commission on Consumer Finance.* Washington, D.C.: U.S. Government Printing Office, Dec. 1972.

National Consumer Law Center. *Credit Discrimination.* 3rd ed. Boston: National Consumer Law Center, 2002.

National Research Council. *Research Laboratories in Industrial Establishments of the United States, 1920–21.* New York: R. R. Bowker, 1920.

Neal, Roger. "Women's Prerogative: To Switch to Sport Utility Vehicles; Advertising Gears Up to Reach Women with Macho, Independent Image of New 4WD Category." *Adweek* (Apr. 7, 1986): n.p.

Nelson, Daniel. *Managers and Workers: Origins of the New Factory System in the United States, 1880–1920.* Madison, Wis.: University of Wisconsin Press, 1975.

———. *Frederick W. Taylor and the Rise of Scientific Management.* Madison, Wis.: University of Wisconsin Press, 1980.

———. "Industrial Engineering and the Industrial Enterprise, 1890–1940." In *Coordination and Information,* edited by Naomi R. Lamoreaux and Daniel M. G. Raff, 35–50. Chicago: University of Chicago Press, 1995.

Nelson, Richard. "The Simple Economics of Basic Scientific Research." *Journal of Political Economy* 67 (June 1959): 297–306.

Nelson, Richard R., and Sidney G. Winter. *An Evolutionary Theory of Economic Change.* Cambridge, Mass.: Harvard University Press, 1982.

Nelson, Richard R., and Gavin Wright. "The Rise and Fall of American Technological Leadership: The Postwar Era in Historical Perspective." *Journal of Economic Literature* 30 (Dec. 1992): 1931–64.

"New Cars: Who Buys Them, How They're Paid For." *U.S. News & World Report* (Mar. 14, 1958): 84–86.

Nickles, Shelley. "'Preserving Women': Refrigerator Design as Social Process in the 1930s." *Technology and Culture* 43 (Oct. 2002): 693–727.

———. "More Is Better: Mass Consumption, Gender, and Class Identity in Postwar America." *American Quarterly* 54 (Dec. 2002): 581–622.

Nicolaides, Becky M. *My Blue Heaven: Life and Politics in the Working-Class Suburbs of Los Angeles, 1920–1965.* Chicago: University of Chicago Press, 2002.

"1953 Survey of Consumer Finances. Part II. Purchases of Durable Goods in 1952 and Buying Patterns for 1953." *Federal Reserve Bulletin* 39 (July 1953): 697–703.

"1955 Survey of Consumer Finances: Purchases of Durable Goods in 1954." *Federal Reserve Bulletin* 41 (May 1955): 465–70.

"1955 Survey of Consumer Finances: Technical Appendix." *Federal Reserve Bulletin* 41 (May 1955): 471–81.

"1958 Survey of Consumer Finances: Purchases of Durable Goods." *Federal Reserve Bulletin* 44 (July 1958): 760–74.

"1959 Survey of Consumer Finances: The Financial Position of Consumers." *Federal Reserve Bulletin* 45 (July 1959): 700–23.

Nishiguchi, Toshihiro. *Strategic Industrial Sourcing: The Japanese Advantage.* New York: Oxford University Press, 1994.

Noll, A. Michael. "Telecommunication Basic Research: An Uncertain Future for the Bell Legacy." *Prometheus* 21 (June 2003): 177–93.

Nonaka, Ikujiro, and Hirotaka Takeuchi. *The Knowledge Creating Company: How Japanese Companies Create the Dynamics of Innovation.* New York: Oxford University Press, 1995.

Norberg, Arthur L. *Computers and Commerce: A Study of Technology and Management at Eckert-Mauchly Computer Company, Engineering Research Associates, and Remington Rand, 1946–1957.* Cambridge, Mass.: MIT Press, 2005.

Norberg, Arthur L., and Judy E. O'Neill. *Transforming Computer Technology: Information Processing for the Pentagon, 1962–1986.* Baltimore: Johns Hopkins University Press, 1996.

Offer, Avner. *The Challenge of Affluence: Self-Control and Well-Being in the United States and Britain Since 1950.* Oxford: Oxford University Press, 2006.

Olmstead, Alan L., and Paul W. Rhode. "Reshaping the Landscape: The Impact and Diffusion of the Tractor in American Agriculture, 1910–1960." *Journal of Economic History* 61 (Sep. 2001): 663–98.

———. "The Red Queen and the Hard Reds: Productivity Growth in American Wheat, 1800–1940." *Journal of Economic History* 62 (Dec. 2002): 929–66.

334

Olney, Martha. *Buy Now, Pay Later: Advertising, Credit, and Consumer Durables in the 1920s.* Chapel Hill: University of North Carolina Press, 1991.

———. "When Your Word Is Not Enough: Race, Collateral, and Household Credit." *Journal of Economic History* 58 (June 1998): 408–31.

O'Shea, James, and Charles Madigan. *Dangerous Company: The Consulting Powerhouses and the Businesses They Save and Ruin.* New York: Times Books, 1997.

Owens, Larry. "Patents, the 'Frontiers' of American Invention, and the Monopoly Committee of 1939: Anatomy of a Discourse." *Technology and Culture* 32 (Oct. 1991): 1076–93.

Pascale, Richard T. "Perspectives on Strategy: The Real Story Behind Honda's Success." *California Management Review* 26 (Spring 1984): 47–72.

Patterson, James T. *Grand Expectations: The United States, 1945–1974.* New York: Oxford University Press, 1996.

Peters, Thomas J., and Robert H. Waterman Jr. *In Search of Excellence: Lessons from America's Best-Run Companies.* New York: Harper & Row, 1982.

Piore, Michael J., and Charles F. Sabel. *The Second Industrial Divide: Possibilities for Prosperity.* New York: Basic Books, 1984.

Porter, Robert H. "A Study of Cartel Stability: The Joint Executive Committee, 1880–1886." *Bell Journal of Economics* 14 (Autumn 1983): 301–14.

Porter, Robert M. *The Rise of Statistical Thinking, 1820–1900.* Princeton: Princeton University Press, 1986.

Posner, Richard A. *Antitrust Law: An Economic Perspective.* Chicago: University of Chicago Press, 1976.

Pratt, Joseph A., Tyler Priest, and Christopher Castaneda, *Offshore Pioneers: Brown & Root and the History of Offshore Oil and Gas.* Houston: Gulf, 1997.

Priest, Tyler. *The Offshore Imperative: Shell Oil's Search for Petroleum in Postwar America.* College Station, Texas: Texas A&M University Press, 2007.

Pugh, Emerson W. *Memories That Shaped an Industry: Decisions Leading to IBM System/360.* Cambridge, Mass.: MIT Press, 1984.

———. *Building IBM: Shaping an Industry and Its Technology.* Cambridge, Mass.: MIT Press, 1995.

Purkayastha, Dev. "Note on the Motorcycle Industry—1975," Case # 9-578-210. Boston: Harvard Business School Press, 1978.

Quinn, James Brian. "The Honda Motor Company." In *The Strategy Process*, edited by Henry Mintzberg, James Brian Quinn, and Sumantra Ghoshal, 293–314. London: Prentice-Hall, 1998.

Rae, John B. *American Automobile Manufacturers: The First Forty Years.* Philadelphia: Chilton, 1959.

Raff, Daniel M. G., and Manuel Trajtenberg. "Quality-Adjusted Prices for the American Automobile Industry: 1906–1940." In *The Economics of New Goods*, edited by Timothy F. Bresnahan and Robert J. Gordon, 71–107. Chicago: University of Chicago Press, 1997.

"Regulation B: Amendment." *Federal Reserve Bulletin* 63 (Jan. 1977): 89–95.

Reich, Leonard S. "Research, Patents, and the Struggle to Control Radio: A Study of Big Business and the Uses of Industrial Research." *Business History Review* 51 (Summer 1977): 208–35.

———. "Industrial Research and the Pursuit of Corporate Security: The Early Years of Bell Labs." *Business History Review* 54 (Winter 1980): 504–29.

————. *The Making of American Industrial Research: Science and Business at GE and Bell, 1876–1926.* New York: Cambridge University Press, 1985.

————. "Lighting the Path to Profit: GE's Control of the Electric Lamp Industry, 1892–1941." *Business History Review* 66 (Summer 1992): 305–34.

Reizenstein, Gail R. "A Fresh Look at the Equal Credit Opportunity Act." *Akron Law Review* 14 (Fall 1980): 215–51.

Riordan, Michael, and Lillian Hoddeson. *Crystal Fire: The Birth of the Information Age.* New York: W. W. Norton, 1997.

Robinson, Edgar Eugene, and Paul Carroll Edwards, eds. *The Memoirs of Ray Lyman Wilbur, 1875–1949.* Stanford, Calif.: Stanford University Press, 1960.

Rogers, Everett M., and Judith K. Larsen. *Silicon Valley Fever: Growth of a High-Technology Culture.* New York: Basic Books, 1984.

Rome, Adam. *The Bulldozer in the Countryside: Suburban Sprawl and the Rise of American Environmentalism.* New York: Cambridge University Press, 2001.

Rose, Mark H. *Interstate: Express Highway Politics, 1941–1956.* Lawrence, Kans.: Regents Press of Kansas, 1979.

————. *Cities of Light and Heat: Domesticating Gas and Electricity in Urban America.* University Park, Penn.: Pennsylvania State University Press, 1995.

————. "Reframing American Highway Politics, 1956–1995." *Journal of Planning History* 2 (Aug. 2003): 212–36.

Roseman, Ellen. "The Consumer Game Car Maker Driving to Win Over." *The Globe and Mail* (Canada) (Apr. 21, 1983): n.p.

Rosenberg, Nathan. *Inside the Black Box: Technology and Economics.* New York: Cambridge University Press, 1982.

————. "Why Do Firms Do Basic Research (With Their Own Money)?" *Research Policy* 19 (Apr. 1990): 165–74.

————. *Exploring the Black Box: Technology Economics and History.* New York: Cambridge University Press, 1994.

Rosenbloom, Richard S. "Leadership, Capabilities, and Technological Change: The Transformation of NCR in the Electronic Era." In *The SMS Blackwell Handbook of Organizational Capabilities*, edited by Constance E. Helfat, 364–92. Malden, Mass.: Blackwell, 2003.

Rosenbloom, Richard S., and William J. Spencer, eds. *Engines of Innovation: U.S. Industrial Research at the End of an Era.* Boston: Harvard Business School Press, 1996.

Ross, Dorothy. *The Origins of American Social Science.* New York: Cambridge University Press, 1991.

Rubenstein, James M. *Making and Selling Cars: Innovation and Change in the U.S. Automotive Industry.* Baltimore: Johns Hopkins University Press, 2001.

Rumelt, Richard P. "The Many Faces of Honda." *California Management Review* 38 (Summer 1996): 103–11.

Russo, Arturo. "Fundamental Research at Bell Laboratories: The Discovery of Electron Diffraction." *Historical Studies in the Physical Sciences* 12, Part 1 (1981): 117–60.

Sabel, Charles F., and Jonathan Zeitlin, eds. *World of Possibilities: Flexibility and Mass Production in Western Industrialization.* Cambridge, U.K.: Cambridge University Press, 1997.

336   Sager, Ira, and Diane Brady. "Big Blue's Blunt Bohemian." *Business Week* (June 14, 1999): 107–108.

Saxenian, AnnaLee. *Regional Advantage: Culture and Competition in Silicon Valley and Route 128*. Cambridge, Mass.: Harvard University Press, 1994.

Schatz, Ronald W. *The Electrical Workers: A History of Labor at General Electric and Westinghouse, 1923–1960*. Urbana: University of Illinois Press, 1983.

Schempf, F. Jay. *Pioneering Offshore: The Early Years*. Houston: Offshore Energy Center, 2007.

Scherer, F. M. *Industrial Market Structure and Economic Performance*. Chicago: Rand McNally, 1970.

Schumpeter, Joseph A. "The Creative Response in Economic History." *Journal of Economic History* 7 (Nov. 1947): 149–59.

———. *Capitalism, Socialism and Democracy*. 3rd ed. New York: Harper & Row, 1950.

———. "Comments on a Plan for the Study of Entrepreneurship." In *Joseph A. Schumpeter*, edited by Richard Swedberg, 406–24. New York: Cambridge University Press, 1991.

———. "The Meaning of Rationality in the Social Sciences." In *Joseph A. Schumpeter*, edited by Richard Swedberg, 327–30. New York: Cambridge University Press, 1991.

Schwartz, Michael. "Markets, Networks, and the Rise of Chrysler in Old Detroit, 1920–1940." *Enterprise and Society* 1 (March 2000): 63–99.

Scranton, Philip. *Proprietary Capitalism: The Textile Manufacture at Philadelphia*. New York: Cambridge University Press, 1983.

———. *Figured Tapestry: Production, Markets, and Power in Philadelphia Textiles, 1885–1941*. New York: Cambridge University Press, 1989.

———. *Endless Novelty: Specialty Production and American Industrialization, 1865–1925*. Princeton: Princeton University Press, 1997.

Serafin, Raymond. "Dealers Hold Key to Sales Success in Segment." *Advertising Age* (Sep. 15, 1986): S-16.

———. "Still Leading Herd: Ford Mustang Gallops into 25th Year." *Advertising Age* (Apr. 10, 1989): 30.

Sloan, Alfred P. Jr. *My Years with General Motors*, edited by John McDonald with Catherine Stevens. Garden City, N.Y.: Doubleday, 1964.

Smith, George David. *The Anatomy of a Business Strategy: Bell, Western Electric and the American Telephone Industry*. Baltimore: Johns Hopkins University Press, 1985.

Snyder, Thomas D. *120 Years of American Education: A Statistical Portrait*. Washington, D.C.: U.S. Department of Education, 1993.

Sobel, Robert. *IBM: Colossus in Transition*. New York: Times Books, 1981.

Sokoloff, Kenneth L., and B. Zorina Khan. "The Democratization of Invention During Early Industrialization: Evidence from the United States, 1790–1846." *Journal of Economic History* 50 (June 1990): 363–78.

Sorensen, Todd, Price Fishback, and Shawn Kantor. "The New Deal and the Diffusion of Tractors in the 1930s." Unpublished paper, Faculty of Economics, University of California, Riverside, 2008.

Speer, J. B. "The Functional Organization of the University." *Journal of Higher Education* 5 (Oct. 1934): 414-21.

Sproul, Robert G. "Opportunity Presented by Budgetary Limitations." *Chronicle of*   **337**
Higher Education 5 (Jan. 1934): 7-13.

Stalk, George W. Jr., and Carl W. Stern. *Perspectives on Strategy from The Boston Consulting Group.* New York: John Wiley & Sons, 1998.

Stigler, George J. "The Theory of Economic Regulation." *Bell Journal of Economics and Management Science* 2 (Spring 1971): 3–21.

Stigler, Stephen M. *The History of Statistics: The Measurement of Uncertainty Before 1900.* Cambridge, Mass.: Harvard University Press, 1986.

Stone, Alan. *Economic Regulation and the Public Interest: The Federal Trade Commission in Theory and Practice.* Ithaca, N.Y.: Cornell University Press, 1977.

Stradling, David. *Smokestacks and Progressives: Environmentalists, Engineers, and Air Quality in America, 1881–1951.* Baltimore: Johns Hopkins University Press, 1999.

Strasser, Susan. *Satisfaction Guaranteed: The Making of the American Mass Market.* New York: Pantheon Books, 1989.

Sturgeon, Timothy J. "How Silicon Valley Came to Be." In *Understanding Silicon Valley,* edited by Martin Kenney, 15–47. Stanford, Calif.: Stanford University Press, 2000.

Sugrue, Thomas J. *The Origins of the Urban Crisis: Race and Inequality in Postwar Detroit.* Princeton: Princeton University Press, 1996.

"Survey of Consumer Finances, Part I. Expenditures for Durable Goods and Investments." *Federal Reserve Bulletin* 33 (June 1947): 647–63.

Survey Research Center, Institute for Social Research. *1960 Survey of Consumer Finances.* Ann Arbor, Mich.: University of Michigan, 1961.

Sutch, Richard C. "Henry Agard Wallace, the Iowa Corn Yield Tests, and the Adoption of Hybrid Corn." NBER Working Paper No. 14141 (2008).

Swedberg, Richard, ed. *Joseph A. Schumpeter: The Economics and Sociology of Capitalism.* New York: Cambridge University Press, 1991.

Taylor, Frederick Winslow. *The Principles of Scientific Management.* New York: W. W. Norton, 1967.

Tedlow, Richard S. *New and Improved: The Story of Mass Marketing in America.* New York: Basic Books, 1990.

———. *The Watson Dynasty: The Fiery Reign and Troubled Legacy of IBM's Founding Father and Son.* New York: HarperBusiness, 2003.

Teece, David J. "Technological Change and the Nature of the Firm." In *Technical Change and Economic Theory,* edited by Giovanni Dosi, Christopher Freeman, Richard Nelson, Gerard Silverberg, and Luc Soete, 256–81. London: Pinter, 1998).

Teece, David J., and Gary Pisano. "The Dynamic Capabilities of Firms: An Introduction." In *Technology, Organization, and Competitiveness: Perspectives on Industrial and Corporate Change,* edited by Josef Chytry, Giovanni Dosi, and David J. Teece, 193–212. New York: Oxford University Press, 1998.

Temin, Peter, with Louis Galambos. *The Fall of the Bell System: A Study in Prices and Politics.* New York: Cambridge University Press, 1987.

Terman, Frederick E. "A Brief History of Engineering Education." *Proceedings of the IEEE* 64 (Sept. 1976): 1403.

Thorelli, Hans B. *The Federal Antitrust Policy: Organization of an American Tradition.* Baltimore: Johns Hopkins University Press, 1955.

338   Tidman, K. R. *The Operations Evaluation Group: A History of Naval Operations Analysis*. Annapolis, Md.: Naval Institute Press, 1984.

Tomlins, Christopher L. *The State and the Union: Labor Relations, Law, and the Organized Labor Movement in America, 1880–1960*. New York: Cambridge University Press, 1985.

Tripsas, Mary, and Giovanni Gavetti. "Capabilities, Cognition, and Inertia: Evidence from Digital Imaging." In *The SMS Blackwell Handbook of Organizational Capabilities*, edited by Constance E. Helfat, 393–412. Malden, Mass.: Blackwell, 2003.

U.S. Board of Governors of the Federal Reserve System. *Consumer Instalment Credit*. Washington, D.C.: U.S. Government Printing Office, 1957.

U.S. Bureau of the Census. *Historical Statistics of the United States: From Colonial Times to the Present*. Washington, D.C.: U.S. Government Printing Office, 1976.

———. "Population Estimates, 2000–2006." http://factfinder.census.gov/servlet/GCT Table?_bm=y&-geo_id=01000US&-_box_head_nbr=GCT-T1&-ds_name=PEP_2006_EST&-_lang=en&-format=US-9&-_sse=on (accessed March 7, 2008).

U.S. Congress, House of Representatives. *Credit Discrimination: Hearings Before the Subcommittee on Consumer Affairs, House Committee on Banking and Currency, Relating to Legislation Prohibiting Sex Discrimination in the Granting of Credit*. 93rd Cong., 2nd sess., part 1. June 20 and 21, 1974.

U.S. Department of Labor. *How American Buying Habits Change*. Washington, D.C.: U.S. Government Printing Office, 1959.

U.S. Department of Labor, Bureau of Labor Statistics. *Study of Consumer Expenditures, Incomes and Savings: Statistical Tables, Urban U.S.-1950*, vol. 18. Philadelphia: University of Pennsylvania, Wharton School of Finance and Commerce, 1957.

———. "Consumer Expenditures and Income: Total United States, Urban and Rural, 1960–61." BLS Report No. 237-93 (Feb. 1965).

———. "Consumer Expenditure Survey, 1998–99." *Bulletin* 955 (Nov. 2001).

U.S. Patent Office. *Annual Reports of the Commissioner of Patents*. Washington, D.C.: U.S. Government Printing Office, 1845–1936.

———. *Names and Addresses of Attorneys Practicing Before the United States Patent Office*. Washington, D.C.: U.S. Government Printing Office, 1883 and 1905.

———. "U.S. Patent Activity: Calendar Years 1790 to the Present." http://www.uspto.gov/web/offices/ac/ido/oeip/taf/h_counts.pdf (accessed March 7, 2008).

Usselman, Steven W. "IBM and Its Imitators: Organizational Capabilities and the Emergence of the International Computer Industry." *Business and Economic History* 22 (Winter 1993): 1–35. Reprinted in *Industrial Research and Innovation in Business*, edited by David E. H. Edgerton, 452–86. London: Edward Elgar, 1996.

———. "Fostering a Capacity for Compromise: Business, Government, and the Stages of Innovation in American Computing," *Annals of the History of Computing* 18 (Summer 1996): 30–39.

———. *Regulating Railroad Innovation: Business, Technology, and Politics in America, 1840–1920*. New York: Cambridge University Press, 2002.

———. "Public Policies, Private Platforms: Antitrust and American Computing." In *Information Technology Policy: An International History*, edited by Richard Coopey, 97–120. Oxford, U.K.: Oxford University Press, 2004.

———. "Learning the Hard Way: IBM and the Sources of Innovation in Early Computing." In *Financing Innovation in the United States*, edited by Naomi R. Lamoreaux and Kenneth L. Sokoloff, 317–63. Cambridge, Mass.: MIT Press, 2007.   339

Vagelos, P. Roy, with Louis Galambos. *Medicine, Science and Merck.* New York: Cambridge University Press, 2004.

Veldman, Hans, and George Lagers, *50 Years Offshore.* Delft, Holl.: Foundation for Offshore Studies, 1997.

Veysey, Laurence R. *The Emergence of the American University.* Chicago: University of Chicago Press, 1965.

Vietor, Richard H. K. *Energy Policy in American Since 1945: A Study of Business-Government Relations* (New York: Cambridge University Press, 1984).

———. *Contrived Competition: Regulation and Deregulation in America.* Cambridge, Mass.: Harvard University Press, 1994.

Waddington, C. H. *OR in World War II: Operations Research Against the U Boat.* London: Elek Science, 1973.

Waller, Spencer Weber. "The Antitrust Legacy of Thurman Arnold." *St. John's Law Review* 78 (Summer 2004): 569–613.

Walsh, Margaret. "Gendering Mobility: Women, Work and Automobility in the United States." *History* 93, no. 311 (July 2008): 376–95.

Wasserman, Elizabeth, and Patrick Thibodeau. "Microsoft, IBM Face Off." *Info World* 21 (June 14, 1999): 30.

Watson, Andrew, Terence Rodgers, and David Dudek. "The Human Nature of Management Consulting: Judgment and Expertise." *Managerial and Decision Economics* 19 (Nov.–Dec. 1998): 495–503.

Watson Thomas J. Jr., and Peter Petre. *Father, Son & Co.: My Life at IBM and Beyond.* New York: Bantam Books, 1990.

Welke, Barbara Young. *Recasting American Liberty: Gender, Race, Law, and the Railroad Revolution, 1865–1920.* New York: Cambridge University Press, 2001.

Whinston, Michael D. "Tying, Foreclosure, and Exclusion." *American Economic Review* 80 (Sep. 1990): 837–59.

Whittington, Richard. *What Is Strategy—and Does It Matter?* 2nd ed. London: Thompson Learning, 2001.

Williamson, Jeffrey G., and Peter H. Lindert. *American Inequality: A Macroeconomic History.* New York: Academic Press, 1980.

Wise, George, *Willis R. Whitney, General Electric, and the Origins of U.S. Industrial Research.* New York: Columbia University Press, 1985.

Wise, T. A. "IBM's $5,000,000,000 Gamble." *Fortune* 74 (Sep. 1966): 118–23, 224, 226, and 228.

———. "The Rocky Road to the Marketplace." *Fortune* 75 (Oct. 1966): 138–43, 199, 201, 205–6, and 211–12.

Witt, John Fabian. *The Accidental Republic: Crippled Workingmen, Destitute Widows, and the Remaking of American Law.* Cambridge, Mass.: Harvard University Press, 2004.

"Wooing Women Buyers: Women Are Involved in 80 Percent of New Vehicle Purchases." *Automotive News* (Aug. 22, 1988): E13.

340   Wright, Gavin. *Old South, New South: Revolutions in the Southern Economy Since the Civil War*. New York: Basic Books, 1986.

Yarrow, Andrew L. *Forgive Us Our Debts: The Intergenerational Dangers of Fiscal Irresponsibility*. New Haven: Yale University Press, 2008.

Yates, JoAnne. *Control Through Communication: The Rise of System in American Management*. Baltimore: Johns Hopkins University Press, 1989.

———. *Structuring the Information Age: Life Insurance and Technology in the Twentieth Century*. Baltimore: Johns Hopkins University Press, 2005.

Zachary, G. Pascal. *Endless Frontier: Vannevar Bush and the Engineering of the American Century*. Cambridge, Mass.: MIT Press, 1999.

INDEX

Knox, H. G., 200
Kodak, 8

Lally, Maureen Ellen, 291
Large firms: Chandler on superiority of, 2;
    cooperation between, 14–15; examination
    of role of, in innovation, 2–4, 26nn8–9;
    movement of inventors into, 4–5, 39–40;
    Schumpeter on rise of, 1, 43–44. *See also*
    In-house R&D
"Learning from Honda" (Mair), 221
"Learning 1, Planning 0" (Mintzberg), 221
Learson, Vin, 277n59
Lécuyer, Christophe, 175, 187n1
Lee, Griff, 192, 208, 209, 210, 215
Leslie, Stuart, 13, 21, 173, 187n1
Lichtenstein, Nelson, 291
Littleton, Joseph, 113n38
Lockwood, T. D., 5, 27n17
Lowen, Rebecca, 187n1
Lucent Technologies, 163
Lucy, Chuck, 106

MacAvoy, Tom, 100, 101, 102
MacPherson v. Buick, 22
Magnolia Oil, 196, 202
Mair, Andrew, 221, 223–24, 226, 230
Marchand, Roland, 127
Marshall, Burke, 266
Marshall, Peter, 213
Mauchly, John, 257, 258–59
Maurer, Robert, 105
May, Elaine Tyler, 289
McCarthy, Tom, 282, 298n1
McCauley, George, 92
McClelland, Bramlette, 198
McCraw, Thomas, 17
MCI, 11, 108
McKinsey & Company, 224, 227, 230
Memento (film), 219–20, 221
Merck, 11, 14
Meyer, Ron, 223
Michelson, Albert, 139
Microsoft, 245, 250, 269, 270–71
Mintzberg, Henry, 221, 225–27
MIT, and Terman, 171, 172–73
Molina, E. C., 122, 123, 124, 125

Monopoly Committee. *See* Temporary
    National Economic Committee
    (TNEC)
Moore, Geoffrey, 285–86, 292
Morison, J. R., 199
Morrill Land Grant Act, 20
Motion picture industry, antitrust case
    against, 253, 274n14
Mowery, David, 7, 20
Munk, W. H., 200
Munn and Company, 48

National Commission on Consumer
    Finance, 291, 292–93
National companies, defined, 63–64
National Patent Planning Commission,
    255–56
NCR, 9
Nelson, Richard, 8
New Deal: farm regulations, 17–18; and
    government prosecution of Glass Trust,
    94–95, 112n24; regulatory initiatives, and
    consulting firms, 16. *See also* Antitrust
    law
"New economies of information," 44
New York City, 40, 72
Nixon, Richard, 288–89, 306n51
Nolan, Christopher, 219, 235n2
Nonaka, Ikujiro, 227
Norberg, Arthur, 21
Northern Telecom, 112n33
Norton Villiers Triumph (NVT), 232
"Note on the Motorcycle Industry—1975"
    (Purkayastha), 221, 223, 230, 238n64

Offshore oil industry, 165, 191–218; areas of
    research by, 198–201; basic approaches
    of, 196–98; first Gulf of Mexico struc-
    tures, 194–95; government regulation of,
    216n4; and Hurricanes Katrina and Rita,
    214–15; impact of World War II on,
    195–96; knowledge of, about Gulf of
    Mexico offshore conditions, 191–93,
    215–16, 218n54; 1950s' developments in,
    201–6; 1960s' innovations in, 206–14; risk-
    taking culture in, 193–94
Olmstead, Alan, 20